The Unity of Science

The Unity of Science

David Bensimon

CRC Press
Taylor & Francis Group
Boca Raton London New York

CRC Press is an imprint of the
Taylor & Francis Group, an **informa** business

First edition published 2022
by CRC Press
6000 Broken Sound Parkway NW, Suite 300, Boca Raton, FL 33487-2742

and by CRC Press
2 Park Square, Milton Park, Abingdon, Oxon, OX14 4RN

CRC Press is an imprint of Taylor & Francis Group, LLC

Library of Congress Cataloging-in-Publication Data

Names: Bensimon, David, author.
Title: The unity of science / David Bensimon.
Description: Boca Raton : Taylor and Francis, 2021. | Includes
bibliographical references and index.
Identifiers: LCCN 2021035071 (print) | LCCN 2021035072 (ebook) | ISBN
9781032112404 (hardback) | ISBN 9781032112411 (paperback) | ISBN
9781003218999 (ebook)
Subjects: LCSH: Science--Philosophy. | Science.
Classification: LCC Q175 .B458 2021 (print) | LCC Q175 (ebook) | DDC
501--dc23
LC record available at https://lccn.loc.gov/2021035071
LC ebook record available at https://lccn.loc.gov/2021035072

ISBN: 9781032112404 (hbk)
ISBN: 9781032112411 (pbk)
ISBN: 9781003218999 (ebk)

DOI: 10.1201/9781003218999

Typeset in Nimbus font
by KnowledgeWorks Global Ltd.

Dedication

—————

To the future enlightenment of my grand-children.

Contents

Preface

The goal of this book is to show that many Natural phenomena and all of our present day technology can be explained and derived from a few basic concepts that have been known for more than a hundred years. These concepts —Newton's Laws of motion and gravitation, Coulomb's law of electrostatic interactions and special relativity (i.e. Maxwell's equations of Electromagnetism), Dirac (or Schrödinger) equation for quantum mechanics —and the concept of entropy for many particle systems are the foundations of Science and Technology. To appreciate their miraculous power and beauty one needs a minimal understanding of Nature's language: Mathematics. Indeed seeing how effective that language is one cannot help but think of E.Wigner's remark about "the Unreasonable Effectiveness of Mathematics in the Natural Sciences". In this book I will exemplify "the miracle of the appropriateness of the language of mathematics" in the description of Nature by using the aforementioned laws to explain various natural phenomena and point to their ramifications in many scientific fields (chemistry, astrophysics, optics, etc.) and technologies (electronics, communication, hydrodynamics, etc.). There are evidently may excellent treatises covering these topics: Feynmann's lectures on Physics and the Landau course on Theoretical Physics are probably the best, but they are aimed at physicists. Since I believe that an understanding of Science should be part of everybody's education, I have tried here to address a larger audience covering topics (how spin and antiparticles arise from Dirac's equation, how Statistical Mechanics can be derived from Information Theory, why the Universe is an ideal black-body, etc.) seldom presented to such an audience. To that purpose I have used a minimal set of mathematical tools (essentially vector calculus and linear algebra) and have derived the various equations as comprehensively as possible in a concise book. These tools will, I hope, be familiar to high school or undergraduate students with basic mathematical training and to laymen willing to learn Nature's language and to make the effort to follow the (sometimes lengthy) mathematical derivations to their ends. It is with them in mind that this book was written.

Acknowledgements

The work presented here is an outcome of courses on Electromagnetism, Quantum Mechanics, Statistical Mechanics and Biophysics that I taught over many years to students with various backgrounds (physicists, chemists, biologists, engineers, etc.). Their questions and their feedback were essential in the development of this book.

My scientific training and philosophy owes a lot to my teachers starting from my high school chemistry teacher, Dr. Levine, to my undergraduate and graduate teachers and mentors: Profs. A. Man, A. Ron, A. Peres, L. Schulman, S. Shtrikman, M. Slatkine, S. Chandrasekhar, S. Shenker, L. Kadanoff and A. Libchaber. My friends and colleagues, Drs. V. Croquette, L. Jullien, B. Shraiman, S. Weiss, M. Volovitch, A. Prochiantz, S. Vriz, B. Gelbart, Y. Rabin, I. Procaccia, E. Domany and R. Phillips have further contributed to my scientific education.

I further thank my friend, Dr. M.Amaral, —who, as a non-physicist with a deep interest in Science, was the personification of the reader for whom this book was written —for his willingness to critically read the material in this book and for his many remarks which helped me clarify its original draft.

Finally, I am grateful to my wife for her constant support and encouragements.

1 The Unity of Science

1.1 INTRODUCTION

I have always found the visit of a Science Museum to be a frustrating experience. Often the purpose of the exhibits is to visually impress the visitor, but even in the best museums (such as the Planetarium in San Francisco) where the goal is to pass on some concepts, the whole experience feels like a magic show: nice tricks that few even try to understand. In my opinion, the essence of my trade has been missed: the beauty of Science is not in the myriad tricks, experiments, devices, etc. that one can design but in the fact that this abundance of phenomena (both natural and artificial) are all explained by or derived from very few basic ideas, laws or concepts. In fact much of our daily experience with natural phenomena or man made artefacts can be explained by four laws or concepts: Newton's laws of motion and gravitation, Maxwell's description of electromagnetic phenomena, Schrödinger's equation for quantum mechanical systems and the concept of entropy for systems with many particles. Using these laws only, we can understand what gases, liquids and solids are, why certain material conduct electricity and others not, why certain compounds (e.g. hydrogen) react with some (e.g. oxygen) but not others (Helium), what gives rises to a rainbow, to lightning or to Aurora Borealis, why is the temperature on Earth in a range that supports life, how do neurons and enzymes work, how old is the Earth or the Universe, etc., etc. Using the same basic laws we can also explain how a motor (electric or thermal) or a refrigerator work, how to generate radio-waves and how to use them to communicate, what principles govern the design and function of a camera or a computer, why is glue sticky, etc., etc. To explain these phenomena and to understand how living systems function, we don't need Einstein's theory of gravitation, Quantum electrodynamics or any "Theory of Everything". The classical concepts, more than a hundred years old, are enough to explain many of the phenomena that we encounter in our daily life. This is in my view the beauty of Science: the simple unifying explanation of Nature and its outcome, technology. My goal in this book is to convey that unity of Science by sketching how myriad fields (chemistry, astronomy, optics, electronics, etc.) are all offshoots of the four basic laws or concepts mentioned above, how some common natural phenomena can be explained by them and what are the principles behind some of our technologies. As a virtual Science Museum, this book is organized around four main chapters dealing with Classical Mechanics, Electromagnetism, Quantum and Statistical Mechanics. The core concepts and some important examples (Black-body radiation, Hydrogen atom, random walk, loaded die, ideal gas, etc.) are introduced in the sections labeled with ◆. The other sections are applications of these core concepts to a variety of natural phenomena and technologies and can be sampled as desired.

To better appreciate the beauty of a poem, one needs to understand the language it was written in. Similarly, to appreciate the beauty of Science one needs to learn and

DOI: 10.1201/9781003218999-1

understand its language: Mathematics. Indeed why "the book of Nature is written in the language of Mathematics" as famously quipped by Galileo Galilei is one of the marvels and mysteries of Science. At the time of Galileo that belief was grounded in Euclidean geometry which while founded on axiomatic statements (such as the sum of the angles in a triangle is 180°) seemed to provide a-priori knowledge on the world. Thus Galileo believed (as many philosophers before and after him) that truths about Nature could be obtained by pure reason, without the need for experimental validation. Experiments are always noisy and yield results that are never as clear cut as a logical argument. However with the invention (some say discovery) of non-Euclidean geometries it became obvious that Mathematics, per se, had nothing to say about natural phenomena without experimental input. In some sense Mathematics is a game like Chess or Go: it has its arbitrary rules (the axioms) and a number of theorems which results from applications of these rules (e.g. the angle bisectors in a triangle meet at a single point; for a given position of the pieces, white wins in 3 moves; etc....). There is therefore no a-priori reason for Mathematics to provide a better description of Nature than Chess or Go. However, it is a fact that the appropriateness of Mathematics in the description of natural phenomena has been in the words of Eugene Wigner - one of the founders of Quantum Mechanics - a "miracle ... that we do not understand nor deserve"[1, 2]. This amazing success of Mathematics is even more noteworthy considering the very small number of problems that we can solve analytically (e.g. the motion in a central force field or the harmonic oscillator). It is the development of approximations, perturbation expansions and numerical methods that have allowed us to expand the power of Mathematics way beyond the very small set of exactly soluble problems.

It is even more remarkable that while Mathematics has been extremely successful at explaining and predicting phenomena at the extremely small or extremely large scale (atoms and molecules or stars and galaxies) it has so far failed miserably at predicting phenomena that occur at our scale, where one finds many of the hardest and unsolved problems in Nature. Thus we cannot predict the weather, a coming earthquake, a crash in the stock market or a future war. From an evolutionary point of view one might have expected the reverse to be true: that evolution would have equipped us with tools to predict a coming flood, a future drought and the behavior of our neighbors rather than the eclipse of the Sun, the return of some comet or the behavior of atoms. In fact the amazing success of Mathematics in the description of physical phenomena relies on one crucial feature of physical (and in particular quantum mechanical) systems at these extreme scales: they are in some sense linear problems. Thus, the force exerted on one atom (or star) by its neighbors is the sum of the individual forces from each neighbor. This property is what makes the mathematical formulation of these systems often soluble[1] and thus useful for understanding them and predicting their behavior. When systems are non-linear (when the effect of the sum of individuals is not the sum of the individual effects, such as for

---

[1]Some classical problems such as the motion of 3 or more bodies interacting via gravitational (or electrostatic) forces can be chaotic, i.e. extremely sensitive to initial conditions and thus unpredictable.

the weather, the stock market, etc.) the predictions become extremely sensitive to uncertainties (initial conditions, noise) and present day Mathematics is much less of a help. Therefore the miracle, mentioned by Wigner, is that Nature at these extreme scales is linear: one could certainly imagine intelligent life arising in a completely non-linear and unpredictable Universe, where some orderly behavior could yet still occur (while the weather is unpredictable, when the sky is cloudy we know that it may rain). Mathematics however would have been much less effective in understanding this Universe. As expressed by Einstein: "the eternal mystery of the world is its comprehensibility ... The fact that it is comprehensible is a miracle"'. It is with this child-like wonder at the comprehensibility of Nature that I would like the reader to approach the subjects of this book.

1.2   THE INVENTION OF MATHEMATICS

Most of the mathematics that is needed to understand common Natural phenomena and our present day technology has been invented and developed in the seventeen and eighteen century: infinitesimal calculus (by Newton, Leibnitz, Bernoulli ..), linear algebra (by Leibnitz, Cramer, Gauss, ...) and probability theory (by Pascal, Bernoulli, Laplace, ...). While it is not my purpose to teach these subjects, some important notions will be reviewed in the following[3].

Infinitesimal calculus is the mathematical study of continuous change and is mainly interested either in the rate of change (differentiation) or in the accumulation of change (integration). This field is essential to the understanding of phenomena that vary with time and/or space, in fact most of physics: electromagnetism, quantum and fluid mechanics, dynamical systems, etc.. It has a long "pre"-history, dating back to the Greeks and the famous paradox of Zeno: Achilles the fast runner can never reach a tortoise since it would take him an infinite series of times to reach the previous positions of the tortoise. Of course, now we know that the sum of this infinite series is finite: if v_1 is Achilles' speed and v_2 the tortoise and if L is their initial distance it will take Achilles a time $T_1 = L/v_1$ to reach the first position of the tortoise, a further time $T_2 = T_1 v_2/v_1$ to reach the second position of the tortoise, etc. The sum: $T = T_1 + T_2 + T_3 + \cdots = T_1/(1 - v_2/v_1) = L/(v_1 - v_2)$, i.e. Achilles catches up with the tortoise in a time T such that $v_1 T = L + v_2 T$ (as known to any schoolchild). Nonetheless, the summation of an infinity of infinitesimally small quantities remained for a long time quite mysterious. Surprisingly, in fact it is another Greek, Archimedes, who was the first to perform what we now call an integration: computing the area of a parabola by summing up (using an ingenious trick), the areas of an infinite number of infinitesimally small triangles, see Fig.1.1.

At the time of Newton the study of curves in space was quite popular, attracting the interests of Pascal, Descartes and Fermat among others. Fermat, in particular, developed a heuristic method quite similar to differentiation to compute the tangent to a curve and determine the maxima and minima of the curve by finding the points were the tangent was parallel to the abscissa (the x-axis), see Fig1.2. It was however Isaac Newton and Gottfried Wilhelm Leibnitz who independently were the true inventors of calculus. Newton came to calculus by thinking about the trajectory of

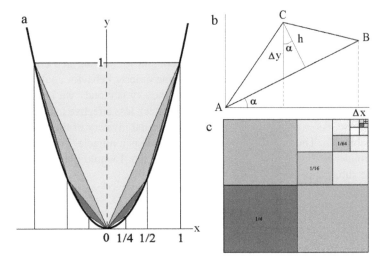

Figure 1.1 (a) Drawing of the construction used by Archimedes to compute the area comprised between the parabola $y = x^2$ and the line $y = 1$. (b) The area of the triangle ABC: $AB \cdot h/2 = \Delta y \cos \alpha \cdot \Delta x/2 \cos \alpha = \Delta x \Delta y/2$. As shown in (b) since the horizontal span Δx of the blue triangle is $1/2$ that of the green one and its vertical span Δy is $1/4$ of the green one, the area of the blue triangle is $1/8$ that of the green, and since there are two blue triangles their joint area is $1/4$ that of the green. Similar argument show that the total area of the red triangles is $1/16$. Hence the total area comprised between the parabola and the line is $1 + 1/4 + 1/16 + 1/64 + ...$ of the area of the green triangle. To compute that sum Archimedes used a clever geometric construction shown in (c). The area $1/4$ of the dark red square is $1/3$ the area of the red squares, similarly the area $1/16$ of the dark green square is $1/3$ the area of the green squares, etc. The sum of all the dark colored squares is thus $1/3$ the area (1) of the big square. Hence the total area of the parabola is $4/3$ of the area (1) of the green triangle in (a).

objects in a plane: their position $(x(t), y(t))$. He reasoned that the tangent to a given trajectory $f(x, y) = 0$ at a time t would be given by the ratio of the velocities (x', y') at that point. He further showed that given $x(0)$ and the tangent $y'/x' = f(x)$ at every point along the abscissa he could reconstruct the curve by integration. Leibnitz at the same time considered curves as we do today, with the ordinate being a function of the abscissa: $y = f(x)$. He introduced the notation of the derivative that we are still using: $dy/dx = \lim_{dx \to 0} (f(x + dx) - f(x))/dx$. He also introduced the inverse operation and its notation the integration, as the area $F(x)$ under a curve $f(x)$, see Fig.1.2, so that $\int f(x)dx = F(x)$ and $dF/dx = f(x)$. Following their lead many other Mathematicians (Bernoulli, Cauchy, etc.) formalized and expanded on these seminal works, which the interested reader can find in many textbooks[4].

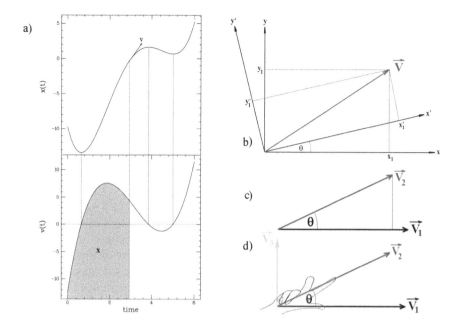

Figure 1.2 (a) The derivative of a curve $x(t)$ (the position versus time) is the tangent to the curve (i.e. the velocity). Conversely, the integral (gray area) of the velocity $v(t)$ yields the position $x(t)$. (b) Expression of the vector $\vec{v}$ in two coordinate systems that differ by a rotation by an angle θ. (c) The scalar (or dot) product of two vectors $a = \vec{v}_1 \cdot \vec{v}_2 = |\vec{v}_1||\vec{v}_2|\cos\theta$. (d) The vector (or cross) product of two vectors: $\vec{v}_3 = \vec{v}_1 \times \vec{v}_2$ is a vector perpendicular to the plane defined by the two vectors and whose magnitude is $|\vec{v}_3| = |\vec{v}_1||\vec{v}_2|\sin\theta$. Its direction is given by the right-hand rule: when the thumb point along $\vec{v}_1$, the index along $\vec{v}_2$, $\vec{v}_3$ points along the bent major.

Linear algebra is the branch of Mathematics concerned with linear equations such as: $a_{i1}x_1 + \cdots + a_{in}x_n = b_i$ ($i = 1, \ldots, n$), linear functions such as $f(x_1, \ldots, x_n) = a_1x_1 + \ldots + a_nx_n$ and their representations through matrices and vector spaces. Linear algebra is central to most fields of Mathematics, Science, Engineering, Economics, etc. It is at the core of Physics and particularly as we shall see Quantum Mechanics.

While the solution of linear equations with two and three unknown was familiar to the Egyptians and Chinese thousands of years ago, the introduction of matrices allowed to solve systematically n linear equations with n unknowns. Thus the following system of linear equations:

$$A_{11}x_1 + A_{12}x_2 + \cdots + A_{1n}x_n = b_1$$
$$A_{21}x_1 + A_{22}x_2 + \cdots + A_{2n}x_n = b_2$$
$$\vdots \qquad\qquad (1.1)$$
$$A_{n1}x_1 + A_{n2}x_2 + \cdots + A_{nn}x_n = b_n$$

which can be concisely written as: $\sum_{j=1}^{n} A_{ij}x_j = b_i$ or $\mathbf{A}x = b$. With $x = (x_1, x_2, ..., x_n)$, $b = (b_1, b_2, ..., b_n)$, vectors of size n and $\mathbf{A}$, an $n \times n$ matrix, the previous system of equations can be written as:

$$\mathbf{A}x \equiv \begin{pmatrix} A_{11} & \cdots & A_{1i} & \cdots & A_{1n} \\ A_{21} & \cdots & A_{2i} & \cdots & A_{2n} \\ & \vdots & & \vdots & \\ A_{n1} & \cdots & A_{ni} & \cdots & A_{nn} \end{pmatrix} \begin{pmatrix} x_1 \\ x_2 \\ \vdots \\ x_n \end{pmatrix} = \begin{pmatrix} b_1 \\ b_2 \\ \vdots \\ b_n \end{pmatrix} \equiv b$$

The solution of this matrix equation is formally similar to the solution of the regular equation $ax' = b$ (namely $x' = a^{-1}b \equiv b/a$): $x = \mathbf{A}^{-1}b$, where the matrix $\mathbf{A}^{-1}$ is the inverse of $\mathbf{A}$. Defining matrix multiplication by the identity: $\mathbf{C} = \mathbf{AB}$ (with $C_{ij} = \sum_k A_{ik}B_{kj}$), the inverse satisfies : $\mathbf{A}^{-1}\mathbf{A} = \mathbf{I}$ ($\mathbf{I}$ is the unity matrix which has ones on its diagonal and zeros elsewhere). For a square matrix (i.e. one solving n equations with n unknowns) there is an inverse and thus a solution if the determinant of $\mathbf{A}$ is non-zero. The computation of the product of two matrices, the inverse of a matrix or its determinant scales as n^3 and is a straightforward but tedious task, see Appendix A.2. However with the advent of computers, these tasks can now be performed with fast and efficient algorithms[5]. As a consequence linear algebra has moved to a central place in computation[6]. It is used to solve linear and differential equations (after their discretization), to compute eigenvalues and eigenvectors in problems ranging from Quantum Mechanics to electromagnetism and heat transfer, to compress images and information, to find optimal solutions to various problems, etc.

As mentioned earlier, most problems in Science are not exactly soluble. Besides perturbation methods about the soluble problem, mathematical methods have also been developed to analyze the qualitative behavior of these solutions (dynamical system theory, fixed points, attractors, etc.). Furthermore the differential equations that describe the studied problem can be discretized (in space and time) to yield a system of coupled linear (or non-linear) equations that can be simulated (or solved) on a computer[5]. Central issues in that area are the stability of the solutions and their computability (how many steps it takes to compute them and how that number scales with the system size). These questions are active areas of research in Science and Technology, but are beyond the scope of this book and will not be treated here.

Vector spaces. As the interest in vector quantities (forces, velocities, etc.) grew the study of vectors and their transformation using matrices was formalised during the 19th century by Caylet, Grassman, Gibbs, etc.. Thus as shown in Fig.1.2, the transformation of a vector $\vec{v} = x_1\hat{x} + y_1\hat{y} \equiv (x_1, y_1)$ from one coordinate system[2] into another rotated by an angle θ: $\vec{v} = x_1'\hat{x}' + y_1'\hat{y}' \equiv (x_1', y_1')$ can be written as a product between the matrix of rotation by an angle θ and the original vector:

$$\begin{pmatrix} x_1' \\ y_1' \end{pmatrix} = \begin{pmatrix} \cos\theta & \sin\theta \\ -\sin\theta & \cos\theta \end{pmatrix} \begin{pmatrix} x_1 \\ y_1 \end{pmatrix} \qquad (1.2)$$

---

[2]the $\hat{}$ in $\hat{x}, \hat{y}$ means they are unit vectors in the x, y directions

A vector in three dimensions (position, velocity, force, torque, etc.) is an object characterised by two quantities: its magnitude $v \equiv |\vec{v}|$ and its direction in 3D space ($\hat{v}$). A vector (e.g. the velocity $\vec{v}$) can be multiplied by a scalar (e.g. the mass) to yield another vector (e.g. the momentum $\vec{p} = m\vec{v}$). The product of two vectors can either be a scalar quantity (e.g. the power $P = \vec{v} \cdot \vec{F}$) or a vector one (e.g. the torque $\vec{\Gamma} = \vec{r} \times \vec{F}$).

The inner (or scalar) product is defined as follows:

$$\vec{v}_1 \cdot \vec{v}_2 = v_{1,x}v_{2,x} + v_{1,y}v_{2,y} + v_{1,z}v_{2,z}$$

Thus the magnitude (or norm) of a vector $v \equiv |\vec{v}| = \sqrt{\vec{v} \cdot \vec{v}}$. The inner product yields the magnitude of the projection of one vector onto another: $\vec{v}_1 \cdot \vec{v}_2 = v_1 v_2 \cos\theta$ (where θ is the angle between the two vectors).

The vector product of $\vec{v}_1$ and $\vec{v}_2$ is a vector $\vec{v}_3$ perpendicular to the plane defined by v_1, v_2 which magnitude is $v_3 = v_1 v_2 \sin\theta$. Its direction is given by the right-hand rule: when the thumb is aligned with $\vec{v}_1$, the index with $\vec{v}_2$ the bent major points to the direction of $\vec{v}_3$, see Fig.1.2. In cartesian coordinates $(x, y, z)^3$: $\hat{x} \times \hat{y} = \hat{z}$; $\hat{y} \times \hat{z} = \hat{x}$; $\hat{z} \times \hat{x} = \hat{y}$. If the vectors are defined in these coordinates $\vec{v}_i = v_{i,x}\hat{x} + v_{i,y}\hat{y} + v_{i,z}\hat{z}$ then the vector product can be written as:

$$\vec{v}_3 = \vec{v}_1 \times \vec{v}_2 = \begin{vmatrix} \hat{x} & \hat{y} & \hat{z} \\ v_{1,x} & v_{1,y} & v_{1,z} \\ v_{2,x} & v_{2,y} & v_{2,z} \end{vmatrix}$$

$$\equiv (v_{1,y}v_{2,z} - v_{1,z}v_{2,y})\,\hat{x} + (v_{1,z}v_{2,x} - v_{1,x}v_{2,z})\,\hat{y} + (v_{1,x}v_{2,y} - v_{1,y}v_{2,x})\,\hat{z} \qquad (1.3)$$

While the concept of a vector arose initially in the context of physical variables in 3D space, it was quickly generalized to vector-spaces in arbitrary dimensions, endowed with addition, scalar multiplication and an inner product: $\vec{v}_1 \cdot \vec{v}_2 = \sum_i v_{1,i}v_{2,i}$, with $\vec{v} = \sum v_i \hat{i}$ (where v_i are scalars and $\hat{i}$ are orthonormal unit vectors $\hat{i} \cdot \hat{j} = \delta_{ij}$). Hilbert generalized the concept to spaces of functions (vector fields): $\psi(x) = \sum_i a_i \eta_i(x)$ endowed with an inner product: $\langle \psi, \phi \rangle = \int dx \psi^*(x)\phi(x)$. Hilbert spaces play a central role in Quantum Mechanics.

Vector calculus, which is the branch of Mathematics dealing with differentiation and integration of vector fields[7], arose from the study of the dynamics and variation of physical vector fields (electric fields, fluid flows, etc.). Considering the operator $\vec{\nabla} = (\partial_x, \partial_y, \partial_z)$ as a vector operator, its application on a scalar field generates a vector field: the gradient of the scalar field that points along the direction of steepest variation of the field. For example the negative gradient of the potential $\Phi(x, y, z)$ yields the force $\vec{F}(x, y, z)$:

$$\vec{F} = -\vec{\nabla}\Phi(x, y, z) \equiv -\partial_x \Phi \hat{x} - \partial_y \Phi \hat{y} - \partial_z \Phi \hat{z}$$

---

^{3}In other orthogonal coordinate systems, see Appendix A.5 similar rules hold. Thus in cylindrical coordinates (ρ, ϕ, z): $\hat{\rho} \times \hat{\phi} = \hat{z}$; $\hat{\phi} \times \hat{z} = \hat{\rho}$; $\hat{z} \times \hat{\rho} = \hat{\phi}$. And in spherical coordinates (r, θ, ϕ): $\hat{r} \times \hat{\theta} = \hat{\phi}$; $\hat{\theta} \times \hat{\phi} = \hat{r}$; $\hat{\phi} \times \hat{r} = \hat{\theta}$.

(where ∂_x is the derivative of $\Phi(x, y, z)$ with respect to x, etc.), which drives the physical system to a potential minima: $\vec{F} = -\vec{\nabla}\Phi = 0$.

One can similarly define the equivalent of the scalar product, the divergence ($\vec{\nabla}\cdot$) of a vector field (such as the flow $\vec{J}$) which yields a scalar field:

$$\vec{\nabla} \cdot \vec{J} = \partial_x J_x + \partial_y J_y + \partial_z J_z$$

The divergence of the flow is usually associated to the time variation of some conserved quantity (e.g. charge or mass) advected by the flow.

The rotational (or curl, $\vec{\nabla}\times$) operator is the equivalent of the vector product:

$$\vec{\nabla} \times \vec{J} = \begin{vmatrix} \hat{x} & \hat{y} & \hat{z} \\ \partial_x & \partial_y & \partial_z \\ J_x & J_y & J_z \end{vmatrix} \equiv (\partial_y J_z - \partial_z J_y)\hat{x} + (\partial_z J_x - \partial_x J_z)\hat{y} + (\partial_x J_y - \partial_y J_x)\hat{z} \quad (1.4)$$

The curl operator measures how much of the flow is rotational (swirling).

One may also consider the integration of vector fields. As we shall see below, in physical applications, the most important integrals are line integrals and surface integrals. A line integral is a scalar obtained by integrating a vector field along a line, for example the work W performed by a force $\vec{F}$ along a path $\vec{r}$ between an initial point r_i and a final one r_f: $W = \int_i^f \vec{F} \cdot d\vec{r}'$. If the force is obtained as the gradient of a potential field, $\vec{F} = -\vec{\nabla}\Phi$ (as is the case for gravitational and electrical forces, see below) then the work performed by the field is the difference between the initial and final potentials[4]:

$$W = \int_i^f \vec{F} \cdot d\vec{r}' = -\int_{r_i}^{r_f} \vec{\nabla}\Phi \cdot d\vec{r}' = \Phi(r_i) - \Phi(r_f) \quad (1.5)$$

If the path closes on itself then Stokes' theorem states:

$$\oint \vec{F} \cdot d\vec{r}' = \int \vec{\nabla} \times \vec{F} \cdot d\vec{S}' \quad (1.6)$$

If $\vec{F}$ is the gradient of a potential one can easily verify that $\vec{\nabla} \times \vec{\nabla}\Phi = 0$, hence no work is performed by going along a closed curve in a potential field.

The second type of integrals often encountered in physics is the surface integral, a scalar obtained by computing the flow $\vec{J}$, passing through a surface: $\int \vec{J} \cdot d\vec{S}'$ (where $d\vec{S}'$ points along the normal to the surface). For a closed surface the divergence theorem states:

$$\oint \vec{J} \cdot d\vec{S}' = \int \vec{\nabla} \cdot \vec{J} d^3 r' \quad (1.7)$$

---

[4]Notice that many textbooks use single, double or triple integral signs when integrating respectively over a line ($d\vec{r}'$), a surface ($d^2\vec{r}$ or $d\vec{S}'$) or a volume ($d^3 r'$ or dV'). To save space and lighten the notation, I will be using a single integral sign in all these cases and use r (or r' if there is an ambiguity) as the integration variable.

Notice that in Eq.1.7 and Eq.1.6, the operator ($\vec{\nabla}$) acts on the integration variable r' (to emphasize that point it is sometimes written as $\vec{\nabla}'$). If the operator acts on a variable that is not integrated it can often be taken out of the integral. Thus for example:

$$\int d^3r'\, \rho(r')\frac{\vec{r}-\vec{r}'}{|\vec{r}-\vec{r}'|^3} = \int d^3r'\, \rho(r')\vec{\nabla}'\frac{1}{|\vec{r}-\vec{r}'|}$$

$$\text{but also :} \quad = -\int d^3r'\, \vec{\nabla}\frac{\rho(r')}{|\vec{r}-\vec{r}'|} = -\vec{\nabla}\int d^3r'\, \frac{\rho(r')}{|\vec{r}-\vec{r}'|}$$

More details can be found in Appendix A.3 and in the many textbooks on calculus[3, 7, 4].

The theory of probability[8, 9] arose from the investigation of games of chance (rolling dice, throwing coins or playing cards) by Cardano, Fermat, Pascal and Huygens. It was however J.Bernoulli and A.de Moivre who put the theory of probability on a sound mathematical basis. In particular Bernoulli was the first to use what would later be called the principle of indifference (or insufficient reason), namely that if we do not have a-priori information on the outcomes of a game of chance the best guess is to assign equal probabilities to all outcomes. He could then show that the average value of a large number of outcomes (e.g. the number of heads in a large number N of tosses of a fair coin) is equal to the expected value (i.e. $N/2$). This is known as the Law of large numbers and the principle of insufficient reason has been reformulated as the maximization of entropy (or missing information, see chapter 5). Both sit at the core of Statistical Mechanics.

2 Classical Mechanics

Physics deals mostly with the prediction of the trajectories of particles (atoms, bullets, stars, etc.). One can therefore date the beginning of modern Science (most of which can be derived as we shall see from basic physical laws) with the publication in 1687 of the first treatise addressing the problem of the trajectories of moving objects: Isaac Newton's "Philosophiae Naturalis Principia Mathematica" (Mathematical Principles of Natural Philosophy[10, 11]).

2.1 ◆ NEWTON'S LAWS OF MOTION

Newton's first law of motion. Newton decided to describe the motion of bodies[12] with respect to certain coordinate systems, known as inertial frames of reference: coordinate systems that are either at rest or moving at constant velocity. In such a frame of reference if the vector sum of the forces acting on a body is null, Newton's first law of motion states that the body will remain at rest or move at constant velocity (an idea formulated earlier by Galileo). In our days of space-probes reaching the confines of the solar system (by their own inertia) we know that law to be true, but in Newton's time the common wisdom (dating back to Aristotle) was that bodies with no force acting on them would come to rest (motion required a mover). Moreover according to Aristotle the velocity of a body would be proportional to the force acting on it, which is indeed the case for a body moving in a viscous medium due to viscous drag (for example a boat in water), but not for a body moving in free space (the famous apple falling from its tree).

Newton's second law in contrast to Aristotle states that the rate of change of a body's momentum $\vec{p} = m\vec{v}$ (or its acceleration $\vec{a} = d\vec{v}/dt$) is proportional to the force $\vec{F}$ acting on that body:

$$\frac{d\vec{p}}{dt} = m\vec{a} = \vec{F} \qquad (2.1)$$

Where m is the inertial mass of the body[1]. In honor of Newton the unit of force in the MKS (meter-kilogram-sec) system is the Newtons (N): a mass of 1kg accelerated

---

[1]Notice that when the mass m is changing with time (as in a rocket which gains momentum by ejecting gases at high speed), one needs to consider the total (constant) mass of the system in which case Eq.2.1 becomes: $d\vec{p}/dt = md\vec{v}/dt + \vec{u}_{rel}dm/dt = \vec{F}$ where $\vec{u}_{rel}$ is the relative velocity of the escaping mass. In the case of a rocket in space $\vec{F} = 0$ and u_{rel} is the exhaust velocity of the gases which is opposite to the rocket's acceleration, hence: $mdv = -u_{rel}dm$. If the initial velocity $\vec{v}_i = 0$, the final velocity of the rocket is: $\vec{v}_f = \vec{u}_{rel}\log m_{init}/m_{final}$, an equation known as Tsiolkovsky rocket equation. Notice that the final velocity depends essentially on the exhaust velocity of the gases. Hence ionic propulsion with exhaust velocities ($\sim$ 50km/s) about 10 times larger than chemical propulsion (used in most rockets) is a better choice in space (where its much lower thrust is less of a problem).

DOI: 10.1201/9781003218999-2

at a rate of 1 m/sec^2 experiences a force of 1 N. In the CGS (centimeter-gram-sec) system the unit of force is the dyne (1 dyne $= 10^{-5}$ N).

In absence of an external force, momentum is conserved and as stated by Newton's first law the body moves at constant velocity. An immediate application of Newton's second law was the computation of the trajectories of projectiles (bullets, arrows, cannon balls, etc.) launched (from position $x_0 = 0; z_0 = 0$) at an angle α to the horizon with an initial velocity $\vec{v}_0$. The force acting on these bodies is their weight w which pulls them downward and thus $\vec{a} = -(w/m)\hat{z} = -g\hat{z}$, (where g is the Earth gravitational acceleration). Experiments by Galileo (and many others since) have shown that g is a constant, independent of the mass, composition, temperature, etc. of the body[2].

The equations of motion of the projectile (in the x, z plane) are then:

$$a_x = \frac{dv_x}{dt} = \frac{d^2x}{dt^2} = 0$$

$$a_z = \frac{dv_z}{dt} = \frac{d^2z}{dt^2} = -g$$

which can be integrated to yield:

$$x = v_x t = v_0 t \cos\alpha \qquad\qquad\qquad (2.2)$$

$$z = v_z t - gt^2/2 = v_0 t \sin\alpha - gt^2/2 = x\tan\alpha - \frac{g}{2v_0^2 \cos^2\alpha}x^2$$

The projectile's trajectory is a parabola (as previously observed by Galileo), see Fig.2.1(a) and its range is: $x = v_0^2 \sin 2\alpha/g$. This was a great achievement and —in that age of wars —with consequential applications: the field of ballistics had been created.

Newton's second law also explained the motion of pendulums (another system studied by Galileo). For a pendulum consisting of a weight w at the end of a rope of length l forming an angle α to the vertical, see Fig.2.1(b) the combined action of the pendulum weight $w = mg$ and the tension in the rope T results in a force $F = -w\sin\alpha$ perpendicular to the rope:

$$ma = ml\frac{d^2\alpha}{dt^2} = F = -w\sin\alpha \approx w\alpha$$

the approximation being valid at small angles. The solution of that equation ($\alpha(t) = \alpha_0 \cos\omega t$) corresponds to angular oscillations of amplitude α_0 at a frequency: $\omega =$

---

[2]This is not obvious and in fact it stood in contradiction with the common belief then, which dated to the Greek philosophers: following Aristotle they thought that heavier objects would fall faster. Galileo's results however confirmed an intuition of the 1st century BC epicurean philosopher, Lucretius who in "de Rerum Natura"[13] claimed that in vacuum all bodies would fall at the same speed. Galileo's observation of the universality of the gravitational acceleration is at the core of Einstein's general theory of relativity (theory of gravitation) which asserts that it is the structure of space itself which is modified by the gravitational field, thus naturally accounting for the composition-independent motion of bodies in such fields.

$\sqrt{w/ml} = \sqrt{g/l}$. The frequency ω depends only on the length of the pendulum and is unaltered if the motion is damped (due to friction with air for example) a property that was used for the manufacture of clocks. Newton similarly considered the motion of a body of mass m under the force $\vec{F}$ of a spring, see Fig.2.1(c), which Hooke had shown a few years earlier to be proportional to the spring deformation $\vec{x}$: $\vec{F} = -k\vec{x}$ (k is known as the spring constant). The result, as for the pendulum, are oscillations at a frequency $\omega = \sqrt{k/m}$.

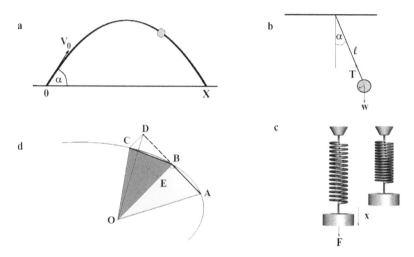

Figure 2.1 (a) A cannon ball fired with a velocity V_0 at an angle α to the horizontal reaches a range $x = v_0 t \sin 2\alpha$. The maximal range is thus obtained when $\alpha = 45°$. (b) A pendulum of length l bearing a weight $w = mg$ oscillates at a frequency $\omega = \sqrt{g/l}$. (c) A spring bearing a mass m and pulled out of equilibrium by a force $F = kx$ will, upon release, oscillate with a frequency $\omega = \sqrt{k/m}$. (d) Newton's proof of Kepler second law: the line joining a planet and the Sun sweeps out equal areas during equal time intervals. Let a body move from point A to point B within a time t. At B a force perpendicular to the trajectory deviates it so that it reaches point C after a time t. Had it not been deviated it will have reached point D. The area of the triangles OBA and OBD are equal since AB=BD and OB is a common base. Similarly the triangle OBD and OBC have equal areas, since they share a common base OB and have similar heights BD=CE (= distance traveled without the perturbation at B, BE is the perpendicular distance traveled during a time t due to the perturbation at B). Hence the surface of triangles OBA and OBC are equal, which is Kepler's second law. When the time interval $t \to 0$ the orbit is continuous and one recovers the result from conservation of angular momentum: $L = mr^2\omega = const$, see text.

Notice that a force is exerted on a body if the direction of its velocity changes (but not its magnitude), as in circular motion of radius r at fixed frequency ($\dot{\theta} = \omega$): $\vec{v} = -\omega r \sin\theta \hat{x} + \omega r \cos\theta \hat{y}$:

$$\vec{F} = md\vec{v}/dt = -m\omega\dot{\theta}r\cos\theta\hat{x} - m\omega\dot{\theta}r\sin\theta\hat{y} = -m\omega^2\vec{r}$$

Where $\vec{r}$ is the vector along the radius ($\vec{r} = \vec{x} + \vec{y} = r\cos\theta\hat{x} + r\sin\theta\hat{y}$). The force $\vec{F}$ is known as the centripetal force and it acts towards the center of the circular motion. When swirling a body at the end of a rope that force is provided by the tension in the rope.

Newton also assumed that a rotating body would keep on rotating unless a force would be exerted to stop it. It was much later understood that torque $\vec{\Gamma} = \vec{r} \times \vec{F}$ plays the same role for angular momentum $\vec{L} = \vec{r} \times \vec{p}$ as force does for momentum:

$$\frac{d\vec{L}}{dt} = \vec{\Gamma} \tag{2.3}$$

For a rotating body: $\vec{L} = I\vec{\omega}$ where I is the moment of inertia about the rotational axis $I = \sum_i m_i r_i^2 = \int \rho(r) r^2 d^3 r$ (r is the distance to the rotational axis and $\rho(r)$ the local density). By convention, the angular velocity vector $\vec{\omega}$ is directed along the rotational axis according to the right-hand rule: when the fingers point in the direction of rotation, the thumb points in the direction of $\vec{\omega}$.

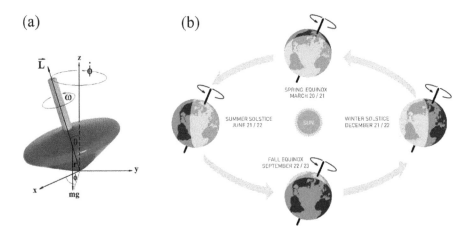

Figure 2.2 (a) A top spins with angular velocity ω at an angle θ to the vertical axis. As a result of the torque applied by its weight ($w = -mg\hat{z}$), the angular momentum $\vec{L} = I\omega\hat{r}$ rotates around the vertical axis at an angular velocity: $\dot{\phi} = mgr/I\omega$. (b) The Earth spins about its axis at an angle $\approx 23.4°$ to the plane of its rotation about the Sun (the ecliptic). Since the angular momentum is a fixed vector, it sometimes point away from the Sun (Winter solstice) and six months later towards the Sun (Summer solstice). The conservation the Earth angular momentum thus results in yearly variation of the Sun's illumination in the Northern and Southern hemispheres which is the cause of the seasons (adapted from ref.[14]).

This law is verified every time one spins a top, see Fig.2.2(a): The torque on the top is due to its down pulling weight ($w = mg$): $\vec{\Gamma} = -\vec{r} \times w\hat{z}$, which leads to a precession of the top momentum $\vec{L} = I\omega\hat{r}$ at an angular velocity[3] $\dot{\phi} = mgr/I\omega$.

In absence of an external torque the angular momentum is conserved. This is used by skaters to increase their spin by bringing their arms and legs closer to their rotating bodies (reducing their moment of inertia I and therefore increasing ω). Conservation of angular momentum is also the reason for the existence of seasons on Earth. The Earth spins on its axis, hence its angular momentum points to a fixed direction in space[4] while the Earth is orbiting the Sun, see Fig.2.2(b). As a result when the rotational axis point away from the Sun it is winter in the Northern hemisphere and Summer in the Southern one and vice-versa six months later when the axis points towards the Sun.

The orbital angular momentum is also conserved in an attractive central force field, where the force is directed along the radial distance between the center of the field and an orbiting body: $\vec{F}$ is parallel to $\vec{r}$ (written as $\vec{F} \parallel \vec{r}$) such as for a planet orbiting the Sun, see below. In that case $\vec{\Gamma} = \vec{r} \times \vec{F} = 0$ and thus $\vec{L} = mr^2\vec{\omega} = \text{const}$: in equal time intervals Δt the line between the Sun and a planet sweeps equal areas $r^2\Delta\theta$, see Fig.2.1(d), an observation first due to Kepler and known as his second law of planetary motion.

Newton's third law of motion states that if a body A exerts a force $\vec{F}_{AB}$ on a second body B, then this second body reacts by exerting an opposite force $\vec{F}_{BA} = -\vec{F}_{AB}$ on the first one. This makes sense since otherwise one could pull himself out of water by pulling on his hair (like the legendary baron Münchausen[5]). One direct consequence of this law of action and reaction is that when two bodies interact their total momentum is conserved:

$$0 = \vec{F}_{AB} + \vec{F}_{BA} = m_B d\vec{v}_B/dt + m_A d\vec{v}_A/dt = d(m_B\vec{v}_B + m_A\vec{v}_A)/dt \qquad (2.4)$$

Hence: $m_B\vec{v}_B + m_A\vec{v}_A = \text{const}$. This law is used and verified by anyone playing billiard or shooting a bullet and feeling the recoil of the rifle.

A generalization of Newton's third law to rotational motion states that if a body A exerts a torque $\vec{\Gamma}_{AB}$ on a body B than this second body reacts by exerting an opposite torque $\vec{\Gamma}_{BA} = -\vec{\Gamma}_{AB}$. Hence by virtue of Eq.2.3, the total angular momentum $\vec{L}_A + \vec{L}_B$ is conserved:

$$0 = \vec{\Gamma}_{AB} + \vec{\Gamma}_{BA} = d(\vec{L}_A + \vec{L}_B)/dt \qquad (2.5)$$

---

[3]Using: $d\vec{L}/dt = -I\omega\dot{\phi}(\sin\theta\sin\phi\hat{x} - \sin\theta\cos\phi\hat{y})$ and $\vec{\Gamma} = -wr(\sin\theta\sin\phi\hat{x} - \sin\theta\cos\phi\hat{y})$

[4]In fact because the earth is not exactly spherical, as for a top the gravitational pull of the Sun and the Moon cause it to precess slightly about its fixed direction, a phenomenon known as the precession of the equinoxes which was explained by Newton.

[5]The real one was a contemporary of Newton.

2.2 ◆ NEWTON'S LAW OF GRAVITATION

Besides inventing calculus and formulating his famous Laws of motion, Newton perhaps greatest contribution to Science was the discovery of the Law of Gravitation[15]. That discovery more than anything else launched the program of modern Science, which gave rise as we shall see to Coulomb's law of electrostatic interactions (modeled on Newton's law of gravitation), to Maxwell's equations, Einstein particular and general theory of relativity and is the main incentive behind the attempts to unify gravitation and quantum mechanics.

As a result of the law of action and reaction, Newton reasoned that the gravitational force between two bodies had to depend on the product of their masses. Newton thus assumed that the "force between two point masses is directly proportional to the product of their masses" and (inspired by Hooke) "inversely proportional to the square of their separation distance". The force F_1 on an object with mass m_1 at position $\vec{r}_1$ exerted by an object with mass m_2 at position $\vec{r}_2$ is "always attractive and acts along the line joining them" ($\hat{r}_{12} = (\vec{r}_1 - \vec{r}_2)/|\vec{r}_1 - \vec{r}_2|$):

$$\vec{F}_1 = -G\frac{m_1 m_2}{|\vec{r}_1 - \vec{r}_2|^2}\hat{r}_{12} \tag{2.6}$$

where $G = 6.67 \ 10^{-11}$ N m^2/kg^2 is known as the gravitational constant. On Earth (mass $M_E = 6 \ 10^{24}$ kg and radius $r \equiv R_E = 6371$ km) an object of mass m experiences a gravitational attraction:

$$\vec{F} = -\frac{GM_E}{R_E^2}m\hat{r} = -mg\hat{r}$$

where the gravitational acceleration on Earth is: $g = GM_E/R_E^2 = 9.8\text{m/sec}^2$.

At that time Newton's theory was revolutionary: how could a force act between bodies that are physically separated in space? What are those spooky actions at a distance? People were familiar with pushing or pulling bodies to which they were connected, not to forces propagating in empty space. Yet, the explanatory power of Newton's law of gravity was so vast and precise that people became quickly convinced of its exactitude. One of its first achievements was to account for the relation between the orbital period T of the planets around the Sun and their distance to it R. Equating the centripetal force exerted by the Sun (of mass M_S) on a planet (of mass m) with the gravitational force led to:

$$m\omega^2 R = m(\frac{2\pi}{T})^2 R = G\frac{mM_S}{R^2} \tag{2.7}$$

from which one deduces that $T \sim R^{3/2}$, a relation that was discovered earlier by Kepler and is known as his third law of planetary motion.

Newton's law of gravity also provided an explanation for the solar eclipse, that terrified generations of humans, as a projection of the Moon's shadow on Earth. Since the plane of rotation of the Moon makes an angle of about 5° with the ecliptic (the orbital plane of the Earth), an eclipse doesn't occur at every new Moon. For an eclipse to occur the Moon must cross the line joining the Sun and the Earth. This understanding and Newton's laws allowed Edmund Halley to be the first to predict and observe the solar eclipse of May 3, 1715. It is the same Halley that also used Newton's laws to predict the return of the comet (bearing his name) in 1759 (14 years after his death). These two precise predictions definitely established the validity of Newton's ideas.

Another success of Newton's theory of gravity was the explanation of the tides as a result of the differential attraction of the Moon (and the Sun) on the Earth's water mass. Since the distance between the centers of the Earth and the Moon ($R_{EM} \sim$ 384,000 km) is much larger than the radius of the Earth, one can compute the force (or pressure) differential on a water mass (m) closest and farthest from the Moon (of mass $M_M = 0.0123 \, M_E$) when compared with a similar mass at the Earth center:

$$\delta F = -[G\frac{mM_M}{(R_{EM} \pm R_E)^2} - G\frac{mM_M}{R_{EM}^2}]\hat{r} \approx \pm \frac{2GmM_M R_E}{R_{EM}^3}\hat{r} \qquad (2.8)$$

where $\hat{r}$ is the unit vector from the Moon's center to the Earth's center. Hence the closest point to the Moon experiences a force differential directed towards the Moon ($-\hat{r}$) while the point further away experiences a force differential away from the Moon ($+\hat{r}$). Consequently, at both these antipodal points the water mass bulges slightly from a spherical shape (which it would adopt in absence of these perturbations). As one rotates with the Earth, one encounters a higher than usual water mass (a tide) twice a day. The differential force responsible for the tides is tiny when compared to the Earth gravitational pull ($\delta F/mg \approx 1.1 \, 10^{-7}$ for the pull of the Moon and $0.52 \, 10^{-7}$ for that of the Sun), yet it is enough to explain the amplitude of the tides (of order $(\delta F/mg)R_E \sim 60cm$), though much variations exist due to the coastline that may channel the tides.

In 1798, a hundred years after Newton exposed his theory of gravitation, Cavendish tested it in an experiment that is still the main technique for measuring the gravitational constant G. In this experiment Cavendish used a torsional pendulum, which consists of two masses at the extremity of a rod held at its center by a thin wire, see Fig.2.3. When two large masses M are brought in the vicinity of these smaller masses the wire is twisted by a measurable angle θ. Knowing the torsional modulus k of the wire (from its natural oscillation frequency $\omega = \sqrt{k/I}$) one can deduce the force acting between the masses. Cavendish could thus test quantitatively the inverse square law and mass dependence postulated by Newton, Eq.2.6, and deduce G (with a precision of 1%!).

Perhaps the greatest success of Newtonian gravity (and calculus) came many years later when the French astronomer and mathematician Le Verrier predicted in 1846 the existence and position of a new planet (Neptune) by mathematically analyzing the perturbations to the orbit of another (Uranus). Notice that this was done analytically with pen and paper, long before the advent of computers that routinely direct probes to the various planets of the solar system using Newton's laws! The planet Neptune was duly discovered by Galle at the predicted position immediately upon reception of le Verrier's note by the Berlin Observatory.

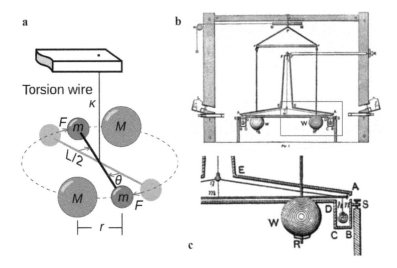

Figure 2.3 (a) Schematic drawing of the torsional balance used by Cavendish to test Newton's theory of gravity. The force between two fixed spheres of mass M and two equal spheres
of mass m exerts a torque Γ on the pendulum which results in its rotation by an angle θ. (b)
Original drawings of Cavendish experiment: two small masses $m = 0.73$ kg are positioned
at the extremity of a rod of length L held at its center by a thin wire. Two large lead balls
($M = 158kg$ labeled W) can be brought in their vicinity and the whole system is held in a
closed box (to minimize external perturbations, such as air currents ...). (c) Enlargement of
the region inside the red rectangle in (b): as a result of the gravitational attraction between the
masses, when the large masses are approached the pendulum twist by an angle θ. Knowing
the torsional constant k of the wire (from its frequency of oscillations ($\omega = \sqrt{k/I}$), where I is
the moment of inertia of the pendulum ($I \approx mL^2/2$)), the torque exerted on the mass can be
computed: $\Gamma = k\theta$.

2.2.1   GRAVITATIONAL POTENTIAL AND ENERGY CONSERVATION

The gravitational force between two bodies is attractive and directed along the line
linking the center of mass of the two bodies. The force $\vec{F}_o$ applied on a body of mass
m at position $\vec{r}_o$ by other bodies of masses m_i at positions $\vec{r}_i$ is the vector sum of the
individual forces between m and the various bodies:

$$\vec{F}_o = -\sum_i \frac{Gmm_i}{|\vec{r}_o - \vec{r}_i|^2}\hat{r}_{oi} \tag{2.9}$$

where the unit vector $\hat{r}_{oi} = (\vec{r}_o - \vec{r}_i)/|\vec{r}_o - \vec{r}_i|$. On Earth the gravitational attraction
between bodies is completely negligible (except obviously the terrestrial one), while
in space the attraction between planets and stars and their subsequent motion can
sometimes be very difficult to predict, since they all move under the influence of the
others. In fact, the motion of 3 bodies (planets or stars) of similar mass and initially

at similar distance from each other is in general impossible to predict accurately over a long time (the motion can be chaotic!). However if one star (or planet) has a mass much larger than all other planets, then the trajectory of these planets is essentially determined by their attraction to that massive star (as is the case for the solar system or the Moons of Jupiter and Saturn). The effects of the other planets can be treated as perturbations to the planet's orbit set by its interaction with the central star. This is how Le Verrier was able to predict the existence of Neptune on the basis of perturbations to the orbit of Uranus. This analysis is simplified by the fact that the gravitational force (a vector) can be derived from a scalar field, the gravitational potential:

$$\Phi_G(\vec{r}_o) = -\sum_i \frac{Gm_i}{|\vec{r}_o - \vec{r}_i|} \leq 0 \qquad (2.10)$$

so that the force exerted on mass m at $\vec{r}_o$ is $\vec{F}_o = -m\vec{\nabla}_o\Phi_G$, where the gradient $\vec{\nabla}_o$ is taken with respect to $\vec{r}_o$. Since the gradient of Φ_G points towards the direction of increase of the potential, the gravitational force drives an object to the lowest point of the gravitational potential. On Earth this is the surface of the planet with $\Phi_G(R_E) = -GM_E/R_E$. Newton's second law for a particle at position $\vec{r}_o$ (with velocity $\vec{v}_o$) can be written as[6]:

$$m\frac{d\vec{v}_o}{dt} = -m\vec{\nabla}_o\Phi_G$$

Scalar multiplication of that equation by $d\vec{r}_o$ yields: $m\vec{v}_o \cdot d\vec{v}_o + m\vec{\nabla}_o\Phi_G \cdot d\vec{r}_o = 0$, which upon integration (see Eq.1.5) yields the following constant of motion:

$$\frac{m\vec{v}_o^2}{2} + m\Phi_G(\vec{r}_o) = \text{const} \qquad (2.11)$$

This is one of the most important results in physics: the first term $(m\vec{v}_o^2/2)$ is known as the kinetic energy (it is associated with the motion of the body), the second term $(m\Phi_G(\vec{r}_o))$ is known as the potential energy (it is associated with the position of the body in space). In the MKS system the units of energy are Joules ($1J = 1$ kg m^2/s^2), in the CGS system they are ergs: 1 erg $= 10^{-7}$J). Eq.2.11 is known as the law of energy conservation: a body acquires kinetic energy when it looses potential energy and vice-versa. This law was first discovered by Emilie du Chatelet around 1740 inspired by experiments of Willem Gravesand, who observed that brass balls dropped from various heights h made indentations in clay that were proportional to the square of their final velocity. In that case Eq. 2.11 can be simplified using the approximation, $h \ll R_E$:

$$\Phi_G(R_E + h) \approx \Phi_G(R_E) + \partial_{r_o}\Phi_G\bigg|_{R_E} h = \Phi_G(R_E) + gh$$

---

[6]Notice that if the inertial mass (the one entering into Newton's second law, Eq.2.1) is equal (or proportional to) the gravitational mass (the one entering into Eq.2.6) then all bodies fall with the same acceleration as discovered by Galileo.

to yield: $m\vec{v}_o^2/2 + mgh = $ const. Hence balls dropped from a height h have a final kinetic energy $mv_o^2/2$ equal to their initial potential energy mgh.

Thus from Newton's laws three constants of motion (conservation laws) can be derived : total energy, total momentum and total angular momentum. A beautiful theorem due to Emmy Noether (1915) relates these conservation laws to the symmetry of the physical system under time and space translations and rotations (in the absence of friction/dissipation): if the interactions don't depend explicitly on time and vary only with the relative distances and orientations between the particles then energy, linear momentum and angular momentum are conserved[12, 16, 15]. That is often the case for physical systems at the very small scale (atoms, molecules) or very large scale (planets, stars) which is one of the reasons that make these systems often predictable. In physical systems at our scale however friction and viscous dissipation invalidate the conditions for Noether's theorem to hold[7]: kinetic energy is lost to heat and bodies do come to rest as argued by Aristotle.

If an external force $\vec{F}_{ex}$ is acting on the system, then $m\vec{v} \cdot d\vec{v} + m\vec{\nabla}\Phi_G \cdot d\vec{r} = \vec{F}_{ex} \cdot d\vec{r}$ and Eq.2.11 becomes:

$$\frac{m}{2}(v_f^2 - v_i^2) + m[\Phi_G(r_f) - \Phi_G(r_i)] = \int_i^f \vec{F}_{ex} \cdot d\vec{r} \equiv W \qquad (2.12)$$

The change in total energy (kinetic + potential) equals the integral, known as the work W, of the component of the external force along the trajectory from the initial point at $\vec{r}_i$ to the final one at $\vec{r}_f$. For rotational motion in presence of an external torque $\vec{\Gamma}_{ex}$ the work is:

$$\int_i^f I\vec{\omega} \cdot d\vec{\omega} = \frac{I}{2}(\omega_f^2 - \omega_i^2) = \int \vec{\Gamma}_{ex} \cdot \vec{\omega}d\theta \equiv W \qquad (2.13)$$

In that case the work is the angular integral of the external torque component along the circular trajectory.

In conclusion, by the end of the eighteen century, Newtonian mechanics was accepted as a very powerful and successful way of studying Nature: it explained many phenomena and made amazingly precise predictions. It has set ever since the standard for a valid scientific theory. In the next chapters we will study those extensions of Newtonian Science: Electromagnetism, Quantum Mechanics and Statistical Mechanics. Together they set the foundations of our present day understanding of much of Nature and Technology.

---

[7] In orbiting planets the dissipation due to tidal waves leads to a decrease of the spining velocity which ultimately results in the planet presenting the same face to the attracting body. This is the case for the Moon orbiting the Earth and for Mercury orbiting the Sun.

3 Electromagnetism

Electromagnetism is the study of electric and magnetic phenomena[17, 18, 19, 20]. In presenting this subject I shall usually follow the historical narrative. First static fields were investigated. The electric field by Coulomb, Cavendish, Gauss, Franklin and others and the magnetic field by Ampère, Biot, Savart, Volta, Galvani and others. One of the great discovery of that time was that electricity and magnetism were actually related phenomena. Since the equivalence of electric and magnetic fields is more clearly displayed by considering their description in different moving frames using the special theory of relativity, in presenting static magnetic fields I shall adopt that point of view even though it differs from the historical narrative. We shall then study the electromagnetic fields generated by time dependent currents (accelerated charges). These investigations are associated with the works of Faraday and Maxwell. The later in particular was the first to formulate a full theory of electromagnetism which unified electricity, magnetism and optics. It predicted the existence of electromagnetic waves which were indeed observed by Hertz and gave rise to modern communication by radio waves[17, 19, 20]. Maxwell's equations were the reason that forced Einstein to reconsider Galilean invariance and propose his special theory of relativity. These equations and the problems they raised were also at the foundation of Quantum Mechanics. Studying and understanding their enormous predictive power is thus of paramount importance for the comprehension of modern Technology and Science. Since that is such a vast field with so many ramifications, we will only study its foundations, use those to explain various natural phenomena (lightning, aurora borealis, the rainbow, etc.) and illustrate their practical applications in our everyday life (motors, generators, electronic circuits, optics, radio, etc.).

3.1 ◆ ELECTROSTATICS

Electricity is a phenomenon known for thousands of years. People had direct experience of it through contacts with electric fish (called by the ancient Egyptians "thunderer of the Nile"). By rubbing certain materials, man also knew that a force could be generated between them. In fact the word electricity comes from the Greek word for amber (elektron) which when rubbed with fur can attract or repel other objects. In particular it repels an other piece of rubbed amber, but it attracts a piece of glass rubbed with silk. The charge left on amber is defined as negative (then called resinous electricity), that left on glass as positive (then called vitreous electricity). Thus, it was known long before it was understood that like-charged objects repel each other, whereas opposite charge objects attract each other. This action at a distance of what we now call charged objects was very intriguing and mysterious to the pre-scientific world.

With the publication in 1687 of Newton's master piece - "Mathematical Principles of Natural Philosophy" (known from its original latin title as *"Principia"*) - a

DOI: 10.1201/9781003218999-3

new area of scientific inquiry opened up: Natural phenomena were to be described in mathematical terms. Influenced by this novel approach (like many of his contemporaries) Coulomb decided to investigate the force between charged objects as was done for the gravitational force. At that time (18th century) the only way to generate charges was by an electrostatic generator that used the friction of a rubber band to transport charges and store them in a so-called Leyden jar (a primitive form of capacitor, see below). People knew that some materials (such as metals) conducted electricity and some didn't (they were insulators like amber). They could charge a Leyden jar by connecting it with metallic wires to a charged one. To evidence the charge in the jar people would use an electroscope, which consisted of a small sphere of non-conducting material (a pith ball) which would be attracted to charged surfaces. Alternatively one could use the repulsion between two charged gold leafs to demonstrate the presence of charges on a conducting material brought into contact with the leaves.

Even though there was no understanding of electrical phenomena, these observations led in 1774 to the first invention of a telegraphic machine. It consisted of 26 insulated wires, each representing a letter of the alphabet connected at their distant end to an electroscope. To send a message, a desired wire was connected momentarily to a charged Leyden jar, whereupon the electroscope at the other end would get charged (the gold leaves or a pith ball would fly out) and in this way messages were transmitted at a distance!

3.1.1 ◆ COULOMB'S LAW

Intrigued by these observations, Coulomb decided in 1784 to measure the force F between two charged spheres using a torsional pendulum similar to the one used at about the same time by Cavendish to study Newton's Law, see Fig.3.1. It consisted of two balanced spheres (of mass m, one of them conducting) a distance $L/2$ from the axis of rotation. The force between the charged conducting sphere on the pendulum and a nearby charged sphere a distance r away yields a torque $\Gamma = FL/2$, which twists the pendulum by an angle $\theta = \Gamma/\kappa$, measured by the deflection of a light ray incident on an attached mirror, where κ is the torsional stiffness of the pendulum[1]

The electrostatic force exerted on one of the spheres of the pendulum by a nearby one is then: $F = 2\kappa\theta/L$. Coulomb observed that same sign charges repel while opposite charges attract each other with a force proportional to the product of the charges (q_1, q_2) on the spheres, inversely proportional to the square of the distance between their centers (r) and directed along the unit vector $\hat{r} = \vec{r}/r$ linking these centers:

$$\vec{F} = k\frac{q_1 q_2}{r^2}\hat{r}$$

The value of the proportionality constant k sets the units of the charge q. In the CGS system, which we will mostly adopt here, $k = 1$ and the charge is measured in

---

[1]The torsional stiffness κ is deduced from the oscillation frequency $\omega_0 = \sqrt{\kappa/I}$ of the pendulum of moment of inertia $I = mL^2/2 : Id^2\theta/dt^2 + \kappa\theta = 0$.

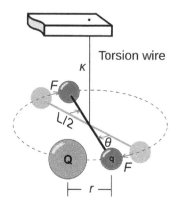 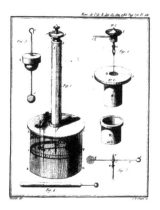

Figure 3.1 (a) Schematic drawing of Coulomb's torsional balance modeled after the one used by Cavendish to test Newton's theory of gravity (Chapter 2). The force between a fixed sphere bearing a charge Q and a sphere bearing a charge q on a balanced torsional pendulum exerts a torsion on the pendulum which results in its rotation by an angle θ. (b) Original drawings of Coulomb's pendulum: the fixed sphere is labeled J (grey) and the charged sphere on the torsional pendulum is labeled g (red). The whole pendulum was held in a conducting box to prevent external interferences (stray charges, air currents, etc.).

so-called electrostatic units (esu) or statCoulomb. Two charges of one esu a distance 1 cm away exert on each other an attractive (or repulsive) force of 1 dyne. In the MKS (or SI) system $k = 1/4\pi\epsilon_0 = 9 \ 10^9$ N m^2/C^2 (ϵ_0 is known as the dielectric constant of the vacuum[2]) and the unit of charge is the Coulomb (C): 1 C = 3 10^9 esu.

The charge of an electron is $e = -4.8 \ 10^{-10}$esu $= -1.6 \ 10^{-19}$C (see below how it was measured). The neutrality of atoms (the fact that matter is rarely charged and even then very weakly) is ensured by the fact that protons (the charged particles in the atom's nucleus) have equal and opposite charge to that of electrons.

The force between an electron and a proton in a Hydrogen atom (radius $r \sim 0.1$nm $= 10^{-8}$cm) is $F = e^2/r^2 \sim (4.8 \ 10^{-10})^2/(10^{-16}) = 23 \ 10^{-4}$ dynes $= 23 \ 10^{-9}$N = 23 nN (this set the scale of the strength (order nN) of the covalent bonds holding the atoms in a molecule together). Notice that this is of the same magnitude as the gravitational force between two masses $m_1 = m_2 = 2$kg held a distance $r = 10$cm apart

---

[2]In contrast with most textbooks on EM, I will be using here esu (and CGS) units. The dielectric constant ϵ_0 and magnetic permeability μ_0 of the vacuum (or luminous ether) were historically introduced when light was supposed to propagate like sound in a medium (the luminous ether) which properties set its speed $c = 1/\sqrt{\epsilon_0\mu_0}$, just like the speed of sound $c_s = 1/\sqrt{\kappa_a\rho}$, is set by the properties of the medium (its adiabatic compressibility $\kappa_a = 1/\gamma P$ and density ρ, see section 5.11.2). With the demise of the theory of the luminous ether we know that these constants have no physical meaning. Only the velocity of light does. The introduction of these constants unduly complicates the exposition of EM and obscures the intimate connection between electric and magnetic fields (addressed in the next section) made by Einstein in his theory of relativity.

$F_g = Gm_1m_2/r^2 \sim 28$ nN! (where $G = 6.67\ 10^{-11}$ N (m/kg)^2 is the gravitational constant).

Charge is a conserved and velocity independent quantity. If this was not the case, the imbalance of charges between the moving electrons and the more static protons in an atom would generate forces that overwhelm the gravitational attraction between bodies.

If there are many charges ($\{q_i\}$) located at positions $\{\vec{r}_i\}$ acting on a charge q at position $\vec{r}$, the total force on the charge q is the vectorial sum of the individual forces:

$$\vec{F}_{tot} = \sum_i \frac{qq_i(\vec{r}-\vec{r}_i)}{|\vec{r}-\vec{r}_i|^3} \tag{3.1}$$

That property of the electrostatic (and also gravitational) forces known as linear superposition is what allows one to understand and predict their action (this is not the case with hydrodynamic forces, which is why the weather is so unpredictable). The electric field $\vec{E}$ of a distribution of charges $\{q_i\}$ at positions $\{r_i\}$ (or charge density $\rho(\vec{r})$ at position[3] $\vec{r}$) is defined by their combined force on a positive test charge q located at position r:

$$\vec{E} \equiv \frac{\vec{F}_{tot}}{q} = \sum_i \frac{q_i(\vec{r}-\vec{r}_i)}{|\vec{r}-\vec{r}_i|^3}$$

$$= \int d^3r' \frac{\rho(\vec{r}')(\vec{r}-\vec{r}')}{|\vec{r}-\vec{r}'|^3} \tag{3.2}$$

This equation can be further developed to yield a function often used in electrostatic problems, the electrostatic potential Φ:

$$\vec{E} = \int d^3r' \frac{\rho(\vec{r}')(\vec{r}-\vec{r}')}{|\vec{r}-\vec{r}'|^3} = -\frac{\partial}{\partial \vec{r}} \int d^3r' \frac{\rho(\vec{r}')}{|\vec{r}-\vec{r}'|}$$

$$\equiv -\frac{\partial \Phi}{\partial \vec{r}} \equiv -\vec{\nabla}\Phi \tag{3.3}$$

Where the potential due to a charge distribution is defined as:

$$\Phi(\vec{r}) \equiv \sum_i \frac{q_i}{|\vec{r}-\vec{r}_i|} = \int d^3r' \frac{\rho(\vec{r}')}{|\vec{r}-\vec{r}'|} + \text{Const} \tag{3.4}$$

The constant (which has no physical consequences) is usually chosen such that the potential at infinity is zero. The electrostatic potential is measured in units of statvolts (statV) in CGS units and in volts (V) in MKS units: 1 statV = 300 V. Since $\vec{E} = -\vec{\nabla}\Phi$, by standard vector calculus identities, see Appendix A.3:

$$\vec{\nabla}\times\vec{E} = -\vec{\nabla}\times\vec{\nabla}\Phi = 0 \tag{3.5}$$

---

[3]The charge in a vicinity ϵ of $\vec{r}$ is $q' = \int_\epsilon \rho(\vec{r})d^3r'$.

The work performed against the electric field by moving a charge between two distinct points $(\vec{r}_1, \vec{r}_2)$ is thus simply proportional to the difference in electrostatic potential between those points[4]:

$$W = \int_1^2 \vec{F}_{ex} \cdot d\vec{r} = -q \int_1^2 \vec{E} \cdot d\vec{r} = q \int_1^2 \vec{\nabla}\Phi \cdot d\vec{r}$$
$$= q[\Phi(\vec{r}_2) - \Phi(\vec{r}_1)] = q\Delta\Phi \tag{3.6}$$

Where $\Delta\Phi$ is the potential difference (or voltage: V). As for the gravitational field, see Eq.2.12, this work correspond to the difference in electric potential energy $\mathcal{E}_e = q\Phi$ between final and initial states. The work is positive if the final state has higher energy than the initial. For example, the work done to ionize an hydrogen atom (bring its electron a large distance away from the proton: $r_2 = \infty$) is: $W = e^2/r \sim 23 \; 10^{-12}$erg $= 23 \; 10^{-19}$J $= 14$ eV (electron-Volt). That energy (a few eV) is the typical energy of a covalent bond (the bond between atoms in a molecule).

No work is done displacing a charge on an equipotential surface, i.e. a surface for which $\Phi(x,y,z) = const$. Consequently the electric field $\vec{E}$ is always perpendicular to the equipotential surfaces. Due to these properties equipotential surfaces play an important role in electrostatic problems. The same is true in Newtonian gravity: the gravitational force is perpendicular to the equipotential surfaces, eg. the surface of the Earth (the sea to be more precise).

Moreover by Stokes' theorem no work is done against (or by) the electric field while moving a charge q along a closed path:

$$W = \oint \vec{F}_{ex} \cdot d\vec{r} = -q \oint \vec{E} \cdot d\vec{r} = -q \int \vec{\nabla} \times \vec{E} \cdot d\vec{S} = 0 \tag{3.7}$$

This is also true for a gravitational field which does no work on a closed path: a stone thrown up with a given velocity returns with the same velocity (neglecting air friction).

3.1.2 APPLICATIONS OF COULOMB'S LAW

3.1.2.1 Lightning

From these considerations we can understand the mechanism of lightning. During a lightning storm strong air-currents in the clouds results in collisions between super-cooled water drops and ice particles, that leave the drops slightly negatively charged and the ice particles slightly positively charged (the same way amber is charged when rubbed with fur or a silk shirt is charged when taken off and rubbed over one's hair). The less dense ice particles are advected towards the top of the cloud leaving the

---

[4]The work against the field requires an external force F_{ex} opposite to the electric force: $F_{ex} = -q\vec{E}$.

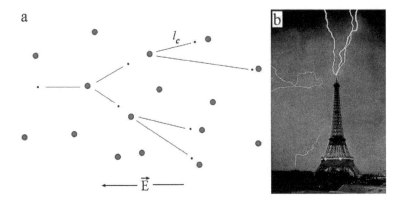

Figure 3.2 (a) Schematic drawing of dielectric breakdown: the process where an electron (small red dot) accelerated by the electric field $\vec{E}$, hits a particle (atom or molecule, blue dot) in the air and ionizes it. The two electrons are accelerated again until the next collision with two atoms, generating 4 electrons, etc. (b) Lightning on the Eiffel Tower (the charges are initially accelerated in the vicinity of the tower's tip where the field is maximal).

bottom of the cloud negatively charged. When the electric field between the top and bottom of the cloud is large enough to ionize the surrounding air an electric discharge occurs which is the lightning bolt. Like the explosion of an atom bomb, lightning is an avalanche process, Fig. 3.2: a spontaneously ionized electron is accelerated by the electric field in the cloud. It hits an atom in the air and ionizes it, i.e. it ejects an other electron; the two electrons are then accelerated again until the next collision with two atoms, generating 4 electrons, etc. For this avalanche to occur the energy gained by the electron between two collisions must be large enough to ionize an atom (a few eV). Once the avalanche has been initiated a path is traced in air where the positively charged ionized gas channels (attracts) the negatively charged electrons and where the lightning current passes.

To be more precise, let the gas density be $\rho = N_A/V_A \simeq 3 \ 10^{19}$ particles/cm^3 (where $N_A = 6 \ 10^{23}$ is Avogadro's number and $V_A = 22.7$ l is the molar volume of a gas in standard conditions, see section 5.5). If the ionization cross section $S = \pi r^2 \simeq 1$ Å^2 (typical cross-section of an atom), then the mean free path between collisions $l_c = 1/S\rho = 3\mu$m. For a typical ionization energy of $W_{ion} \sim 3$eV this corresponds to an electric field of $E_b = W_{ion}/e \ l_c \approx 10^6$ V/m, which is known as the dielectric breakdown field in air. Hence, the potential difference in a cloud or between it and the ground may reach many million volts!

3.1.2.2 The Field of a Dipole

Consider two opposite charges ($\pm q$) at positions $\pm d\hat{z}$. This charge configuration is known as a dipole (its orientation sets the z-axis). To compute the electric field of a dipole at a distance $r \gg d$ consider first the potential it generates :

$$\Phi(\vec{r}) = \frac{q}{|\vec{r} - d\hat{z}|} - \frac{q}{|\vec{r} + d\hat{z}|} \tag{3.8}$$

In the far field $r \gg d$, we may expand the denominators to first order[5] in d/r:

$$\begin{aligned}
|\vec{r} \pm d\hat{z}|^{-1} &= 1/\sqrt{x^2 + y^2 + (z \pm d)^2} \\
&= r^{-1}(1 \pm 2zd/r^2)^{-1/2} \\
&\approx r^{-1}(1 \mp zd/r^2)
\end{aligned} \tag{3.9}$$

We obtain:

$$\Phi(\vec{r}) = \frac{2qd\hat{z} \cdot \vec{r}}{r^3} = \frac{\vec{p} \cdot \vec{r}}{r^3} = \frac{p \cos\theta}{r^2} \tag{3.10}$$

Where the dipole moment[6] $\vec{p}$ is defined as: $\vec{p} \equiv 2qd\,\hat{z}$. Its units are esu·cm, though to characterize the dipole moments of molecules one also uses the Debye (1D = 10^{-10} esu·Å = 10^{-18} esu·cm). Using the expression of the $\vec{\nabla}$ operator in spherical coordinates (r, θ, ϕ), see Appendix A.3:

$$\vec{\nabla}\Phi \equiv \frac{\partial\Phi}{\partial r}\hat{r} + \frac{1}{r}\frac{\partial\Phi}{\partial\theta}\hat{\theta} + \frac{1}{r\sin\theta}\frac{\partial\Phi}{\partial\phi}\hat{\phi} \tag{3.11}$$

the electric field becomes:

$$\vec{E} = -\vec{\nabla}\Phi = \frac{2p\cos\theta}{r^3}\hat{r} + \frac{p\sin\theta}{r^3}\hat{\theta} \tag{3.12}$$

Notice that the attractive (or repulsive) component of the field, the component along $\hat{r}$, is maximal when $\theta = 0$ (or π), i.e. along the dipole axis. The energy $\mathcal{E}_e$ of a dipole in an external electric potential $\Phi(\vec{r})$ (or field $\vec{E}$) is then[7]:

$$\mathcal{E}_e = q\Phi(\vec{r} + \vec{d}) - q\Phi(\vec{r} - \vec{d}) = 2q\vec{d} \cdot \vec{\nabla}\Phi = -\vec{p} \cdot \vec{E} = -pE\cos\theta \tag{3.13}$$

Where θ is the angle between the orientations of the dipole and the external field. Thus to minimize their energy, dipoles tend to align parallel to the external field $\vec{E}$ ($\theta = 0$) or (in absence of an external field) parallel to each other, possibly forming chains of dipoles aligned head to tail. The torque aligning the dipole is then: $\vec{\tau} = \vec{r} \times \vec{F} = \vec{p} \times \vec{E}$.

The long range attractive interaction between dipoles is crucial to our understanding of the forces between molecules[21]. These interactions are always present

---

[5]Using the Taylor expansion $(1 + x)^\alpha = 1 + \alpha x + \alpha(\alpha - 1)x^2/2! + ...$

[6]The dipole moment and the momentum, like the energy and magnitude of the electric field, share the same symbol ($\vec{p}$ or E), but should not be confused. That is unfortunate, but a consequence of historical conventions and the finite number of letters.

[7]Using the Taylor expansion: $\Phi(\vec{r} + \vec{d}) \approx \Phi(\vec{r}) + d_x\partial_x\Phi + d_y\partial_y\Phi + d_z\partial_z\Phi = \Phi(\vec{r}) + \vec{d} \cdot \vec{\nabla}\Phi$

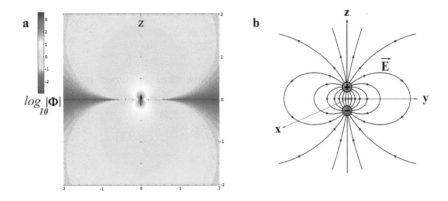

Figure 3.3 (a) The electric potential $\Phi(r,\theta)$ around a dipole $\vec{p} = p\hat{z}$ (arrow at the center). The color coding represent the logarithmic strength of the potential ($\log_{10}|\Phi|$). (b) The electric field lines (the lines parallel to $\vec{E}$) of a dipole.

whether the molecules are charged or not, whether they present permanent dipoles or not. In the latter more general case, an instantaneous asymmetric distribution of charges in one molecule creates a dipole which field $\vec{E}_o$ induces in nearby molecules a similarly oriented dipole ($\vec{p}_i \sim \vec{E}_o$) (see section 3.1.6). Since the dipole field $\vec{E}_o$ decreases as r^{-3}, the energy of this induced dipole in the inducing field ($-\vec{p}_i \cdot \vec{E}_o$) decreases as r^{-6}. These weak interactions, first described by van der Waals and bearing his name, are long range and attractive. They stand in contrast to the strong covalent bonds ensuring the cohesion of molecules (see section 4.10), which result from electron sharing and which decay exponentially over much shorter distances (of order of 1 Angstrom: $O(\text{Å})$). The van der Waals attraction[21] is responsible for the properties of gases and liquids (see section 5.10.2), the wetting properties of surfaces, the instability of emulsions (such as mayonnaise where the attraction between water molecules is stronger than between water and oil molecules possibly causing the emulsion to separate), the adhesion between materials as, for example, between the Gecko's feet and a vertical wall, etc., etc.

3.1.2.3 Piezo-Electricity

The previous small exercise with electric dipoles has interesting practical consequences. Certain materials known as ferroelectrics (for example Barium titanate, $BaTiO_3$) have small permanent electric dipoles which - according to our preceding analysis - will tend to align parallel to each other generating a strong electric field in the material and an electric potential difference $\Delta\Phi$ across it. Small mechanical deformation of the crystalline material affect the dipole moment (the distance d between the charges) and thus the potential across the material. This effect known as piezo-electricity is used either to generate electricity (in electric gas lighters or wrist-watches) or to precisely move an object (as in scanning probe microscopes) by applying an electric field on a piezo-electric material thus affecting its extension.

Because changes in temperature cause a crystal to expand or contract (thus altering the distance between charges), ferro-electric materials also display pyro-electricity, i.e. a change in the voltage across the material as a function of temperature. This property is being used to make infrared cameras (and non-contact thermometers), where changes in temperature on a pixel (picture element) as small as $0.1°C$ is converted into a detectable voltage change on that pixel.

3.1.3 ◆ GAUSS'S THEOREM

Electricity is evidently fundamental to modern civilization and the control of static electric fields is an essential part of modern technology (for example in electron microscopes or old TV tubes, see below). For that purpose Eq.3.4 is all one needs to compute the electric field of a given charge distribution. There are however various ways to compute that sum (or integral) which we will now review.

Consider a charge q and a closed surface S surrounding that charge at a distance $r(\theta, \phi)$. Since the electric field $\vec{E}$ is directed along the radius and decays as $1/r^2$ (a property also shared by the gravitational field), Gauss proved that:

$$\oint \vec{E} \cdot d\vec{S} = 4\pi q$$

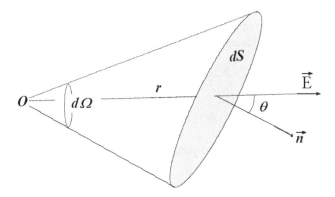

Figure 3.4 A surface element dS whose normal $\vec{n}$ makes an angle θ with the electric field $\vec{E}$ originating from a charge q at point O. The solid angle $d\Omega$ spanned by dS is given by $d\Omega = dS \cos\theta / r^2$. Therefore: $\vec{E} \cdot d\vec{S} = EdS \cos\theta = Er^2 d\Omega = qd\Omega$, which integral yields Gauss's theorem.

To convince oneself of the validity of Gauss's theorem one may write $\vec{E} \cdot d\vec{S} = \vec{E} \cdot \hat{n}dS = E\cos\theta dS$, see Fig.3.4. However $\cos\theta dS$ is the area perpendicular to the radius vector r which defines the solid angle $d\Omega$ spanned by $d\vec{S}$: $\cos\theta dS/r^2 = d\Omega$ and therefore: $\int \vec{E} \cdot d\vec{S} = \int Er^2 d\Omega = q \int d\Omega = 4\pi q$. Because of linear superposition, the field due to many charges $\{q_i\}$ will obey:

$$\oint \vec{E} \cdot d\vec{S} = 4\pi \sum_i q_i \tag{3.14}$$

As we shall see below, Gauss's theorem is often used to compute the electric field when the charge distribution is constant on some symmetric surfaces (planes, spheres, cylinders, etc.). Using the divergence theorem from vector calculus, see Appendix A.3:

$$4\pi q = \oint \vec{E} \cdot d\vec{S} = \int \vec{\nabla} \cdot \vec{E} \, dV \tag{3.15}$$

Since $q = \int \rho dV$, one obtains Coulomb's law:

$$\vec{\nabla} \cdot \vec{E} = 4\pi \rho \tag{3.16}$$

or using the relation $\vec{E} = -\vec{\nabla}\Phi$, Poisson's equation[8]:

$$\nabla^2 \Phi = -4\pi \rho \tag{3.17}$$

The solution of this equation is in fact Eq.3.4, as can be checked using the identity[9]:

$$\nabla_r^2 \frac{1}{|\vec{r} - \vec{r}'|} \equiv -4\pi\delta(\vec{r} - \vec{r}') \tag{3.18}$$

Using Eq.3.17 and taking care not to count the charges twice[10], the energy $\mathcal{E}_e$ stored in the electric field generated by a charge distribution (q_i or $\rho(r)$), is:

$$\mathcal{E}_e = \frac{1}{2} \sum_{i \neq j} \frac{q_i q_j}{|\vec{r}_i - \vec{r}_j|}$$
$$= \frac{1}{2} \int \rho(r) \, \Phi(r) \, d^3r = -\frac{1}{8\pi} \int \Phi \, \nabla^2 \Phi \, d^3r \tag{3.19}$$

Integration by part of the last integral yields:

$$\mathcal{E}_e = -\frac{1}{8\pi} \left\{ \oint \Phi \, \vec{\nabla}\Phi \cdot d\vec{S} - \int (\vec{\nabla}\Phi)^2 d^3r \right\} \tag{3.20}$$

---

[8] In the gravitational context, Poisson's equation reads: $\nabla^2 \Phi_G = 4\pi G\rho$, where ρ is the mass density.

[9] The Dirac delta function $\delta(x)$ is defined as being zero except at $x = 0$ where it diverges such that $\int dx\delta(x) = 1$. Consequently the units of $\delta(x)$ are the inverse of the units of x and $\delta(ax) = \delta(x)/|a|$. The Dirac delta function is often viewed as the limit of a Gaussian distribution centered at $x = 0$ with vanishing variance.

[10] A term such as $q_1 q_2/(|\vec{r}_1 - \vec{r}_2|)$ also appears as $q_2 q_1/(|\vec{r}_2 - \vec{r}_1|)$ which is identical.

Since the field Φ vanishes at infinity, considering a closed surface S far away from the charges the first integral on the right vanishes[11]. We are therefore left with:

$$\mathcal{E}_e = \frac{1}{8\pi} \int (\vec{\nabla}\Phi)^2 d^3 r = \frac{1}{8\pi} \int |\vec{E}|^2 d^3 r \qquad (3.21)$$

The local energy density of a static electric field is thus: $e_e(r) = \frac{1}{8\pi}|\vec{E}(r)|^2$.

3.1.4 APPLICATIONS OF GAUSS'S THEOREM

3.1.4.1 The Capacitor

When the distribution of charges is symmetric and uniform Gauss's theorem is often the simplest way to compute the electric field. In particular in charged conducting materials, since like charge repel each other they will distribute themselves on the surface such as to cancel the average force on them, i.e. $\vec{F}_{tot} = 0$ which implies that inside a conductor $\vec{E} = 0$. Since $\vec{E} = -\vec{\nabla}\Phi$ the surface S of a conductor is an equipotential surface $\Phi|_S =$Const.

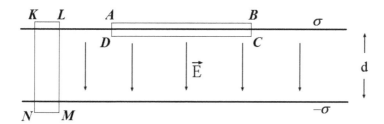

Figure 3.5 A capacitor consists of two conducting surfaces of area S separated by a small gap d. The surfaces are charged with equal and opposite charge densities σ and $-\sigma$. Application of Gauss's theorem in the element KLMN (which encloses a charge equal to zero) shows that the field outside the capacitor is null. Similarly application of Gauss's theorem in the element ABCD yields: $ES = 4\pi q = 4\pi\sigma S$, i.e. $E = 4\pi\sigma$.

For example consider two large flat conducting planes a small distance d from each other bearing a constant but opposite surface charge density $\sigma = Q/S$, see Fig.3.5. Since both plates have equal and opposite charges, there is no field outside them. Between the two planes however application of Gauss's theorem in a box enclosing one of the surfaces yields: $ES = 4\pi\sigma S$, i.e. $E = 4\pi\sigma$ and the potential difference between the plates is: $\Delta\Phi = Ed = 4\pi Qd/S = Q/C$ where $C = S/4\pi d$ is known as the capacitance of the plates. Notice that in the CGS system capacitance is

---

[11]Far from the charges $\Phi(r) \sim Q_{tot}/r$ and $\nabla\Phi \sim Q_{tot}/r^2$, hence the integral over S will scale as $1/r \to 0$

measured in cm, while in the MKS system capacitance is measured in Farad (F): 1 F = 1 C/V (= 9 10^{11} cm). A capacitor is an element that stores electric energy:

$$\mathcal{E} = \int \Delta\Phi dQ = \frac{Q^2}{2C} = \frac{(ES/4\pi)^2}{2C} = S d \frac{E^2}{8\pi} \tag{3.22}$$

Thus the energy stored per unit volume is $E^2/8\pi$, in agreement with or previous result, Eq.3.21.

The two flat parallel plates configuration is the simplest example of a capacitor, but any two conducting plates not necessarily flat separated by a small gap will form a capacitor which can store charge. The Leyden jar mentioned earlier consisted of a glass bottle whose inner and outer surfaces were covered with a conducting sheet. The inner surface was charged by an electrostatic generator while the outer surface was grounded (thus ensuring that its charge counter-balances the charge deposited inside the jar). Capacitors are ubiquitous components of every electrical circuit. By storing charge they buffer the circuit against sudden changes in the voltage.

3.1.4.2 Measuring the Charge of the Electron

The fact that the electric field between the two plates of a capacitor is a constant has allowed Robert Millikan [22, 23] to precisely measure in 1909 the charge of an electron. He proceeded in the following way: he sprayed small charged oil-droplets and let them fall between the plates of a capacitor, see Fig.3.6.

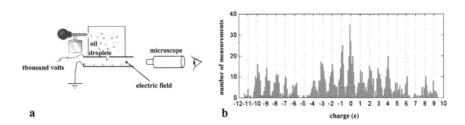

Figure 3.6 (a) Scheme of Millikan's oil drop experiment. Charged oil-droplets are introduced between the plates of a capacitor and their movement is observed with a microscope. Millikan deduced the charge on the droplet from the dependence of its final velocity on the electric field . (b) Histogram of the charge of a colloidal droplet in a recent experiment similar to Millikan's original study. Adapted from ref.[23] copyright (2008) by the American Physical Society.

Due to friction with air (of known viscosity η and density ρ_{air}) the droplets (of density ρ_{oil}) reached a constant drift velocity v (upward or downward) that he could measure. Equating the viscous drag on a droplet (the Stokes' force: $F_{Stokes} = 6\pi\eta r v$, see section 5.11.3) with the balance of electrostatic (qE) and gravitational force[12] ($w_{eff} = (4\pi/3)(\rho_{oil} - \rho_{air})r^3 g$) yields: $F_{Stokes} = \pm qE - w_{eff}$. Studying the same droplet with different values of electric field E and thus final drift velocity v allowed Millikan to deduce the droplet radius r and charge q. Repeating the experiment on different droplets, Millikan noticed that the droplets' charges were quantized: they were an integer multiple of an elementary value, see Fig.3.6, the charge of the electron $e = -1.6 \ 10^{-19}$ C, which he got with only a 1% error-bar!

3.1.4.3 The Electrostatic Lens

The simple capacitor can also be used to focus electron beams in applications such as cathode ray tubes (TV sets) or electron microscopes. The principle of an electrostatic lens is the following. Electrons emitted by some hot cathode are accelerated in a voltage difference V_0, to a velocity v_1 such that: $mv_1^2/2 = eV_0$. As they enter a capacitor consisting of two curved grids separated by a small distance d and a voltage difference $V \ll V_0$ their velocity increases so that their velocity v_2 at the exit from the capacitor satisfies: $mv_2^2/2 = mv_1^2/2 + eV = e(V + V_0)$, see Fig.3.7(a). As a consequence of the longitudinal acceleration of the electron (the acceleration along the capacitor field $\vec{E}$), their propagation direction is altered (like the light rays that we will address later in this book they are "refracted"). Since the transverse component of the velocity is not altered by the capacitor field: $v_\perp = v_1 \sin\alpha = v_2 \sin\beta$ and therefore:

$$\frac{\sin\alpha}{\sin\beta} = \frac{v_2}{v_1} = \sqrt{1 + V/V_0} \tag{3.23}$$

If the capacitor is a thin spherical shell of radius $r \gg d$ and in the small angle limit ($\alpha, \beta \ll 1$ it is easy to show that a parallel electron beam will be focused at a distance f such that: $f \sin(\alpha - \beta) = r \sin\alpha$, i.e. $f = r/(1 - \beta/\alpha) \simeq 2rV_0/V$. The fact that f is independent of the incident angle α means that all the rays in a wide electron beam will cross at the same point O, a distance f from the capacitor. This point is the focus of this simple electrostatic lens, see Fig.3.7(b). In practice many electric field configurations whose equipotential surfaces (surfaces on which $\Phi =$constant) are curved can be used to focus an electron beam, see for example Fig.3.7(c). Such configurations are used in Transmission Electron Microscopes (TEM) to illuminate a sample with an electron beam and project its image on a camera.

In old TV sets capacitors were not only used to focus the electron beam on a fluorescent screen but also (by applying a transverse electrostatic force) to steer it horizontally and vertically across the screen. Finally by modulating the voltage difference between the electron emitting hot cathode and a nearby grid, the amount of electrons emitted from the cathode (the beam intensity, i.e. the brightness of the image) could be controlled. This modulation of the emitted electron current by a charged grid was also the working principle of vacuum tubes (triodes), which served

---

[12]We took into account the small Archimedes force on the droplet which reduces its effective weight $w = \rho_{oil}V$ by an amount $\rho_{air}V$ where $V = 4\pi r^3/3$ is the volume of the droplet.

]

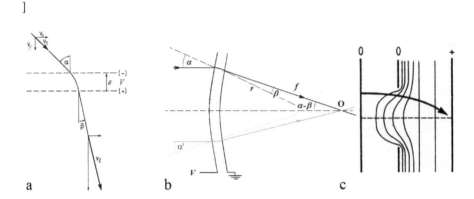

Figure 3.7   a): Change in the direction of propagation of an electron beam as it is accelerated between the plates of a capacitor. (b): Examples of an electron lenses: a thin spherical capacitor which can focus the electron beam at a distance f satisfying: $f \sin(\alpha - \beta) = r \sin \alpha$. (c) Example of a field configuration which equipotential surfaces are curved and that can also be used to focus an electron beam.

as voltage amplifiers in many electronic devices (radio, radars, TV sets, etc.) before the invention of the transistor, see Fig.3.8.

3.1.5   SOLVING POISSON'S EQUATION

Often in electrostatic problems, the potential (or charge) on certain surfaces is controlled and one needs to determine the resulting electric field[19, 20]. That is the case for example when designing an electrostatic lens, see Fig.3.7. The problem then is to solve Eq.3.17 with $\rho = 0$, known as the Laplace (or homogeneous Poisson) equation:

$$\nabla^2 \Phi = 0 \qquad\qquad (3.24)$$

When the potential is given on a surface $\Phi|_S = \Phi_0$, the problem is known as a Laplace problem with Dirichlet boundary condition. When the charge is given on the surface, by Gauss's theorem this is equivalent to setting the electric field perpendicular to the surface, i.e. the normal derivative of the potential $\partial_n \Phi|_S$. This type of boundary condition is known as Neumann boundary condition.

   In cases where the boundary condition depends on all three coordinates, one needs to solve Eq.3.24 in 3D. In most cases this is done numerically. There are however a few geometries where Eq.3.24 can be solved analytically. It is interesting to consider these cases for various reasons. First they can be used as test grounds for numerical algorithms. Second, the Laplace equation and its generalization, the Helmholtz equation ($\nabla^2 \Phi = \epsilon \Phi$) are encountered in a great variety of contexts (electromagnetic radiation, quantum mechanics, heat equation, etc.), so that their solutions in one field can be relevant in a totally different area. Because of their relevance, the solutions

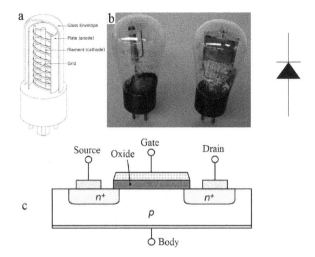

Figure 3.8 (a –b) The vacuum tube was used as a voltage amplifier before the invention of the transistor. (a) A hot cathode emitted a current of electrons that were collected on a cold anode. The voltage difference between the cathode and a grid placed between it and the anode controlled the intensity of the current (and voltage) between cathode and anode. In absence of a grid the device works as a diode (symbol shown on the right) which allows current to flow in only one direction: from the hot cathode to the cold anode. To prevent collisions with the air and corrosive reactions at the electrodes the tube (b) was held in vacuum, hence its name (adapted from ref.[24, 25]). (c) Advances in the understanding of the conductance of semiconductors allowed the development in 1947 of the transistor which in some implementation (the Field Effect Transistor, FET) is similar to a vacuum tube: a voltage on the gate controls the density of charge carriers beneath it and thus controls the current flowing between the source and the drain, see section 5.9.2.4.

of the Helmholtz equation in rectangular, cylindrical and spherical coordinates (geometries) are described in Appendix A.5 .

If due to the boundary conditions being independent of the z-coordinate, the potential depends on two coordinates only (x, y), there are ingenious means to solve the Laplace equation based on the theory of analytic functions (functions of the complex variable $z = x + iy$). While this is beyond the scope of this book, the interested reader can have a glimpse at these methods in Appendix A.7.1.

3.1.5.1 The Method of Images

When point charges are located in the vicinity of grounded conducting surfaces the so-called method of images is a trick often used to solve electrostatic problems[20]. The idea is to introduce a charge of opposite sign on the other side of the conducting surface such as to ensure that the potential on the surface is null (it is grounded).

For example consider finding the field $\vec{E}$ due to a charge q located on the z-axis a distance d from a flat infinite grounded plane located at $z = 0$. Introducing a charge $-q$ at $z = -d$ ensures that the potential on the plane is indeed zero, as can be verified:

$$\Phi(x,y,0) = \frac{q}{\sqrt{x^2 + y^2 + d^2}} - \frac{q}{\sqrt{x^2 + y^2 + d^2}} = 0 \tag{3.25}$$

Since the two charges (q and its image charge $-q$) form a dipole of moment $\vec{p} = 2qd\hat{z}$, the field in the upper half space ($z > 0$) is simply that of a dipole, Eq.3.12.

The method of images can also be used if a charge q is placed a distance r from the center of a grounded sphere of radius r_0, see Fig.3.9. In that case we have to determine the charge q' and position r' of the image charge so that:

$$\Phi(r_0,\theta,\phi) = \frac{q}{\sqrt{r_0^2 + r^2 + 2rr_0 \cos\theta}} + \frac{q'}{\sqrt{r_0^2 + r'^2 + 2r'r_0 \cos\theta}} = 0 \tag{3.26}$$

Which we may rewrite as:

$$\Phi(r_0,\theta,\phi) = \frac{q}{r\sqrt{1 + (r_0/r)^2 + 2(r_0/r)\cos\theta}} + \frac{q'}{r_0\sqrt{1 + (r'/r_0)^2 + 2(r'/r_0)\cos\theta}} = 0 \tag{3.27}$$

The potential is null on the sphere if an image charge $q' = -qr_0/r$ is placed at a distance $r' = r_0^2/r$ from the center of the sphere along the line joining it to the charge q.

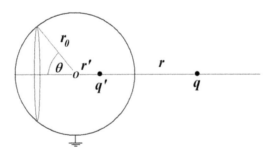

Figure 3.9 A charge q is placed a distance r from the center O of a grounded sphere of radius r_0. The potential outside the sphere is the same as would be generated by the charge q and an image charge $q' = -qr_0/r$ placed a distance $r' = r_0^2/r$ from O.

3.1.6 ELECTRIC FIELD IN MATTER: DIELECTRICS

In the previous discussions we have considered the electric field arising from isolated charges in vacuum. Often though it is of interest to know the field arising from charges embedded in a material environment (liquid, gas or solid) which environment may itself be responsive to the electric field (though non-conducting: in a conducting material charges would move to the surface).

Matter is made of atoms and molecules which although electrically neutral can polarize in the presence of an external electric field: their charge distribution may become anisotropic due to the opposite forces felt by positive and negative charges. They will thus develop an induced dipole $\vec{p}$ and generate a potential $\Phi = \vec{p} \cdot \vec{r}/r^3$, see Eq.3.10. For a distribution of induced dipoles with density $\vec{P}$ (the vector sum of individual dipoles per unit volume) and in the presence of charges with density ρ the total potential is the sum of the contributions from the charges (which define the field $\vec{E}$) and the dipoles (permanent or induced by the field $\vec{E}$):

$$
\begin{aligned}
\Phi(\vec{r}) &= \int d^3 r' \frac{\rho(\vec{r}\,')}{|\vec{r} - \vec{r}\,'|} + \int d^3 r' \frac{\vec{P}(\vec{r}\,') \cdot (\vec{r} - \vec{r}\,')}{|\vec{r} - \vec{r}\,'|^3} \\
&= \int d^3 r' \left(\frac{\rho(\vec{r}\,')}{|\vec{r} - \vec{r}\,'|} + \vec{P}(\vec{r}\,') \cdot \vec{\nabla}' \frac{1}{|\vec{r} - \vec{r}\,'|} \right) \\
&= \int d^3 r' \frac{\rho(\vec{r}\,') - \vec{\nabla}' \cdot \vec{P}(\vec{r}\,')}{|\vec{r} - \vec{r}\,'|}
\end{aligned}
\tag{3.28}
$$

where $\vec{\nabla}'$ stands for the divergence with respect to coordinate $\vec{r}\,'$. Eq.3.17 then becomes:

$$
\vec{\nabla} \cdot \vec{E} = -\nabla^2 \Phi = 4\pi(\rho - \vec{\nabla} \cdot \vec{P})
\tag{3.29}
$$

where we used Eq.3.18 to replace $\nabla^2 |\vec{r} - \vec{r}\,'|^{-1}$ by $-4\pi\delta(\vec{r} - \vec{r}\,')$. Assuming that the induced dipole density $\vec{P} = \chi\vec{E}$ (where χ is known as the electric susceptibility of the material), Eq.3.29 can be rewritten as:

$$
\vec{\nabla} \cdot \vec{D} = 4\pi\rho
\tag{3.30}
$$

where the displacement field $\vec{D}$ is defined as:

$$
\vec{D} \equiv \vec{E} + 4\pi\vec{P} = (1 + 4\pi\chi)\vec{E} \equiv \epsilon\vec{E}
\tag{3.31}
$$

$\epsilon \equiv 1 + 4\pi\chi$ is known as the dielectric constant of the material. Gauss's theorem becomes:

$$
\oint \vec{D} \cdot d\vec{S} = 4\pi q
\tag{3.32}
$$

At the interface between two media of different dielectric constants, Gauss's theorem yields:

$$
(\vec{D}_2 - \vec{D}_1) \cdot \hat{n}_{21} = 4\pi\sigma
\tag{3.33}
$$

where $\hat{n}_{21}$ is the unit vector normal to the interface directed from medium 1 to medium 2 and σ is the surface charge density at the interface (see section 3.1.4.1

above). If $\sigma = 0$, Eq.3.33 expresses the continuity of the normal component of the displacement field $D_n = \epsilon E_n$ across the interface between the two materials.

Similarly from Eq.3.5: $0 = \oint \vec{\nabla} \times \vec{E} \cdot d\vec{S} = \oint \vec{E} \cdot d\vec{l}$. Choosing a path $\vec{l}$ close to and parallel to the interface, one obtains:

$$(\vec{E}_2 - \vec{E}_1) \times \hat{n}_{21} = 0 \tag{3.34}$$

This equation expresses the continuity of the tangential electric field E_t across the interface. Eqs.3.33 and 3.34, are useful when computing the reflection and transmission of light passing between two medium (such as from air to glass), see section 3.5.1.

The work δW performed against the electric field when changing the charge distribution by an amount $\delta \rho(\vec{r})$ is:

$$\delta W = \int d^3 r' \Phi(\vec{r'}) \delta \rho(\vec{r'})$$

By Coulomb's law in a dielectric medium, Eq.3.30: $\delta \rho = \vec{\nabla} \cdot \delta \vec{D}/4\pi$:

$$\delta W = \frac{1}{4\pi} \int d^3 r' \Phi \, \vec{\nabla} \cdot \delta \vec{D} = \frac{1}{4\pi} \left(-\int d^3 r' \vec{\nabla} \Phi \cdot \delta \vec{D} + \int d^3 r' \vec{\nabla} \cdot (\Phi \delta \vec{D}) \right) \tag{3.35}$$

where we used the vector calculus identity $\vec{\nabla} \cdot (\Phi \delta \vec{D}) = \vec{\nabla} \Phi \cdot \delta \vec{D} + \Phi \vec{\nabla} \cdot \delta \vec{D}$. Far from the charges by Stokes' theorem, see Appendix A.3:

$$\int d^3 r' \vec{\nabla} \cdot (\Phi \delta \vec{D}) = \oint \Phi \, \delta \vec{D} \cdot d\vec{S'} \to 0$$

As $\vec{E} = -\vec{\nabla} \Phi$:

$$\delta W = \frac{1}{4\pi} \int d^3 r' \vec{E} \cdot \delta \vec{D} = \frac{\epsilon}{4\pi} \int d^3 r' \vec{E} \cdot \delta \vec{E} = \frac{\epsilon}{8\pi} \int d^3 r' \delta(\vec{E} \cdot \vec{E}) \tag{3.36}$$

If the work $\delta W > 0$, it increased the energy contained in the electric field. Thus in presence of a dielectric medium the total energy stored in the field is:

$$\mathcal{E}_e = \frac{\epsilon}{8\pi} \int d^3 r' \vec{E} \cdot \vec{E} = \frac{1}{8\pi} \int d^3 r' \vec{D} \cdot \vec{E} \tag{3.37}$$

The previous considerations are particular relevant to the case of a capacitor where the medium between the charged plates has dielectric constant $\epsilon > 1$. By Gauss's theorem the displacement field is then: $\vec{D} = 4\pi\sigma = 4\pi Q/S$. The potential between the plates is: $V = Ed = 4\pi Qd/\epsilon S$. Therefore, in comparison with the case where the medium between the plates is air, the capacitance $C = Q/V = \epsilon S/4\pi d$ is increased by a factor ϵ which can be very large. The energy stored in such a capacitor:

$$\mathcal{E}_e = \frac{1}{8\pi} \int d^3 r' \vec{D} \cdot \vec{E} = \frac{\epsilon S d}{8\pi} E^2 = \frac{\epsilon S}{8\pi d} V^2 = \frac{CV^2}{2}$$

is also increased by a factor ϵ. Increasing the dielectric constant between the plates of a capacitor is a very practical way of storing more charge and better buffering against changes in voltage. Thus dielectric constants $\epsilon > 10000$ have been achieved in polycrystalline ceramics of Calcium Copper Titanate ($CaCu_3Ti_4O_{12}$).

3.2 ◆ MAGNETOSTATICS

Magnetism like electricity has been known since ancient times, when people noticed that magnetite (lodestone) has the power to attract iron. Aristotle attributes to Thales (6th century BC) the first scientific discussion of magnetism. Mention of magnetism can also be found as early as the 4th century BC in the Chinese literature. The Chinese were indeed the first to use the property of magnets to point to the North in order to develop in the 11th century the lodestone compass. It took about a century for the invention to arrive in Europe were it revolutionized shipping and sea-trade, and was instrumental in the discovery of America. The unusual properties of the lodestone led the ancients to attribute a "spirit" to the stone. Indeed a full scientific explanation of magnets would have to wait the development of Quantum and Statistical Mechanics to explain how atoms could carry a magnetic moment (through their spin - see section 4.6) and how an ensemble of interacting spins could give rise to a global magnetic moment (see section 5.10.1).

The understanding of magnetism came not from a study of magnetite (the lodestone is a magnetized piece of magnetite), but from the discovery in 1819 by Oersted that current passing through a wire could deflect a nearby magnetic needle. This observation was revolutionary: it showed that two distinct phenomena, electricity and magnetism, were actually related. It immediately led to the development of an instrument (the galvanometer) that measured the amount of current flowing in a given coil from the deflection of a small permanent magnet in its vicinity. The following investigations by Ampère of this phenomenon led him to formulate his theory of electro-dynamics relating magnetic fields and forces to the currents passing in interacting wires. These experiments were made possible by the discovery of chemically generated electricity and the development of the battery.

3.2.1 ◆ VOLTA'S BATTERY AND OHM'S LAW

Until the end of the 18th century, currents were mostly generated by the momentary discharge of a capacitor (the Leyden jar). In 1800, Alessandro Volta observed that when two plates (one made of copper (Cu) and the other of zinc (Zn)) were immersed in a solution of brine or sulphuric acid a constant current would flow in a wire connecting these two plates. The current flow implied the existence of a force, i.e an electric potential difference between the two plates, the unit of which (the Volt) has been named in the honor of its discoverer. Nowadays we understand the functioning of Volta's battery as being an example of an oxido-reduction reaction. An oxidation reaction is taking place at the Zn anode: $Zn \rightarrow Zn^{++} + 2e^-$ and a reduction reaction is happening at the cathode: $Cu^{++} + 2e^- \rightarrow Cu$. Thus, electrons are transferred from the anode to the cathode, i.e. an electrical current is generated, see Fig.3.10.

For its development of the first battery, Volta was honored throughout Europe. Napoleon who was present at the demonstration by Volta of his battery at the French Academy, was so impressed that he made him knight of the Legion of Honor, Count of Lombardy and gave him a gold medal and a lifetime allowance to celebrate his contributions to Science! At that time scientists were admired by politicians who

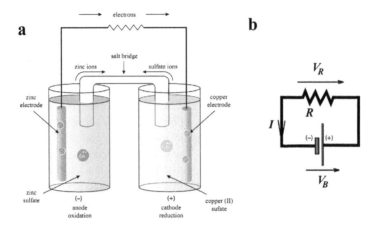

Figure 3.10 (a) Volta's battery. Negatively charged electrons are transferred to the Zn anode from a solution of $ZnSO_4$ where an oxidation of Zn occurs: $Zn \rightarrow Zn^{++} + 2e^-$. They are transferred from the Cu cathode to a solution of $CuSO_4$ where a reduction of Cu occurs: $Cu^{++} + 2e^- \rightarrow Cu$. The electrons are recycled via a wire connecting the anode with the cathode (adapted from ref.[26]). (b) The commonly used symbol for a battery with voltage V_B. When connected to a resistance R, a positive current I flows from the cathode to the anode (opposite to the flow of the negatively charged electrons) generating a voltage $V_R = IR$ opposite to V_B.

attended sessions of the Academy!!

Volta's invention of the battery launched the field of electrochemistry (the use of electricity to catalyze chemical reaction): within a few months of his invention, using a voltaic cell as a current source Carlisle and Nicholson discovered the phenomenon of water electrolysis by which they showed that water is made of hydrogen and oxygen. The battery also allowed for the first investigations of electricity: the behavior of conducting materials. It led Georg Simon Ohm to formulate in 1827 his famous law that the current I (which MKS units are Amperes ($1A = 1$ C/s)) is related to the voltage V across a conductor by:

$$V = IR \tag{3.38}$$

Where R is known as the resistance. Its units in the MKS system are named after Ohm: $1 \, \Omega = 1V/1A$, while in the CGS system the unit of resistance is sec/cm: $1\Omega = 10^{-11}/9$ sec/cm.

The electrical resistance of a material is now understood to result from the scattering of the charge carriers (e.g. the electrons) by the atoms of the material (by their collective thermal vibrations (known as phonons) or by impurities). To see how this comes about, let τ be the mean time between collisions and $\langle \vec{v} \rangle$ the mean velocity of the electrons of mass m. By Newton's Law: $q\vec{E} = \vec{F} = d\vec{p}/dt \sim m \langle \vec{v} \rangle / \tau$. Hence the mean velocity of the charge carriers is:

$$\langle \vec{v} \rangle = \frac{\tau q}{m} \vec{E} \equiv \mu_e \vec{E} \tag{3.39}$$

where $\mu_e \equiv \tau q/m$ is known as the charge mobility. For copper $\mu_e = 1.35 \ 10^4$ cm^2/statV·sec $= 45$ cm^2/V·sec. For a wire of given length l, cross-section S and a density of free electrons n_e, the current is[13]:

$$I = n_e q S \langle v \rangle = n_e q S \mu_e E = n_e \mu_e q \frac{S}{l} V \equiv gV \equiv V/R \tag{3.40}$$

which is Ohm's law. The quantity $\sigma_R = n_e \mu_e q = n_e q^2 \tau/m$ is known as the conductivity of the material. In CGS units it has dimensions of sec^{-1}, while in MKS it is Siemens/m $(= \Omega m)^{-1}$. For good conducting materials (such as copper, silver, etc.) : $10^{17} < \sigma_R < 10^{18}$ sec^{-1} ($10^7 < \sigma_R < 10^8$ Siemens/m). Insulating materials such as wood or air have very low conductivity (they have very few free electrons that don't travel far): $\sigma_R \sim 10^{-5}$ sec^{-1}. The conductance is defined as $g = I/V = \sigma_R S/l$ while the resistance is: $R \equiv 1/g = l/\sigma_R S$. The resistance increases with the length l of the wire and decreases with its cross-section S and conductivity σ_R (it increases with the resistivity $\rho_R \equiv 1/\sigma_R$). On can intuitively understand Ohm's law: the longer the wire the more the electrons are scattered and the higher the resistance, while the larger its cross section the more current can flow, the higher the conductance. As a result of electron scattering electric energy is dissipated: the material is heated (this is how electric heaters work and why lamps are hot). The power dissipated is:

$$P_{diss} = \frac{dW}{dt} = \frac{dVq}{dt} = V\frac{dq}{dt} = IV = I^2 R = V^2/R \tag{3.41}$$

At fixed voltage (which is usually the case with electrical appliances that are powered by batteries or by the electric grid) the smaller the resistance the larger the dissipated power. Nowadays, Ohm's law is appreciated as a particular example of a more general principle (the dissipation-fluctuation theorem, see section 5.11.3). In a system near thermodynamic equilibrium the response (e.g. the current I in an electrical circuit, the velocity $\vec{v}$ of a particle in a viscous medium) is proportional to the driving force (e.g. the voltage $V = IR$ or the hydrodynamic drag: $\vec{F} = \gamma \vec{v}$). The dissipation-fluctuation theorem states that the coefficient of proportionality (e.g. the resistance R, the drag coefficient γ) sets the dissipation of energy in the system and is correlated to the fluctuations at equilibrium of the force (the larger the dissipation, the larger the fluctuations, i.e. the noise).

For a system of charges in motion, one defines the current density as: $\vec{J} = \rho \vec{v}$ (where ρ is the local charge density and $\vec{v}$ the local average velocity of the charges).

---

[13] The voltage difference across the wire V can be derived from Eq.3.6: $V = El$

The total current through a closed surface $\vec{S}$ is: $I = \int \vec{J} \cdot d\vec{S}$. Charge conservation implies that $I = -dQ/dt$, namely:

$$I = \int \vec{J} \cdot d\vec{S} = \int dV \vec{\nabla} \cdot \vec{J} = -dQ/dt = -\int dV \frac{\partial \rho}{\partial t} \tag{3.42}$$

from which one obtains the equation of continuity for the charge density ρ:

$$\frac{\partial \rho}{\partial t} + \vec{\nabla} \cdot \vec{J} = 0 \tag{3.43}$$

The equation of continuity is the stronger local form of a conservation law. It states that any local change in a conserved quantity is due to its transport. Such equation can be formulated in any context where a quantity is conserved (mass, energy, probability, etc.). For example in hydrodynamics, where mass is conserved, the equation of continuity is very similar to Eq.3.43:

$$\partial \rho_m / \partial t + \vec{\nabla} \cdot \rho_m \vec{v} = 0$$

where ρ_m is the fluid density. If the fluid is incompressible ($\rho_m = $ const), the equation of continuity is $\vec{\nabla} \cdot \vec{v} = 0$: the flow that gets in must get out.

In similitude with incompressible hydrodynamics, in magnetostatics the charge density does not depend explicitly on time ($\partial \rho / \partial t = 0$). Therefore $\vec{\nabla} \cdot \vec{J} = 0$, i.e. the total current through a closed surface is null: whatever current gets in must get out. This is the type of current configurations that Ampère studied. These investigations relied on Volta's invention of the battery, which delivered a constant current I for extended times. In particular Ampère studied the force between two very long parallel wires a distance r apart and bearing currents I_1, I_2. He observed that if the currents were flowing parallel to each other the wires would attract, while they would repel if the current were flowing in anti-parallel directions. He further noticed that the force per unit length between the wires dF_{12}/dl was given by:

$$dF_{12}/dl = k \frac{2I_1 I_2}{r} \tag{3.44}$$

In the MKS system $k = \mu_0/4\pi = 10^{-7}$ N/A^2, where μ_0 is known as the vacuum magnetic permeability[14]. In the CGS system $k = 1/c^2$ (where c is the speed of light). To understand how the speed of light enters into a magnetostatic problem, one has to resort to the special theory of relativity. Even though, that connection was made by Einstein about 80 years after Ampère's discovery, we shall deviate from the historical narrative and adopt his point of view since it reveals a profound connection between electrostatics and magnetostatics.

---

[14] As already mentioned, in contrast to the compressibility and density of air which set the sound velocity, the permeability μ_0 has no physical reality. It was introduced at a time when people believed in the existence of a medium (the luminous ether characterized by its dielectric constant ϵ_0 and magnetic permeability μ_0) which set the velocity of electromagnetic(EM) waves, i.e. light, propagating in it. The Michelson-Morley experiments and Einstein theory of relativity, see Appendix A.6, have quashed this belief.

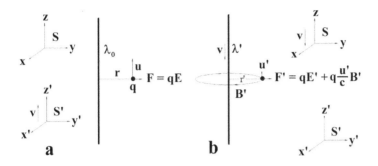

Figure 3.11 Derivation of the Lorentz force from electrostatics and special relativity. (a) In frame S a particle with charge q is moving at velocity $\vec{u} = u\hat{z}$ and experiences the repulsive force from a static wire with charge density λ_0. (b) In a frame S' moving with velocity $\vec{v} = v\hat{z}$, the wire appears to be moving with velocity v, charged with a density $\lambda' = \lambda_0/\sqrt{1 - v^2/c^2}$ and thus generating a current $I' = \lambda'v$. Special relativity predicts that the force experienced by the particle in frame S' can be interpreted as resulting from an electric repulsion $qE' = 2\lambda'q/r'$ and a magnetic repulsion $qu'B'/c$ with $B' = 2I'/r'c$ (i.e. $\vec{B}' = -\vec{v}\times\vec{E}'/c$). The magnetic repulsion in frame S' is the same as predicted by Ampère's Law.

3.2.2 ◆ ELECTROSTATICS, MAGNETOSTATICS AND RELATIVITY: THE LORENTZ FORCE

Consider a charge q moving in a reference frame S with velocity $\vec{u} = u\hat{z}$ parallel to and a distance r from a long wire bearing a charge density $\lambda_0 = Q/L_0$. Using Gauss's theorem one can show (see Eq.A.77) that the electric field of such a wire is: $E = 2\lambda_0/r$. The (repulsive) force on the charge is therefore $F = qE = 2q\lambda_0/r$, see Fig.3.11.

Now consider the description of that system by an observer in an inertial frame S' moving parallel to the wire with velocity $\vec{v} = v\hat{z}$. To this observer, the wire appears to move in the opposite direction with velocity $\vec{v} = -v\hat{z}'$ and the charge with velocity u'. From Einstein's special theory of relativity (see section A.6), we know that for this observer the distance between the charge q and the wire is unchanged, being perpendicular to the direction of motion: $r' = r$. However the length of the wire L' appears shorter than its rest frame length L_0: $L' = L_0\sqrt{1 - v^2/c^2}$. Hence to an observer in S', the charge density on the wire is increased: $\lambda' = \lambda_0/\sqrt{1 - v^2/c^2}$ and there is a current $I' = \lambda'v$. The velocity of the particle in that frame of reference is (Eq.A.69):

$$u' = \frac{u - v}{1 - uv/c^2}$$

and the force on the particle is, see Eq.A.76:

$$F' = \sqrt{1 - u'^2/c^2}\,F_0 = \frac{\sqrt{(1 - u^2/c^2)(1 - v^2/c^2)}F_0}{(1 - uv/c^2)}$$

where F_0 is the force acting on the particle in its **own** frame of reference (one moving with velocity u with respect to the wire), i.e.:

$$F_0 = \frac{2q\lambda_0}{r\sqrt{1-u^2/c^2}}$$

Notice that in frame S: $F = \sqrt{1-u^2/c^2}F_0 = 2q\lambda_0/r$ as expected from electrostatics. In the reference frame S', the force can be written as:

$$F' = \frac{2q\lambda_0\sqrt{1-v^2/c^2}}{r(1-uv/c^2)} = \frac{2q\lambda'(1-v^2/c^2)}{r(1-uv/c^2)}$$

$$= \frac{2q\lambda'}{r'}(1+u'v/c^2) = \frac{2q\lambda'}{r'} + \frac{qu'}{c}\frac{2\lambda'v}{r'c}$$

$$= qE' + \frac{qu'}{c}B' \tag{3.45}$$

The first term in Eq.3.45 represents the electrostatic repulsion (measured by an observer in the moving frame S') between a charge q and a wire with charge density λ', generating an electric field $\vec{E}' = (2\lambda'/r')\hat{r}'$ (in cylindrical coordinates (r',ϕ',z')). The second term corresponds (for the moving observer) to a repulsion force increasing linearly with the particle velocity u'. This is the magnetic repulsion (discovered by Ampère) between a charge q moving with velocity $\vec{u}' = u'\hat{z}'$ parallel to a wire bearing a current $\vec{I}' = -\lambda'v\hat{z}'$ in the opposite direction and generating at a distance r' from it a magnetic field:

$$\vec{B}' = -\frac{2I'}{r'c}\hat{\phi}' \tag{3.46}$$

where $\hat{\phi}'$ is the unit vector in the azimuthal direction. Eq.3.46 can be also be written as $\oint r'd\hat{\phi}' \cdot \vec{B}' = 4\pi I'/c$. The magnetic field direction is given by the right hand rule: if the thumb points along the current, the bent fingers point along the field. The magnetic force (also known as the Lorentz force), $\vec{F}'_m$ becomes:

$$\vec{F}'_m = q\frac{\vec{u}'}{c} \times \vec{B}' \tag{3.47}$$

Electric and magnetic fields are thus relative (and related) concepts for describing in different moving frames the same interaction between charged particles. Comparing the expressions for E' and B', one notices that in the frame S' moving at velocity $\vec{v} = v\hat{z}$: $\vec{B}' = -\vec{v}\times\vec{E}'/c$. Electric and magnetic fields have the same units, even though in the CGS system these are called Gauss (G) for the magnetic field and statV/cm for the electric field. In the MKS system the units of $\vec{B}$ are Tesla: $1\ T = 10^4\ G$.

While the values of the electric and magnetic fields will be different for observers in different inertial frames, the observed physical phenomena (e.g. charge attraction or repulsion) remain the same: only their interpretation differs. However, even though electric and magnetic fields are associated with relative descriptions of the same phenomena, there is a fundamental difference between them: while electric charges are commonly observed no one has ever observed a magnetic charge, a magnetic monopole. In Nature magnets always come as dipoles, with a north and a south pole. In that sense magnetic fields are derived from the more fundamental electric fields. The consequences of this fact are multiple. First, since there are no magnetic charges according to Gauss's theorem:

$$\vec{\nabla} \cdot \vec{B} = 0 \tag{3.48}$$

In other words, the magnetic field lines must close on themselves. The second consequence is that magnetic fields do no work: $W_m = 0$. Indeed, from Eq.3.47:

$$W_m = \int \vec{F}_m \cdot d\vec{r} = \frac{q}{c} \int (\frac{d\vec{r}}{dt} \times \vec{B}) \cdot d\vec{r} = \frac{q}{c} \int (d\vec{r} \times \frac{d\vec{r}}{dt}) \cdot \vec{B} = 0 \tag{3.49}$$

where we used the vector identity $\vec{A} \cdot (\vec{B} \times \vec{C}) = \vec{C} \cdot (\vec{A} \times \vec{B})$, see Appendix A.3. The Lorentz force experienced by a moving charge in a magnetic field is a generalization of Ampères's force law for the interaction between electric currents, Eq.3.44. Indeed, the force on a wire carrying a current I_2 in the magnetic field $\vec{B}_1$ generated by a wire carrying a current I_1 can be computed from Eq.3.47 by noticing that $q\vec{u} = \lambda_2 \vec{u} dl = I_2 d\vec{l}$:

$$dF_m = \frac{1}{c} I_2 d\vec{l} \times \vec{B}_1 = \frac{1}{c} \frac{2I_1 I_2}{rc} d\vec{l} \times \hat{\phi} = -\frac{2I_1 I_2}{rc^2} dl \, \hat{r} \tag{3.50}$$

where we used Eq.3.46 for the magnetic field $\vec{B}_1$ a distance r from a wire carrying a current I_1. This equation is Ampère's force law, Eq.3.44.

3.2.3   APPLICATIONS OF THE LORENTZ FORCE

3.2.3.1   Aurora Borealis

The Lorentz force provides an explanation for the beautiful phenomena of the Aurora Borealis observable at the Earth's poles, see Fig.3.12. This phenomena is due to the excitation of atoms in the upper atmosphere (80km above ground) by charged particles emitted from the sun, the so-called solar wind. These particles moving at a speed of about 400km/s are deflected upon their encounter with the Earth magnetic field to the poles where they are decelerated by their collisions with the atoms in the atmosphere (mostly Oxygen and Nitrogen). As we shall see later on, see section 4.12.3, the colors of the Aurora are due to the light emitted by the excited atoms as they return to their ground state (green light for oxygen, red light for nitrogen).

To study the motion of a particle of charge q and velocity $\vec{v}$ in a magnetic field $\vec{B}$, we shall align the z-axis along the magnetic field lines (a valid local approximation), i.e. $\vec{B} = B_0 \hat{z}$. The Lorentz-force, Eq.3.47, acting on the particle is then: $\vec{F} = (q/c)\vec{v} \times \vec{B} = (qB_0/c)(v_y \hat{x} - v_x \hat{y})$. From Newton's law, $F = m\vec{a} = md\vec{v}/dt \equiv m\dot{\vec{v}}$, thus:

$$\begin{aligned} m\dot{v}_x &= (qB_0/c)v_y \\ m\dot{v}_y &= -(qB_0/c)v_x \\ m\dot{v}_z &= 0 \end{aligned} \tag{3.51}$$

Figure 3.12 Curtains of Aurora Borealis over a Northern forest (reprinted from ref.[27]).

Defining the cyclotron frequency $\omega_c = qB_0/mc$ the solution of these equations is: $v_x(t) = v_0 \sin(\omega_c t + \phi)$; $v_y(t) = v_0 \cos(\omega_c t + \phi)$; $v_z =$ const. From this solution it is clear that:

1. The energy of the particle is unaffected by the magnetic field ($m\vec{v}^2/2 =$ const), as expected since the magnetic field does no work.
2. The velocity of the particle along the field line is constant, hence the deflection of the solar wind to the poles.
3. The particle spirals about the magnetic field line at a frequency ω_c and with a radius $\rho = v_0/\omega_c$.

The deflection of the solar wind to the poles by the Earth magnetic field has allowed for Life on Earth to develop without interference from the highly energetic radiation emitted by the Sun. In fact, the possibility of solar storms emitting high energy particles is one of the main hazards on a trip to Mars.

The confinement of moving charged particles in the vicinity of magnetic field lines which explains the observation of Aurora Borealis at the poles only, is a property that is being used in Nuclear Fusion Reactors, known as Tokamaks. As in the Sun's core which is getting its energy from the fusion of Hydrogen nuclei into alpha particles (Helium nuclei), see section 4.1.1.1, a fusion reactor is trying to fuse two isotopes of Hydrogen (Deuterium and Tritium) into Helium. As a result of that fusion a portion Δm of the initial mass is converted into energy (using Einstein celebrated formula: $E = mc^2$, see appendix A.6). In a Tokamak the nuclei of the Hydrogen isotopes have to be heated to millions of degrees so that their kinetic energy can overcome their electrostatic repulsion potential. At such extremely high temperatures it is a daunting challenge to confine the hot plasma of charged electrons and nuclei in a reactor (without cooling the plasma or melting the reactor). In Tokamaks, the idea is to confine this highly energetic ionized matter along the closed magnetic field lines generated by very strong external magnets, as the solar wind is channeled by the Earth magnetic field to the poles. However due to electro- and thermo-dynamic instabilities, the goal of sustained fusion is still a serious challenge, not to mention the issue of the damage to the reactor structure (and the induced radioactivity) arising from the hot neutrons generated during the fusion reactions which being electrically neutral cannot be confined by electro-magnetic fields.

3.2.3.2   The Particle Accelerator

For non-relativistic velocities ($v \ll c$) the cyclotron frequency, $\omega_c = qB_0/mc$, is independent of the velocity of the particle. This has important practical applications. For example the first particle accelerator (built in 1932 by Lawrence at UC Berkeley) consisted of a chamber in the form of a disk each half of which was held at an opposite alternating potential (see Fig.3.13). A magnetic field perpendicular to the chamber was used to bend the trajectory of the electrons emitted at the center of the chamber and accelerated between its electrodes.

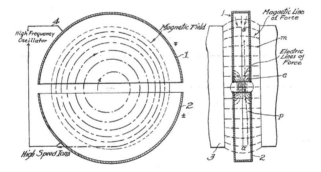

Figure 3.13   The original drawings of Lawrence's patent for the cyclotron. A conducting disk-like chamber was placed between the poles of a magnet. Electrons emitted at the center of the chamber were accelerated by a voltage on each half of the chamber alternating at twice the orbital (cyclotron) frequency of the electrons.

The fact that the cyclotron frequency does not depend on the velocity of the particle allows to accelerate them coherently at twice the cyclotron frequency (i.e. each time they pass the gap between the two half of the disk). As they are accelerated the radius of their motion around the magnetic field lines increases until they escape from the chamber. This very simple particle accelerator is used today to generate high energy proton or ion beams for cancer radiation therapy. In modern particle accelerators where magnetic fields are used to bend the trajectory of the very fast

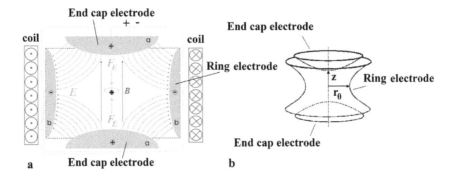

Figure 3.14   The Penning trap is a configuration of electric and magnetic fields that can trap particles in an orbital motion about the magnetic field lines so that their mass to charge ratio can be estimated accurately by a precise measurement of their cyclotron frequency. (a) Cut through a Penning trap with cylindrical symmetry displaying the electrodes (in blue) generating the electric field and the coils (in red) generating the magnetic field (adapted from ref.[28]). (b) 3D view of the electrodes in a Penning trap.

moving particles ($v \lesssim c$) into a more or less circular orbit, relativistic effects are important: the effective mass of the particle increases with the velocity, see Appendix A.6. Consequently, the cyclotron frequency varies with the particle velocity which requires the synchronization of the pulse of electric acceleration with the cyclotron frequency, hence the name of these accelerators: synchrotrons.

The Lorentz force is also used in electron microscopes (and old cathode tube TV sets) together with electrostatic lenses to control (bend) the path of electron beams. In analytical chemistry magnetic fields are used in a mass spectrometer to separate ionized molecules according to their mass to charge ratio (m/q). Thus in so called MALDI-TOF spectrometer, the molecules to be analyzed (organics molecules, proteins, sugars, etc.) are vaporized by a UV laser pulse that charges the molecules and their solvent (the matrix, hence the acronym MALDI for Matrix Assisted Laser Desorption Ionization). The molecules are then accelerated and deflected by a combination of electric and magnetic fields. Their time of flight (TOF) can be measured and related to their velocity v which by energy conservation satisfies: $mv^2/2 = qV$. Thus knowing the acceleration voltage V the ratio m/q can be determined and since q is a small integer number of electron charges (typically $q = \pm e$), the mass of the detected molecule can be inferred. Alternatively the deflection of the molecules by the magnetic field can be measured and related again to their mass to charge ratio.

3.2.3.3 The Penning trap

However the most accurate method to measure the mass to charge ratio is through the use of a Penning trap, see Fig.3.14, an instrument that earned his inventor, H.G. Dehmelt, the 1989 Nobel prize in Physics for its applications in precision atomic clocks. This trap consists of a combination of electric and magnetic fields that force charged particles into circular orbits the frequency of which is related to the cyclotron frequency. The measurement of their orbital frequency allows for a very precise determination of their mass to charge ratio. As a simple approximation of the Penning trap consider the following configuration of charges: charges $+2Q$ on the z-axis at position $\pm d$ and charges $-Q$ on the x and y-axes at position $\pm d$. One can show that the electric field $\vec{E}$ generated in the vicinity of the origin by this charge configuration is: $\vec{E} = (6Q/d^3)(\vec{x} + \vec{y} - 2\vec{z})$. In a Penning trap this charge configuration is supplemented with a constant magnetic field along the z-axis: $\vec{B} = B_0 \hat{z}$, which role is to confine the motion of the particles in the xy plane, while the confinement (of a positively charged particle) along the z-axis is due to the repulsion from the positive charges at $(0, 0, \pm d)$. In the xy-plane the motion of the particle obeys:

$$m\ddot{x} = \frac{qB_0}{c}\dot{y} + \frac{6Qq}{d^3}x$$

$$m\ddot{y} = -\frac{qB_0}{c}\dot{x} + \frac{6Qq}{d^3}y \tag{3.52}$$

Which solution is an epicycle (like the trajectory of the moon about the sun) characterized by two frequencies (ω_1 ; ω_2) the sum of which is the cyclotron frequency $\omega_c = qB_0/mc$ (check[15] for yourself!). Therefore by measuring these frequencies, one can determine very precisely the ratio m/q. The Penning trap can also be used to trap ions at very low temperatures in order to investigate their atomic transitions (see section 4.3.2.1) and use those as time standards (atomic clocks).

3.2.4 ◆ AMPÈRE'S LAW

In our previous derivation of the magnetic field of a current line, Eq.3.46, we saw that $\int r d\vec{\phi} \cdot \vec{B} = 4\pi I/c$, where the integration is around a circle of radius r. This result can be expressed in a more general form analogous to Gauss's theorem in electrostatics:

$$\oint \vec{B} \cdot d\vec{r} = \frac{4\pi}{c}I = \frac{4\pi}{c}\int \vec{J} \cdot d\vec{S} \tag{3.53}$$

i.e. the integral of the magnetic field around a closed loop (not necessarily a circle) is proportional to the current passing through the area delimited by that loop. Eq.3.53 is known as Ampère's law for the magnetic field (not to be confused with Ampère's force law, Eq.3.44). Since from Stokes' theorem, see Appendix A.3:

$$\oint \vec{B} \cdot d\vec{r} = \int \vec{\nabla} \times \vec{B} \cdot d\vec{S} \tag{3.54}$$

---

[15]Define $z \equiv x + iy$ and derive its dynamic equation: $\ddot{z} + i\omega_c\dot{z} - \alpha^2 z = 0$ with $\alpha^2 = 6Qq/md^3$, then look for a solution $z(t) = z_1 e^{-i\omega_1 t} + z_2 e^{-i\omega_2 t}$

One may cast Ampère's law into a differential form (valid for any current distribution):

$$\vec{\nabla} \times \vec{B} = \frac{4\pi}{c} \vec{J} \tag{3.55}$$

This equation is the equivalent of Gauss's equation for electrostatic fields, Eq.3.16. If $\vec{r}$ denotes the position where the magnetic field is measured and $\vec{r}'$ the position of the currents (the sources of the magnetic field), then:

$$\begin{aligned} \vec{B}(\vec{r}) &= \frac{1}{c} \int d^3r' \, \vec{J}(\vec{r}') \times \frac{\vec{r} - \vec{r}'}{|\vec{r} - \vec{r}'|^3} \\ &= \frac{I}{c} \int \frac{d\vec{l} \times (\vec{r} - \vec{r}')}{|\vec{r} - \vec{r}'|^3} \end{aligned} \tag{3.56}$$

where the last equation, known as the law of Biot-Savart is obtained when the current is flowing in a wire and the 1D integration ($d\vec{l}$) is along the wire. The magnetic field $\vec{B}$ in magnetostatics plays the role of the electric field $\vec{E}$ in electrostatics (compare Eq.3.56 with Eq.3.2). In analogy with the introduction of the electric potential in electrostatic problems, it is convenient to rewrite Eq.3.56 differently such as to introduce the magnetic vector potential $\vec{A}$:

$$\begin{aligned} \vec{B} &= \frac{1}{c} \int d^3r' \, \vec{J}(\vec{r}') \times \frac{\vec{r} - \vec{r}'}{|\vec{r} - \vec{r}'|^3} \\ &= \frac{1}{c} \vec{\nabla} \times \int d^3r' \, \frac{\vec{J}(\vec{r}')}{|\vec{r} - \vec{r}'|} \equiv \vec{\nabla} \times \vec{A} \end{aligned} \tag{3.57}$$

The magnetic field is thus obtained as the rotational of the vector potential: $\vec{B} = \vec{\nabla} \times \vec{A}$. The vector potential $\vec{A}$ plays a similar role in magnetostatics to the one played by the potential Φ in electrostatics ($\vec{E} = -\vec{\nabla}\Phi$). The real measurable fields, the ones acting on charged particles are the electric and magnetic fields ($\vec{E}$ and $\vec{B}$). However, the potential fields (Φ and $\vec{A}$) are often simpler to compute. Notice though that while Φ is defined up to a constant, $\vec{A}$ is defined up to the gradient of a function (since the rotational of a gradient is zero, $\vec{\nabla} \times \vec{\nabla}\Psi = 0$). That freedom of choice (known as gauge invariance) plays a major role in Quantum Mechanics. From these considerations we may write:

$$\vec{A} = \frac{1}{c} \int d^3r' \, \frac{\vec{J}(\vec{r}')}{|\vec{r} - \vec{r}'|} + \vec{\nabla}\Psi \tag{3.58}$$

Notice that this equation linking the components A_i with J_i is valid only in cartesian coordinates: $i = x, y$ or z. Since by definition: $\vec{\nabla} \cdot \vec{\nabla} \times \vec{A} = 0$, Eq.3.58 is consistent with Eq.3.48 expressing the absence of magnetic charges. Using the relation $\vec{B} = \vec{\nabla} \times \vec{A}$, Eq.3.55 can be cast as an equation for the vector potential $\vec{A}$:

$$\vec{\nabla} \times \vec{B} = \vec{\nabla} \times \vec{\nabla} \times \vec{A} = \vec{\nabla}(\vec{\nabla} \cdot \vec{A}) - \nabla^2 \vec{A} = \frac{4\pi}{c} \vec{J} \tag{3.59}$$

In magnetostatics charge conservation implies $\vec{\nabla} \cdot \vec{J} = 0$ and therefore $\vec{\nabla} \cdot \vec{A} = 0$, which given a time independent current distribution of sources $\vec{J}$ leads to a Poisson's equation for the vector field $\vec{A}$:

$$\nabla^2 \vec{A} = -\frac{4\pi}{c} \vec{J} \tag{3.60}$$

Thus given a distribution of currents the three components of the vector potential can be computed (analytically or more often numerically) just like the electrostatic potential can be computed for a given charge distribution (see Appendix A.5).

3.2.5 APPLICATIONS OF AMPÈRE'S LAW

3.2.5.1 Magnetic Field of a Current Segment

Let us use Biot-Savart's Law, Eq.3.56 to compute the contribution to the magnetic field at a distance r from the axis of a segment of length l carrying a current I[20], see Fig.3.15:

$$\vec{B}(\vec{r}) = \frac{I}{c} \int \frac{d\vec{l} \times (\vec{r} - \vec{r}\,')}{|\vec{r} - \vec{r}\,'|^3} = \frac{I\hat{\phi}}{c} \int \frac{dz \sin\alpha}{|\vec{r} - \vec{r}\,'|^2} = \frac{I\hat{\phi}}{c} \int \frac{dz \sin\alpha}{r^2 + z^2}$$

Using the relations: $z = r\cot\alpha$; $dz = -rd\alpha/\sin^2\alpha$; $r^2 + z^2 = r^2/\sin^2\alpha$, we can replace the integration over z by an integration over α:

$$\vec{B}(\vec{r}) = -\frac{I\hat{\phi}}{c} \int_{\alpha_1}^{\alpha_2} \frac{rd\alpha}{\sin^2\alpha} \sin\alpha \frac{\sin^2\alpha}{r^2} = -\frac{I\hat{\phi}}{rc} \int_{\alpha_1}^{\alpha_2} d\alpha \sin\alpha$$

$$= \frac{I\hat{\phi}}{rc} (\cos\alpha_2 - \cos\alpha_1) \tag{3.61}$$

One can check that for an infinite wire ($\alpha_1 = \pi$, $\alpha_2 = 0$) one recovers our previous result: $B = 2I/rc$. The previous result can be used to compute the magnetic field at the center of a square loop ($\alpha_1 = 3\pi/4$, $\alpha_2 = \pi/4$; ...): $B = 4\sqrt{2}(I/rc)$ and other variants (octagonal loop, etc.).

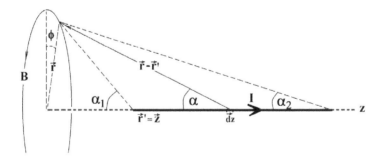

Figure 3.15 Current I is flowing through a segment of length l along the z–axis. The magnetic field at position r is computed using Biot-Savart's law by integrating over $r' = z = r\cot\alpha$ from α_1 to α_2 (with $dr' = dz = -rd\alpha/\sin^2\alpha$ and $|\vec{r} - \vec{r}\,'| = r/\sin\alpha$).

3.2.5.2 Magnetic Field of a Current Loop

To demonstrate the utility of the vector potential $\vec{A}$ in computing the magnetic field $\vec{B}$, consider the field of a current I flowing in a circular loop of radius a, lying in the xy-plane, see Fig.3.16. In polar coordinates the current density can be written as: $J_\phi = I\delta(\cos\theta')\delta(|\vec{r}\,'|-a)/a$. In the cartesian coordinates required to compute the vector potential $\vec{A}$ from Eq.3.58:

$$\vec{A}(\vec{r}) = \frac{1}{c} \int d^3r' \frac{\vec{J}(\vec{r}\,')}{|\vec{r}-\vec{r}\,'|} \tag{3.62}$$

the current density is: $\vec{J}(\vec{r}\,') = -J_\phi \sin\phi'\,\hat{x} + J_\phi \cos\phi'\,\hat{y}$. Since the problem has azimuthal symmetry, we may without loss of generality choose the observation point in the xz-plane (i.e. $\phi = 0$). In that case the denominator becomes:

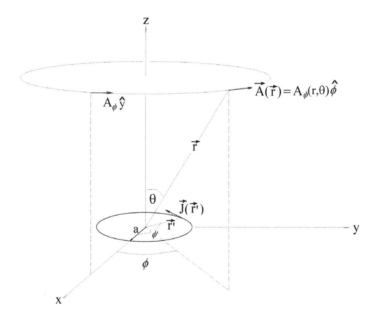

Figure 3.16 A current with density $\vec{J} = I\hat{\phi}\,\delta(\cos\theta')\delta(|\vec{r}\,'|-a)/a$ is flowing in a loop of radius a. The magnetic vector potential $\vec{A}(\vec{r})$ at position $\vec{r}$ is computed from Eq.3.58. Notice that $\vec{A}$ has the same symmetry as the current density, i.e. $\vec{A} = A_\phi(r,\theta)\hat{\phi}$.

$$|\vec{r}-\vec{r}\,'| = \{r^2 + r'^2 - 2rr'(\cos\theta\cos\theta' + \sin\theta\sin\theta'\cos\phi')\}^{1/2} = \sqrt{r^2 + a^2 - 2ar\sin\theta\cos\phi'}$$

where we used the fact that $r' = a$ and $\theta' = \pi/2$. Since the denominator is symmetric with respect to $\phi' \rightarrow -\phi'$, the component of $\vec{J}$ along x will not contribute (it changes sign as $\phi' \to -\phi'$) and thus we are left with:

$$A_\phi(r,\theta) = A_y(r,\theta,\phi=0) = \frac{1}{c}\int r'^2 dr' d\cos\theta' \frac{d\phi' J_\phi \cos\phi'}{\sqrt{r^2 + a^2 - 2ar\sin\theta\cos\phi'}}$$

$$= Ia/c \int_0^{2\pi} \frac{d\phi' \cos\phi'}{\sqrt{r^2 + a^2}}[1 + \frac{ar\sin\theta\cos\phi'}{r^2 + a^2} + ...]$$

$$\simeq \frac{I\pi a^2}{c}\frac{\sin\theta}{r^2} \tag{3.63}$$

Where we have assumed that $r \gg a$ (far field approximation). Because of the azimuthal symmetry: $A_\phi(r,\theta) = A_y(r,\theta,\phi=0)$ and thus we may compute the components of the magnetic fields $\vec{B}$ (see Appendix A.3):

$$B_r = \frac{1}{r\sin\theta}\frac{\partial}{\partial\theta}(A_\phi \sin\theta) = 2\frac{I\pi a^2}{c}\frac{\cos\theta}{r^3} \equiv \frac{2m\cos\theta}{r^3} \tag{3.64}$$

$$B_\theta = -\frac{1}{r}\frac{\partial}{\partial r}(rA_\phi) = \frac{I\pi a^2}{c}\frac{\sin\theta}{r^3} \equiv \frac{m\sin\theta}{r^3} \tag{3.65}$$

which has the same form as the electric field of a dipole, Eq.3.12, where $\vec{m} = (IS/c)\hat{z}$ is the magnetic dipole moment of a small current loop of area $S = \pi a^2$. In atoms the magnetic dipole is often the result of the motion of electrons (with charge e) around the nucleus with a period T (frequency $\omega = 2\pi/T$), generating a current $I = e/T$. We can relate the magnetic dipole to the angular momentum of the electron by writing:

$$\vec{m} = (e\pi a^2/Tc)\hat{z} = em\omega a^2/2mc\hat{z} = (e/2mc)L_z\hat{z} \tag{3.66}$$

where the angular momentum is: $\vec{L} = \vec{r} \times \vec{p} = m\omega a^2\hat{z}$ (with momentum $\vec{p} = m\omega a\hat{\phi}$). We shall revisit that point when studying the QM of an atom in a magnetic field, sections 4.6 and 4.11.3.

Notice that the vector potential can be written as:

$$\vec{A} = \frac{\vec{m}\times\vec{r}}{r^3} = \frac{m\sin\theta}{r^2}\hat{\phi} \tag{3.67}$$

A current flowing through a small loop will produce a magnetic dipole which will tend to align the other magnetic dipoles (e.g. permanent magnets) in its vicinity. As we shall see below, this property has found many applications in our daily life. Notice, that in presence of a distribution of magnetic dipoles with density $\vec{M}(r)$ the total vector potential is obtained as the sum:

$$\vec{A}(r) = \int d^3r' \vec{M}(r') \times \frac{(\vec{r}-\vec{r}')}{|\vec{r}-\vec{r}'|^3} = \int d^3r' \vec{M}(r') \times \vec{\nabla}' \frac{1}{|\vec{r}-\vec{r}'|} \tag{3.68}$$

3.2.5.3 The Solenoid

When the distribution of currents is symmetric and uniform Ampère's law, Eq.3.53, is often the simplest way to compute the magnetic field. Consider for example an

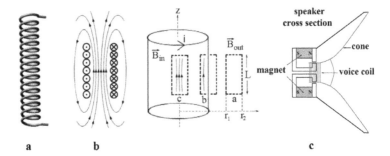

Figure 3.17   a) A solenoid is a made by wrapping a long wire around a cylinder. The current flowing through a solenoid creates a magnetic field that is very large in the solenoid and very weak outside. (b) In the limit of infinitely long solenoid, the magnetic field $\vec{B}$ is aligned along the solenoid $z-$ axis. Its value inside and outside the solenoid (B_{in}, B_{out}) can be computed by applying Ampère's law on the loops a, b and c. (c) A voice coil is a solenoid which interacts with the field of a permanent magnet when a current is flowing through it. In the present example the coil is used to move the membrane of a speaker as it gets in and out of the field of the magnet.

infinitely long solenoid of radius a carrying a current I and characterized by n coils per unit length. Because of symmetry the magnetic field $\vec{B}$ will be aligned along the solenoid (z-)axis. Outside the solenoid, there being no currents, the field along the closed loop (a) shown in Fig.3.17 satisfies: $B_{out}(r_1)L - B_{out}(r_2)L = 0$ (where L is the length of the loop along z) which implies that $B_{out}(r) = $ const $= 0$ (since as $r \to \infty$ the field must go to zero). For a loop (c) inside the solenoid by the same argument $B_{in}(r) = $ const, while for a loop (b) enclosing nL coils Ampère's law states that : $(B_{in} - B_{out})L = 4\pi I n L/c$, hence:

$$B_{in} = 4\pi n I/c \tag{3.69}$$

The magnetic field generated by a solenoid can be used as an actuator to attract or repel a magnetic object along its axis. Since the current in a solenoid can be tuned within a very short time, it has many applications such as a pneumatic valve, the starter solenoid of a car or the voice coil of a speaker. In the latter case, a current through the coil produces a magnetic field which reacts with a permanent magnet fixed to the speaker's frame, thereby moving the cone of the speaker. By applying an audio waveform to the voice coil, the cone will reproduce the sound pressure waves corresponding to the input signal, i.e. it will reproduce the original sound. The fast response time of a voice coil is also used in order to rapidly position the read/write head of hard-disk drives. The head itself is often a microscopic coil in which a current creates a magnetic field that aligns small magnetic dipoles in the disk, a few nm below. The same coil is used to read the magnetic field of these oriented dipoles, via the currents generated in it when "flying" a few nm above the disk (see section 3.3).

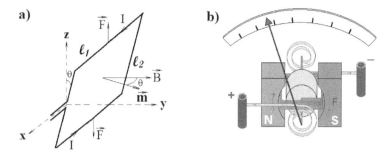

Figure 3.18 a) A flat rectangular loop (area $S = l_1 l_2$) bearing a current I and allowed to swivel about the x-axis is placed in a magnetic field $\vec{B} = B\hat{y}$. It experiences a torque $\vec{\Gamma} = Fl_2 \sin\theta\,\hat{x} = (Il_1 l_2 B \sin\theta/c)\,\hat{x} = \vec{\mathbf{m}} \times \vec{B}$ where $\mathbf{m} = IS/c$ is the magnetic moment of the loop directed along the normal to the surface of the loop and making an angle θ with the magnetic field. b) D'Arsonval's galvanometer consists of a flat solenoid such as shown in (a) placed between the poles of a permanent magnet parallel to the field lines. When current flows in the solenoid, the torque tends to align the dipole moment of the current loop along the magnetic field, namely to align the loop perpendicular to the field lines. The induced torque is balanced by the deformation of a torsional spring coupled to the solenoid and measured by the deflection of a dial, which is therefore proportional to the current flowing in the loop (adapted from ref.[29]).

3.2.5.4 D'Arsonval's Galvanometer

An other common use of the field generated by a solenoid is in the measurement of the current flowing through the solenoid in a so-called D'Arsonval galvanometer, see Fig.3.18. This instrument consists of a flat solenoid (with n coils of area S) placed between the poles of a permanent magnet, parallel to the field lines. When a current flows in the solenoid a magnetic dipole $\vec{\mathbf{m}} = (nIS/c)\hat{n}$ is generated along the normal $\hat{n}$ to the plane of the solenoid. To minimize its energy: $E = -\vec{\mathbf{m}} \cdot \vec{B}$ that dipole tends to align with the magnetic field, thus generating a torque $\vec{\tau} = \vec{\mathbf{m}} \times \vec{B}$ on the loop. A torsional spring linked to the solenoid counter-balances that torque. The distortion of the balancing spring is indicated by an appropriate dial which deflection is thus proportional to the current in the solenoid.

3.2.6 MAGNETIC FIELDS IN MATTER: PERMEABILITY AND PERMANENT MAGNETS

In the previous discussions we have considered the magnetic field arising from the movement of charges (currents) in vacuum (or in very thin conducting wires). Just as we discussed the interaction of static electric fields with dielectric materials we will now discuss the behavior of magnetic fields in permeable and magnetizable materials. The major difference between magnetic and electric fields however is that the former do no work on moving charges. Therefore they cannot induce a magnetic dipole in a material the way the electric field does. However, magnetic fields do interact with existing magnetic dipoles in the materials. These microscopic magnetic dipoles $\vec{m}$ are a purely quantum phenomena: as we shall see below (see Chap.4) they arise from the spin and angular momentum of the particles. The density of these microscopic dipoles set the magnetic dipole moment density $\vec{M}(r)$ which contributes to the total vector potential $\vec{A}$, see Eq.3.68, in addition to the contribution of a current distribution $\vec{J}$:

$$\vec{A}(\vec{r}) = \frac{1}{c} \int d^3r' \frac{\vec{J}(\vec{r}')}{|\vec{r}-\vec{r}'|} + \int d^3r' \vec{M}(\vec{r}') \times \vec{\nabla}' \frac{1}{|\vec{r}-\vec{r}'|}$$
$$= \int d^3r' \frac{\vec{J}/c + \vec{\nabla}' \times \vec{M}}{|\vec{r}-\vec{r}'|} \qquad (3.70)$$

where we used the vector calculus identity

$$\vec{M} \times \vec{\nabla}\psi = \psi\vec{\nabla} \times \vec{M} - \vec{\nabla} \times (\psi\vec{M})$$

(with $\psi \equiv 1/|\vec{r}-\vec{r}'|$, see Appendix A.3). By virtue of Eq.A.7 the volume integral over $\vec{\nabla} \times (\psi\vec{M})$ can be turned into a surface integral. Far away from the dipole moments, the contribution of that surface integral is negligible. These considerations justify Eq.3.70.

We have seen that as a result of charge conservation ($\vec{\nabla} \cdot \vec{J} = 0$): $\vec{\nabla} \times \vec{B} = -\nabla^2\vec{A}$, see Eq.3.59. Using the mathematical identity $\vec{\nabla} \cdot \vec{\nabla} \times \vec{M} = 0$ and Eq.3.18, one then obtains:

$$\vec{\nabla} \times \vec{B} = \frac{4\pi}{c}\vec{J} + 4\pi\vec{\nabla} \times \vec{M}$$

which we can rewrite as:

$$\vec{\nabla} \times \vec{H} = \vec{\nabla} \times (\vec{B} - 4\pi\vec{M}) = \frac{4\pi}{c}\vec{J} \qquad (3.71)$$

The field $\vec{H}$ is sometimes called the magnetic field strength[16] in contrast with the field $\vec{B}$ which is called the magnetic induction (or flux density). In most circumstances (e.g. non-magnetic materials) the average magnetization in the sample is small and proportional to the field ($\vec{B}$ or $\vec{H}$) and it is customary to write:

$$\vec{B} = \vec{H} + 4\pi\vec{M} = \mu\vec{H} \qquad (3.72)$$

with the magnetic permeability of matter: $\mu \sim 1$. In ferromagnetic materials on the other hand, which can have a large magnetic dipole even in absence of a magnetic field, the relation between the fields $\vec{H}$ and $\vec{B}$ is non-linear: $\vec{B} = \vec{F}(\vec{H})$. At large fields the magnetization saturates $\vec{M} \to \vec{M}_{max}$ (all the microscopic dipoles are aligned with

---

[16]The units of H are called Oersted instead of Gauss but here we will not make that distinction.

the field). This is how a strong magnet is usually made: by aligning its intrinsic microscopic dipoles in a strong external field and then turning off the external field and relying on the internal dipole field in the material to keep the dipoles aligned against the thermal agitation which will tend to disorient them, see section 5.10.1. Alternatively to demagnetize a given material one may heat it so as to disorganize its internal dipoles and create an average internal field that is close to zero.

3.3 ◆ ELECTROMAGNETIC INDUCTION

Until now we have dealt with time independent distributions of charges and currents. When these distributions become time dependent new phenomena are observed that have many fundamental as well as practical consequences. We will see that these new phenomena lead to a formulation of electromagnetism by Maxwell that unified electricity, magnetism and optics and predicted the existence of radio-waves (or Hertzian waves as they were called in the 19th century). The unification of so disparate phenomena (the first in physics) was an intellectual revolution that spurred the theory of relativity, quantum mechanics and quantum electrodynamics, the unification of electromagnetism and the weak and strong interactions (in the so-called standard model of particle physics), and is still the driving force behind the efforts to develop a "Theory of Everything" that would unify these forces with gravity.

3.3.1 ◆ FARADAY'S LAW

Electromagnetic induction was discovered independently by Michael Faraday and Josef Henry in 1831. They observed that electric currents could be generated in two different ways: either by moving a conducting loop in a nonuniform magnetic field or by moving a magnet across a stationary conducting loop. The remarkable equivalence of these two phenomena was one of the driving forces that led Einstein to develop his theory of relativity.

Consider first the displacement at velocity $\vec{v} = v_x\,\hat{x}$ of a conducting square loop in a inhomogeneous magnetic field perpendicular to the loop $\vec{B} = B\hat{z}$, see Fig.3.19. The Lorentz force on charges in the conducting wire is : $\vec{F} = (q/c)\vec{v} \times \vec{B}$. Since the field $\vec{B}$ is nonuniform the force on the charges in the right leg of the loop will be different from the force on the charges in the left leg of the loop. As a consequence there is a force difference $\Delta\vec{F}$:

$$\Delta\vec{F} = \frac{q}{c}\vec{v} \times (\vec{B}(x+w/2) - \vec{B}(x-w/2)) = \frac{q}{c}wv_x\frac{dB}{dx}(-\hat{y}) \tag{3.73}$$

which results in a movement of charges around the loop, i.e. a work:

$$q\mathcal{E}_{emf} = \Delta Fl = -\frac{q}{c}wlv_x\frac{dB}{dx} = -\frac{q}{c}\frac{dx}{dt}\frac{d(wlB)}{dx} = -\frac{q}{c}\frac{d\Phi_B}{dt} \tag{3.74}$$

Where $\Phi_B = \oint \vec{B} \cdot d\vec{S} = wlB$ is the magnetic flux through the loop. $\mathcal{E}_{emf} = -(1/c)d\Phi_B/dt$ is known as the electromotive force: it is equivalent to a potential driving the current around the loop. Notice that this induced current flows such as

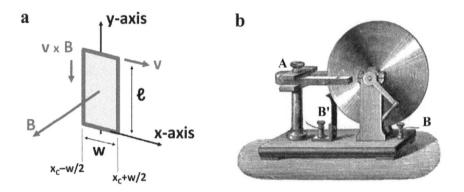

Figure 3.19 (a) Schematics of Faraday's first observation: by moving a conducting loop across a nonuniform magnetic field one can generate a current in the loop. (b) A conducting disk (such as the one used by Faraday) rotated in the magnetic field of magnets A also generates a current due to the differential of the Lorentz force on the charges in the spinning disk. Wires in contact (B, B') with the disk at its periphery and its center close the current loop.

to generate a magnetic field that counters the change in the inducing magnetic flux. In the example shown in Fig.3.19, if the field increases as the loop moves to the right the current induced in the loop flows clockwise which results in a magnetic field which opposes the increasing flux. If on the other hand the field decreases as the loop moves to the right, the induced current runs counter-clockwise and generates a magnetic field in the same direction as the decreasing flux. This is a general stability principle, sometimes called Le Chatelier's principle: a system responds to a perturbation in such a way as to reduce its effect.

To demonstrate the value of his discovery, Faraday invented a continuous current generator (at that time only batteries could supply a constant current). To that purpose he spun a disk of copper between the poles of a permanent magnet, see Fig.3.19. The field B on the charges in the disk is (to a first approximation) constant, but the velocity of the charges varies with their distance from the axis: $v = \omega r$. Hence the Lorentz force on a charge q at a distance r from the axis is in this case: $F = qB\omega r/c$. The electromotive force (work per unit charge) due to this force is therefore:

$$\mathcal{E}_{emf} = \frac{1}{q} \int F dr = \frac{1}{2c} B\omega r^2 \tag{3.75}$$

In the experiments just described the magnetic field is fixed and a current is induced in a moving loop by the action of the Lorentz force on free charges in the loop. Faraday also noticed that an induced current would be generated in a **fixed** solenoid if a magnet was moved across it or if the current in a nearby solenoid was varied. This experiment is equivalent to what an observer in the moving frame of the loop described above would see: a current flowing in the loop when the magnetic flux passing through the loop varies. This motivated Einstein to formulate his relativity principle: physical phenomena are independent of the moving frame in which they are observed. Their interpretation only may vary. There is no way to tell if you are at rest or moving. When the loop is at rest there is no Lorentz force on the charges since the solenoid is not moving, however (as Einstein pointed out later) the force exerted on the charges is a result of the electric field induced in their reference frame by the changing magnetic field. Faraday noticed that just as in the previous experiments, the electromotive force was given by:

$$\mathcal{E}_{emf} = -\frac{1}{c}\frac{d\Phi_B}{dt} \tag{3.76}$$

So it is the relative motion of magnet and conducting loop that matters for the generation of an induced current, not their absolute velocity. The phenomenon observed in different moving frames is the same: a current is flowing in the loop, even if the interpretation of why that is so may differ. It is this remarkable observation of Faraday that spurred Einstein to develop his theory of relativity. Indeed we have already used that theory to demonstrate the equivalence of magnetic and electric fields when deriving the equation of magnetostatics from the electrostatics field of a moving charged wire. In the context of Faraday's law, this equivalence is clearer if we rewrite Eq.3.76 in a differential form. On the one hand the electromotive force driving the current in a closed loop, i.e. the work per unit charge, is:

$$\mathcal{E}_{emf} = \oint \vec{E} \cdot d\vec{l} = \int \vec{\nabla} \times \vec{E} \cdot d\vec{S} \tag{3.77}$$

On the other hand: $\Phi_B = \int \vec{B} \cdot d\vec{S}$. Thus Eq.3.76 can be written as a differential equation relating time varying electric and magnetic fields:

$$\vec{\nabla} \times \vec{E} = -\frac{1}{c}\frac{\partial \vec{B}}{\partial t} \tag{3.78}$$

As an example consider an observer moving with the loop shown in Fig.3.19. For this observer, while the loop is at rest the magnetic field varies with time: $\vec{B} = B(x+vt)\hat{z}$. While for him the charges in the wire are at rest and there is no Lorentz force due to the magnetic field, there is however an electric field, E_y satisfying Eq.3.78:

$$\partial_x E_y = -\frac{1}{c}\partial_t B = -\frac{1}{c}\frac{dx}{dt}\partial_x B = -\frac{v_x}{c}\partial_x B \tag{3.79}$$

Namely $E_y = -(v/c)B$ and more generally[17]: $\vec{E} = \vec{v} \times \vec{B}/c$. For this observer the differential in electric field on the right and left legs of the loop will result in an electromotive force:

$$\mathcal{E}_{emf} = (E_y(x+w/2) - E_y(x-w/2))l = wl\partial_x E_y = -\frac{1}{c}\partial_t \Phi_B \tag{3.80}$$

---

[17]Notice that for a moving charged wire we derived a similar relation: $\vec{B} = -\vec{v} \times \vec{E}/c$. Both exemplify the relationship between electric and magnetic fields in a frame S' moving at velocity $\vec{v}$ with respect to a frame S in which there is electric but no magnetic field (charged wire) or magnetic but no electric field (moving loop).

which is the same result as the one reached by an observer at rest (for whom the magnetic field is constant but nonuniform and the loop is moving).

As we shall now show Faraday's law has had an enormous impact on our daily life, which is fitting since when Faraday was asked shortly after the formulation of his law: "What is the use of it?", he answered: "what is the use of a new-born baby?".

3.3.2 APPLICATIONS OF FARADAY'S LAW

3.3.2.1 Inductance

Consider a solenoid made of N loops (which length $l \gg a$, where a is the loop radius) and carrying a time varying current I. According to Ampère's law the magnetic field inside the solenoid is, see Eq.3.69: $B_{in} = 4\pi n I/c$ $(n = N/l)$, while according to Faraday's Law the variation in magnetic field induces an electromotive force (i.e. potential difference) in one loop: $V_1 = \mathcal{E}_{emf} = -(1/c)d\Phi_B/dt$. The potential across the N loops of the solenoid is thus: $V_N = NV_1$. Therefore, a varying current will result in an induced voltage at the solenoid ends (opposite to the current as in a capacitor) given by:

$$V_N = -\frac{N}{c}\frac{d\Phi_B}{dt} = -\frac{NS}{c}\frac{dB_{in}}{dt} = -\frac{S}{l}\frac{4\pi N^2}{c^2}\frac{dI}{dt} = -L\frac{dI}{dt} \tag{3.81}$$

$L = (4\pi N^2/c^2)(S/l)$ is known as the inductance of the solenoid. Its units in CGS are $\sec^2/\text{cm}$. In the MKS system inductance is measured in Henry (1Hy is the inductance that results in a voltage of 1V for a change of current of 1A/s). Just as electric energy is stored in a capacitor, magnetic energy is stored in an inductor (a solenoid). Consider the energy $\mathcal{E}_m$ stored in a solenoid carrying a current I_f (generating a field B_f):

$$\mathcal{E}_m = \int V_N dQ = -\int V_N I dt = L\int_0^{I_f} I dI = LI_f^2/2 = \frac{L}{2}(\frac{lcB_f}{4\pi N})^2 = Sl\frac{B_f^2}{8\pi} \tag{3.82}$$

Since Sl is the volume of the solenoid the energy stored per unit volume is $\mathbf{e}_m = B^2/8\pi$. As we shall see this is a general result for the energy density of magnetic fields. Notice the similarity with the energy density of electric fields, Eq.3.22.

3.3.2.2 The Electric Transformer

As mentioned above when Faraday moved a solenoid carrying a direct current (DC) inside a fixed one, he observed in the later the generation of a measurable current, see Fig.3.20(a). This observation is the core principle of today's electric transformers. These are built by wounding two solenoids around a common core of iron (which is used to canalize the magnetic field and limit losses), see Fig.3.20(b). An alternating current (AC) flowing in the primary solenoid (with N_p windings) generates an electromotive force (primary voltage) across it: $V_p = -N_p d\Phi/dt$ while across the secondary solenoid (with N_s windings) the induced electromotive force (secondary voltage) is: $V_s = -N_s d\Phi/dt = (N_s/N_p)V_p$ (the ratio of secondary to primary voltage is equal to the ratio of secondary to primary windings). If the losses are minimal

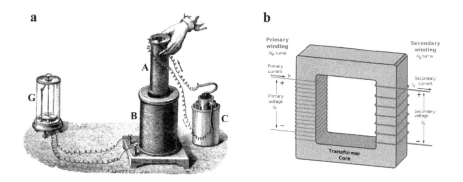

Figure 3.20 (a) The original drawings of Faraday's inductor: by displacing a solenoid A carrying a constant current (supplied by a battery C) inside a fixed solenoid B, Faraday observed a current flowing in solenoid B (measured by the galvanometer G). (b) The modern transformer principle: two solenoids are wound around an iron core (to canalize the magnetic field). The AC voltage in the primary winding is related to the AC voltage in the secondary winding as the ratio of the number of loops in the two solenoids (adapted from ref.[30]). Thus can a 220/110 V transformer be achieved with a ratio 2:1 in the number of loops in the primary and secondary circuits.

(in practice on the order of 2%) the power input to the transformer is equal to the power output, i.e. $V_p I_p = V_s I_s$ and thus $I_p/I_s = N_s/N_p$. The losses themselves are mostly due to Faraday's law: the changing flux through conducting surfaces creates induction currents that heat the surfaces because of their intrinsic electric resistivity (i.e. Ohm's law). To reduce these dissipative currents the surface perpendicular to the magnetic field is sliced by stacking layers of thin steel laminations, so that the changing flux (and thus the induction current) through each layer is minimal.

3.3.2.3   Electric Motor and Alternating Current (AC) Generator

We have seen that Faraday invented a DC current generator by spinning a conducting disk between the poles of a permanent magnet. It is interesting that the first alternating current(AC) generator and motor was invented by an Hungarian physicist, Anyos Jedlik, four years before Faraday's explanation of their working principle! In the first dynamo he generated a current by spinning a conducting loop between the poles of a permanent magnet, see Fig.3.21(a). As the magnetic flux through the loop varies periodically, the induced electro-motive force, Eq.3.76 also varies cyclically giving rise to an alternating current. This application is of immense relevance today as all the currents fed into our homes are still generated in essentially the same way! Jedlik also invented the first DC motor, see Fig.3.21(b). To that end he used a constant current in a fixed loop (the stator) to generate a magnetic field. He then induced a torque on a rotating loop (the rotor) by having a second constant current flow in it and

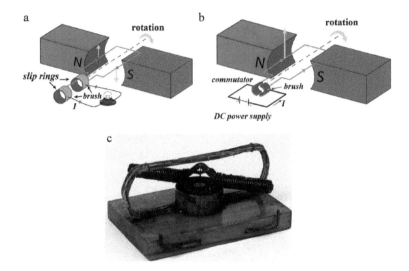

Figure 3.21 a) Schematics of an AC-generator: a loop rotates between the poles of a permanent magnet, the alternating current generated in the loop by the oscillating magnetic flux is picked up by brushes in contact with conducting slip rings. (b) Schematics of a DC motor: a constant current is flowing in a loop set in a constant magnetic field. The direction of the current flow is alternated such as to ensure that the torque on the loop always points in the same direction (adapted from ref.[31]). (c) the first DC motor build by Jedlik (in 1827 and still working!): a fixed external loop (the stator) carrying a constant current generates a constant magnetic field. A freely rotating loop (the rotor) carrying also a constant current is spun by the magnetic torque while a commutator ensures that the torque always points in the same direction.

interact with the magnetic field of the stator. With the help of a commutator —switching the current in the rotor so that the torque always points in the same direction —he made the rotor turn continuously.

3.3.2.4 Electronic Circuits

With batteries, generators and transformers as sources of electric energy, capacitors and inductors to store this energy, resistors to dissipate it, diodes to rectify it and amplifiers (vacuum tubes or transistors) to amplify its modulations (the signals), all of today's electronic circuits can be designed and understood! This is evidently well beyond the scope of this book, but a few examples can be worked out to give a glimpse of the field of electrical engineering.

Consider first the RC-circuit shown in Fig.3.22(a). When it is connected to the battery the voltage across the resistance $V_R = IR$ and across the capacitor V_C sum up to equal the battery voltage V_0. Since $I = dQ/dt = CdV_C/dt$ one obtains:

$$V_R + V_C = RC\frac{dV_C}{dt} + V_C = V_0$$

Which solution is: $V_C = V_0(1 - \exp(-t/RC))$. The capacitor is charged within a typical time $\tau = RC$. If the input voltage is oscillating $V_0(t) = V_{osc}e^{-i\omega t}$ then using Fourier transforms (see appendix A.4), one obtains:

$$(-i\omega RC + 1)\tilde{V}_C(\omega) = V_{osc}$$

Defining the cutoff frequency $\omega_c = 1/RC$ yields

$$|\tilde{V}_C(\omega)| = \frac{V_{osc}}{\sqrt{(\omega/\omega_c)^2 + 1}}$$

At low frequencies ($\omega < \omega_c$): $\tilde{V}_C(\omega) = V_{osc}$ whereas at high frequencies $\tilde{V}_C(\omega) = iV_{osc}\omega_c/\omega$. When the output of that circuit is the voltage across the capacitance V_c, the circuit functions as a low-pass frequency filter: it damps signals whose frequency is above ω_c and let signals pass undisturbed if their frequency is below ω_c.

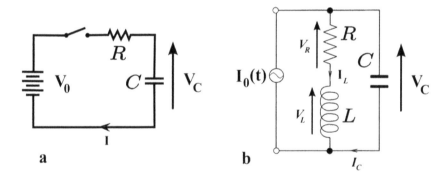

Figure 3.22 (a) Schematics of a RC circuit: the capacitor is charged by a battery when the switch is closed. (b) A RLC circuit fed by an oscillating current $I_0(t)$. The voltage across the capacitor is maximal when the current oscillation frequency $\omega = \omega_0 \equiv 1/\sqrt{LC}$.

Next consider the RLC circuit shown in Fig.3.22(b). When it is connected at $t = 0$ to a current source I_0, the current is split between the capacitor branch with amplitude $I_C = CdV_c/dt \equiv C\dot{V}_c$ and the inductor-resistor branch with amplitude $I_L = I_0 - I_C = I_0 - C\dot{V}_c$. The equality of the voltage across both branches: $V_C = V_R + V_L = I_L R + LdI_L/dt$ yields:

$$LC\frac{d^2V_C}{dt^2} + RC\frac{dV_C}{dt} + V_C = RI_0 + L\frac{dI_0}{dt}$$

If I_0 is constant ($dI_0/dt = 0$), introducing the frequency $\omega_0 = 1/\sqrt{LC}$ and damping $\gamma = R/2L$, one can write the solution of the previous equation as:

$$V_C = RI_0(1 - e^{-\gamma t}\cos\omega t) + Be^{-\gamma t}\sin\omega t$$

with $\omega^2 = \omega_0^2 - \gamma^2$ and $B = I_0(1 - \gamma RC)/\omega C$ chosen such that at $t = 0$: $V_C = 0$ and $I_L = 0$ (the voltage on a capacitor and the current in an inductor cannot change abruptly).

If the current I_0 is oscillating (as it is in a radio receiver, see section 3.4.2.2): $I_0(t) = I_{rec}e^{-i\omega t}$ ($\omega = 2\pi f$), then using Fourier transforms (see appendix A.4), one obtains:

$$\tilde{V}_C(\omega) = \frac{(R - i\omega L)I_{rec}}{1 - LC\omega^2 - iRC\omega} = \frac{(R - i\omega L)I_{rec}Q}{Q[1 - (\omega/\omega_0)^2] - 2i\omega/\omega_0} \equiv A(\omega)(R - i\omega L)I_{rec}$$

Where we define the quality factor $Q = \omega_0/\gamma$ and the transfer function $A(\omega) = |A|e^{i\phi_a}$ with:

$$|A(\omega)|^2 = \frac{Q^2}{Q^2(1 - (\omega/\omega_0)^2)^2 + 4(\omega/\omega_0)^2} \tag{3.83}$$

$$\tan\phi_a = \frac{2\omega/\omega_0}{Q[1 - (\omega/\omega_0)^2]} \tag{3.84}$$

At the resonance of the circuit : $\omega = \omega_0$, the amplitude of the transfer function is maximal $|A| = Q/2$ (its phase $\phi_a = \pi/2$). If ω is slightly away from resonance, the transfer function decreases rapidly, $|A(\omega)|^2$ reaching half its maximal value when $(\omega/\omega_0 - 1) = \delta\omega/\omega_0 \equiv \delta f/f_0 = 1/Q$. A circuit with high Q has low damping at resonance $f = f_0$, i.e. low dissipation. Such a circuit responds maximally to frequencies that are tuned close to its resonance frequency f_0. It acts as a band-pass frequency filter around f_0.

In Amplitude Modulation (AM) radio, the audio signal $A(t)$ modulates the amplitude of the radiation emitted at frequency f_0 known as the carrier frequency (530 kHz< f_0 < 1700 kHz): $E(t) = A(t)e^{-i\omega_0 t}$ ($\omega_0 = 2\pi f_0$), see section 3.4.2.2. Given the finite frequency of audio signals (typically less than 10 kHz), in the frequency domain the emitted signal occupies a bandwidth $\delta f = \pm 10$kHz around f_0. In an AM radio receiver, see Fig.3.23 the resonance frequency of the receiving RLC circuit is tuned to the carrier frequency of the emitting station by varying the capacitor C. To reject the signal emitted by other stations while limiting the damping of the desired audio signal, the value of the quality factor is chosen such that: $\delta f = f_0/Q \leqslant 10$ kHz (typically $Q \sim 50$). The output $V_C(t)$ of the RLC circuit is thus a signal at frequency ω_0 modulated mostly by the audio signal from the desired station. The output voltage is rectified by a diode (that transmits only positive voltages, see Fig.3.8 and section 5.9.2.3) and low-pass filtered to recover the audio signal (in the headphone shown in Fig.3.23) while filtering out the carrier frequency.

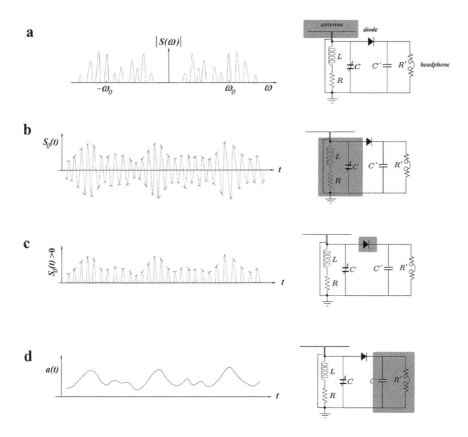

Figure 3.23 Working principle of the simplest AM-radio. (a) The antenna receives the signal emitted by many stations, each occupying a certain frequency range in Fourier space, as shown on the left. (b) The desired audio-signal modulating the amplitude of the carrier frequency ω_0 (shown in red in (a)) is extracted by the RLC circuit shown on the right which resonant frequency is tuned to ω_0 by varying the capacitor C. (c) The extracted signal $S_0(t)$ is rectified by the diode which lets pass only the positive half of the signal. (d) The rectified signal is low-pass filtered by the RC circuit made of a capacitor and a headphone, which filters out the high frequency carrier wave, restoring the audio signal $a(t)$.

3.4 ♦ MAXWELL'S EQUATIONS

By the mid-nineteenth century, much was known about electric and magnetic fields. One knew of Coulomb's law (as expressed by Gauss's theorem), Eq.3.16:

$$\vec{\nabla} \cdot \vec{E} = 4\pi\rho$$

and the absence of magnetic charges, Eq.3.48:

$$\vec{\nabla} \cdot \vec{B} = 0$$

One knew of Faraday's law, Eq.3.78:

$$\vec{\nabla} \times \vec{E} = -\frac{1}{c}\frac{\partial \vec{B}}{\partial t}$$

and Ampère's law, Eq.3.55:

$$\vec{\nabla} \times \vec{B} = \frac{4\pi}{c}\vec{J}$$

James Clerk Maxwell however pointed out that these equations violated charge conservation

$$\frac{\partial \rho}{\partial t} + \vec{\nabla} \cdot \vec{J} = 0$$

Indeed from Coulomb's and Ampère's equations one obtains:

$$\frac{\partial \rho}{\partial t} + \vec{\nabla} \cdot \vec{J} = \frac{1}{4\pi}\vec{\nabla} \cdot \frac{\partial \vec{E}}{\partial t} + \frac{c}{4\pi}\vec{\nabla} \cdot \vec{\nabla} \times \vec{B} = \frac{1}{4\pi}\vec{\nabla} \cdot \frac{\partial \vec{E}}{\partial t} \neq 0 \qquad (3.85)$$

Rather then questioning charge conservation, Maxwell proposed in 1861 to modify Ampère's law such as to satisfy charge conservation, namely by adding on its right hand side a term $(1/c)(\partial \vec{E}/\partial t)$, which he called the displacement current. The resulting four equations describing electromagnetic fields with charge and current sources are known as Maxwell's equations:

$$\vec{\nabla} \cdot \vec{E} = 4\pi\rho \qquad (3.86)$$

$$\vec{\nabla} \times \vec{E} = -\frac{1}{c}\frac{\partial \vec{B}}{\partial t} \qquad (3.87)$$

$$\vec{\nabla} \cdot \vec{B} = 0 \qquad (3.88)$$

$$\vec{\nabla} \times \vec{B} = \frac{4\pi}{c}\vec{J} + \frac{1}{c}\frac{\partial \vec{E}}{\partial t} \qquad (3.89)$$

Away from the sources (i.e. in regions where $\rho = \vec{J} = 0$), a remarkable feature of Maxwell's equations is their symmetry with respect to the electric and magnetic fields: just as a time dependent magnetic field creates an electric field (by Faraday's law) so a time varying electric field generates a magnetic one (by Maxwell's modification of Ampère's law). This mutual generation of time dependent electric fields by magnetic ones (and vice-versa) results in the self-sustained propagation of electro-magnetic waves. In the language of Science (i.e. Mathematics) this can best be seen by taking the rotational of the second (or fourth) of Maxwell's equations:

$$\vec{\nabla} \times \vec{\nabla} \times \vec{E} = -\frac{1}{c}\vec{\nabla} \times \frac{\partial \vec{B}}{\partial t} = -\frac{1}{c^2}\frac{\partial^2 \vec{E}}{\partial t^2}$$

Using the mathematical identity:

$$\vec{\nabla} \times \vec{\nabla} \times \vec{E} = \vec{\nabla}(\vec{\nabla} \cdot \vec{E}) - \nabla^2 \vec{E} = -\nabla^2 \vec{E}$$

where the last equality is due to the absence of charges, $\vec{\nabla} \cdot \vec{E} = 0$, one derives the wave equation for an electro-magnetic (EM) field::

$$\nabla^2 \vec{E} - \frac{1}{c^2}\frac{\partial^2 \vec{E}}{\partial t^2} = 0 \tag{3.90}$$

A generic solution is $\vec{E}(\vec{r},t) = \vec{E}_0(\vec{k} \cdot \vec{r} \pm \omega t)$, which describes a wave of amplitude E_0 propagating at velocity c and characterized by frequency f and wavelength λ, such that $\omega \equiv 2\pi f = 2\pi c/\lambda \equiv kc$ where $k \equiv 2\pi/\lambda$ is known as the wave-vector. More generally, any electric or magnetic field $\vec{E}(\vec{r},t)$ can be described in terms of its Fourier modes $\tilde{E}(\vec{k},\omega)$, see appendix A.4:

$$\vec{E}(\vec{r},t) = \frac{1}{4\pi^2}\int d^3k d\omega \tilde{E}(\vec{k},\omega)e^{i(\vec{k}\cdot\vec{r}-\omega t)}\delta(k-\omega/c) \tag{3.91}$$

We will have more to say about this wave-equation in the following, but when first obtained by Maxwell it initiated a scientific and technological revolution. First, it predicted the existence of a new phenomenon: electromagnetic waves. Second, it showed that these waves propagated at a velocity c that experimentally was very close to the velocity of light. Hence Maxwell predicted that light was an electromagnetic phenomenon! This unified electricity, magnetism and optics!

Yet it took more than 20 years for Heinrich Hertz to demonstrate in 1886-87 the existence of these waves by building a radiating dipole (see below) and an antenna (a simple coil) that produced a current upon receiving the emitted radiation. By measuring the wavelength and frequency of the waves he could show that their velocity was indeed the velocity of light (in his honor the frequency unit is the Hertz: $1Hz = 1sec^{-1}$). Amazingly when asked about the uses of his discovery he answered: "It's of no use whatsoever[...] this is just an experiment that proves Maestro Maxwell was right —we just have these mysterious electromagnetic waves that we cannot see with the naked eye. But they are there."!!

It took about ten more years for Guglielmo Marconi to develop these so-called Hertzian waves (nowadays known as radio-waves) as a means for the transmission of information: thus was wireless communication born. The first and simplest implementation used the transmission of electromagnetic pulses (originating from sparks) to generate a Morse code. This type of wireless telegraph became famous and widespread after the Titanic signaled its collision with an iceberg using a Marconi wireless transmitter.

Maxwell's equations with sources can be solved using the same type of approach that we used to solve Coulomb and Ampère's equations for static electric and magnetic fields, namely via the introduction of the auxiliary vector and scalar potentials: $\vec{A}$ and Φ. From the absence of magnetic charges (Maxwell's third equation): $\vec{\nabla} \cdot \vec{B} = 0$, one deduces that:

$$\vec{B} = \vec{\nabla} \times \vec{A} \tag{3.92}$$

Similarly from Faraday's equation: $\vec{\nabla} \times \vec{E} = -(1/c)(\partial \vec{B}/\partial t)$ one deduces that:

$$\vec{E} = -\vec{\nabla}\Phi - \frac{1}{c}\frac{\partial \vec{A}}{\partial t} \tag{3.93}$$

An important property of these potentials is their so-called gauge invariance: one can modify them without altering the underlying physics, i.e. the value of the real fields $\vec{E}$ and $\vec{B}$. Indeed for any arbitrary function $f(\vec{r},t)$, we can define new potentials: $\Phi' = \Phi + (1/c)\partial f/\partial t$ and $\vec{A}' = \vec{A} - \vec{\nabla}f$, then by virtue of Eq.3.92 and Eq.3.93: $\vec{E}' = \vec{E}$ and $\vec{B}' = \vec{B}$. That freedom in the choice of a gauge (the function f) is often used to bring the equations for the potentials $(\vec{A}, \Phi)$ into a form that is mathematically simpler to solve. Thus inserting the expression for $\vec{E}$ into Coulomb's law (Maxwell's first equation) yields:

$$-\vec{\nabla} \cdot \vec{E} = \nabla^2\Phi + \frac{1}{c}\frac{\partial}{\partial t}\vec{\nabla} \cdot \vec{A} = -4\pi\rho \tag{3.94}$$

Similarly inserting the exressions for $\vec{B}$ and $\vec{E}$ into the fourth of Maxwell's equations yields:

$$\vec{\nabla} \times \vec{B} = \vec{\nabla} \times \vec{\nabla} \times \vec{A} = \vec{\nabla}(\vec{\nabla} \cdot \vec{A}) - \nabla^2\vec{A} \tag{3.95}$$

$$= \frac{4\pi}{c}\vec{J} + \frac{1}{c}\frac{\partial \vec{E}}{\partial t} = \frac{4\pi}{c}\vec{J} - \vec{\nabla}(\frac{1}{c}\frac{\partial \Phi}{\partial t}) - \frac{1}{c^2}\frac{\partial^2 \vec{A}}{\partial t^2} \tag{3.96}$$

which upon rearranging yields:

$$\nabla^2\vec{A} - \frac{1}{c^2}\frac{\partial^2 \vec{A}}{\partial t^2} = -\frac{4\pi}{c}\vec{J} + \vec{\nabla}(\vec{\nabla} \cdot \vec{A} + \frac{1}{c}\frac{\partial \Phi}{\partial t}) \tag{3.97}$$

The Lorentz gauge is a choice of f such that:

$$\vec{\nabla} \cdot \vec{A} + \frac{1}{c}\frac{\partial \Phi}{\partial t} = 0 \tag{3.98}$$

which then yields the following equations for the potential fields:

$$\nabla^2\Phi - \frac{1}{c^2}\frac{\partial^2 \Phi}{\partial t^2} = -4\pi\rho \tag{3.99}$$

$$\nabla^2\vec{A} - \frac{1}{c^2}\frac{\partial^2 \vec{A}}{\partial t^2} = -\frac{4\pi}{c}\vec{J} \tag{3.100}$$

This gauge is often chosen if both the charge and current distributions vary with time. If the charge distribution is time independent (i.e. if $\partial\rho/\partial t = -\vec{\nabla}\cdot\vec{J} = 0$) then one often choose the Coulomb gauge: $\vec{\nabla}\cdot\vec{A} = 0$ and Φ is time independent, which implies :

$$\nabla^2\Phi = -4\pi\rho \tag{3.101}$$

$$\nabla^2\vec{A} - \frac{1}{c^2}\frac{\partial^2\vec{A}}{\partial t^2} = -\frac{4\pi}{c}\vec{J} \tag{3.102}$$

Notice that the equation for the vector potential (and in the Lorentz gauge also for the scalar potential Φ) is a wave-equation with sources. To solve such an equation it is often convenient to use the Fourier transforms of $\vec{A}(\vec{r},t)$, $\vec{J}(\vec{r},t)$ (see appendix A.4):

$$\vec{A}(\vec{r},t) = \frac{1}{\sqrt{2\pi}}\int \tilde{A}(\vec{r},\omega)e^{-i\omega t}d\omega$$

$$\vec{J}(\vec{r},t) = \frac{1}{\sqrt{2\pi}}\int \tilde{J}(\vec{r},\omega)e^{-i\omega t}d\omega \tag{3.103}$$

The equation for the Fourier mode $\tilde{A}(\vec{r},\omega)$ then becomes:

$$\nabla^2\tilde{A} + k^2\tilde{A} = -\frac{4\pi}{c}\tilde{J} \tag{3.104}$$

In absence of sources this equation is known as the Helmholtz equation (see appendix A.5). It is often encountered when computing the eigenmodes (the standing waves) of a particular cavity (and in Quantum Mechanics when solving Schrödinger's equation for bound systems, see below). With sources the solution of the preceding equation is:

$$\tilde{A}(\vec{r},\omega) = \frac{1}{c}\int \frac{\tilde{J}(\vec{r}\,',\omega)e^{ik|\vec{r}-\vec{r}\,'|}}{|\vec{r}-\vec{r}\,'|}d^3r' \tag{3.105}$$

Hence:

$$\vec{A}(\vec{r},t) = \frac{1}{\sqrt{2\pi}}\int \tilde{A}(\vec{r},\omega)e^{-i\omega t}d\omega$$

$$= \frac{1}{\sqrt{2\pi}c}\int \frac{\tilde{J}(\vec{r}\,',\omega)e^{i\omega|\vec{r}-\vec{r}\,'|/c-i\omega t}}{|\vec{r}-\vec{r}\,'|}d^3r'\,d\omega$$

$$= \frac{1}{2\pi c}\int \frac{\vec{J}(\vec{r}\,',t')e^{i[|\vec{r}-\vec{r}\,'|/c-(t-t')]\omega}}{|\vec{r}-\vec{r}\,'|}d^3r'\,d\omega dt' \tag{3.106}$$

Where we used the inverse Fourier transform (see appendix A.4) to replace $\tilde{J}$ by $\vec{J}$:

$$\tilde{J}(\vec{r}\,',\omega) = (1/\sqrt{2\pi})\int \vec{J}(\vec{r}\,',t')e^{i\omega t'}dt'$$

With the help of the Fourier relation:

$$\delta(t'-t+|\vec{r}-\vec{r}\,'|/c) = \frac{1}{2\pi}\int d\omega e^{i[|\vec{r}-\vec{r}\,'|/c-(t-t')]\omega} \tag{3.107}$$

and using the identity $f(t) = \int dt' \, f(t')\delta(t' - t)$, we obtain:

$$\vec{A}(\vec{r},t) = \frac{1}{c} \int \frac{\vec{J}(\vec{r}\,',t-|\vec{r}-\vec{r}\,'|/c)}{|\vec{r}-\vec{r}\,'|} d^3 r' \qquad (3.108)$$

Notice that Eq.3.108 is equivalent to Eq.3.62 taking into account causality. In other words, because an electromagnetic perturbation propagates with velocity c, the field at a point $\vec{r},t$ will be determined by the current density $\vec{J}(\vec{r}\,',t')$ at an earlier time: $t' = t - |\vec{r}-\vec{r}\,'|/c$. Thus, the light we receive from far away stars informs us about their state when it was emitted, many years ago.

3.4.1 ◆ ENERGY CONSERVATION AND RADIATED POWER

With the help of Maxwell's equation we may now revisit the problem of the energy stored in and radiated by an electromagnetic field. Since the magnetic field does no work, for a given charge distribution ρ (in a volume $V = \int d^3 r'$), the work dW performed by the EM fields displacing the charges by an infinitesimal distance $\delta\vec{r}$ is only due to the interaction of the charges with the electric field:

$$dW = \int d^3 r' \, \vec{F} \cdot \delta\vec{r}\,' = \int d^3 r' \, \rho\vec{E} \cdot \delta\vec{r}\,'$$

Since $\vec{v} = \delta\vec{r}/dt$ the power $P = dW/dt$ is:

$$\frac{dW}{dt} = \int d^3 r' \, \rho\vec{E} \cdot \vec{v} = \int d^3 r' \, \vec{E} \cdot \vec{J}$$

where the local current density $\vec{J}$ is simply related to the local velocity of the charges: $\vec{J} = \rho\vec{v}$. Using the fourth of Maxwell's equations, Eq.3.89, to express the current density $\vec{J}$ as a function of $\vec{E}$ and $\vec{B}$ ($\vec{J} = (c/4\pi)\vec{\nabla}\times\vec{B} - (1/4\pi)\partial\vec{E}/\partial t$) one obtains:

$$\frac{dW}{dt} = \frac{c}{4\pi}\int d^3 r' \, \vec{E}\cdot\vec{\nabla}\times\vec{B} - \frac{1}{4\pi}\int d^3 r' \, \vec{E}\cdot\partial\vec{E}/\partial t$$

Using the vector identity: $\vec{\nabla}\cdot(\vec{E}\times\vec{B}) = \vec{B}\cdot\vec{\nabla}\times\vec{E} - \vec{E}\cdot\vec{\nabla}\times\vec{B}$ and Faraday's law $\vec{\nabla}\times\vec{E} = -(1/c)\partial\vec{B}/\partial t$ to replace: $\vec{E}\cdot\vec{\nabla}\times\vec{B} = -\vec{\nabla}\cdot(\vec{E}\times\vec{B}) - (1/c)\vec{B}\cdot\partial\vec{B}/\partial t$ one obtains:

$$\frac{dW}{dt} = -\frac{c}{4\pi}\int d^3 r' \, \vec{\nabla}\cdot(\vec{E}\times\vec{B}) - \frac{1}{4\pi}\int d^3 r' \, (\vec{E}\cdot\partial\vec{E}/\partial t + \vec{B}\cdot\partial\vec{B}/\partial t)$$

$$= -\frac{c}{4\pi}\int d\vec{S}'\cdot\vec{E}\times\vec{B} - \frac{1}{8\pi}\frac{d}{dt}\int d^3 r' \, (|\vec{E}|^2 + |\vec{B}|^2) \qquad (3.109)$$

where we used the divergence theorem, Appendix A.3, to replace the volume integral by an integral over the surface S' bounding the integration volume. The previous equation can be rewritten as:

$$\frac{d}{dt}[W + \frac{1}{8\pi}\int d^3 r' \, (|\vec{E}|^2 + |\vec{B}|^2)] + \frac{c}{4\pi}\int d\vec{S}'\cdot\vec{E}\times\vec{B} = 0 \qquad (3.110)$$

This equation is an energy conservation law, of the same form as the charge conservation law ($dQ/dt + \int d\vec{S} \cdot \vec{J} = 0$). The first term on the left represents the change in energy: the work done by the EM fields on the charges and the energy stored in the electromagnetic fields, see Eq.3.21 with energy density:

$$\epsilon = (|\vec{E}|^2 + |\vec{B}|^2)/8\pi \tag{3.111}$$

The second term on the left of Eq.(3.110) is the radiated power. In other words to generate the electric and magnetic fields and maintain their radiation one has to inject into the system a power $P_{ex} = -dW/dt$.

The so called Poynting vector $\vec{S}_P$ is defined as the instantaneous energy radiated per unit area and unit time (it is measured in erg/sec·cm^2 or W/m^2):

$$\vec{S}_P = \frac{c}{4\pi}\vec{E} \times \vec{B} \tag{3.112}$$

The radiation intensity I is the time averaged value of the Poynting vector: $I = \langle \vec{S}_P \rangle \cdot \hat{k}$. It is the average power radiated per unit area. For a plane wave propagating along the z-axis ($\hat{k} = \hat{z}$):

$$\vec{E} = E_0 \hat{x}\cos(kz - \omega t) = Re\{E_0 \hat{x}e^{i(kz-\omega t)}\} \equiv Re\{\vec{E}_c\}$$
$$\vec{B} = B_0 \hat{y}\cos(kz - \omega t) = Re\{B_0 \hat{y}e^{i(kz-\omega t)}\} \equiv Re\{\vec{B}_c\}$$

Where $\vec{E}_c$ and $\vec{B}_c$ are the complex representations of the electric and magnetic fields[18] (with $B_0 = E_0$ to satisfy Faraday's law, Eq.3.78). The Poynting vector is:

$$\vec{S}_P = \frac{cE_0^2}{4\pi}\hat{z}\cos^2(kz - \omega t)$$

and the radiation intensity:

$$I = \langle \vec{S}_P \rangle \cdot \hat{z} = \frac{cE_0^2}{8\pi} = \frac{c}{8\pi}Re\{\hat{k} \cdot \vec{E}_c \times \vec{B}_c^*\} \tag{3.113}$$

where $\vec{B}_c^*$ is the complex conjugate of $\vec{B}_c$. For the spherical waves generated by radiation sources, because of energy conservation, the radiation intensity decays as $1/r^2$. The average power radiated per unit angle is thus:

$$\frac{dP}{d\Omega} = Ir^2 = \frac{c}{8\pi}Re\{r^2\hat{k} \cdot \vec{E}_c \times \vec{B}_c^*\}$$

In the following we shall discuss a few applications of Maxwell's equations and electro-magnetic radiation.

---

[18] $Re\{\vec{E}_c\}$ (or $Im\{\vec{E}_c\}$) stand for the real (or imaginary) part of the complex field $\vec{E}_c$. The electric and magnetic fields are real quantities, but often it is simpler to work with their complex form (i.e. using $\vec{E}_c$ instead of $E_0\hat{x}\cos(kz - \omega t)$). In the following we shall use the notation $\vec{E}$ and $\vec{B}$ for both representations, with the implicit understanding that if the fields are complex only their real part is relevant (i.e. measured).

3.4.2 APPLICATIONS OF MAXWELL'S EQUATIONS

3.4.2.1 The Radiating Dipole

As we have seen, a time varying current is a source of electromagnetic waves. If we assume that the current is oscillating at a fixed frequency: $J(\vec{r}, t) = J(\vec{r})e^{-i\omega t}$ then according to Eq.3.108:

$$
\begin{aligned}
\vec{A}(\vec{r}, t) &= \frac{1}{c} \int \frac{\vec{J}(\vec{r}\,', t - |\vec{r} - \vec{r}\,'|/c)}{|\vec{r} - \vec{r}\,'|} d^3 r' \\
&= \frac{1}{c} \int \frac{\vec{J}(\vec{r}\,')e^{-i\omega(t - |\vec{r} - \vec{r}\,'|/c)}}{|\vec{r} - \vec{r}\,'|} d^3 r' \\
&= \frac{1}{c} e^{-i\omega t} \int \frac{\vec{J}(\vec{r}\,')e^{ik|\vec{r} - \vec{r}\,'|}}{|\vec{r} - \vec{r}\,'|} d^3 r'
\end{aligned}
\tag{3.114}
$$

where the wavevector $k = 2\pi/\lambda = \omega/c$. Looking for the field $\vec{A}(\vec{r}, t)$ far away from the sources (i.e. $r \equiv |\vec{r}| \gg |\vec{r}\,'|$), we may expand: $|\vec{r} - \vec{r}\,'| \approx r - \hat{r} \cdot \vec{r}\,'$, where $\hat{r} = \vec{r}/r$ is the unit vector in the direction $\vec{r}$. This approximation and the assumption that the wavelength $\lambda \ll r$, define what is known as the far-field approximation. Eq.(3.114) then becomes:

$$
\vec{A}(\vec{r}, t) \approx \frac{1}{c} \frac{e^{ikr - i\omega t}}{r} \int \vec{J}(\vec{r}\,')e^{-ik\hat{r} \cdot \vec{r}\,'} d^3 r'
\tag{3.115}
$$

If the wavelength λ of the emitted radiation is larger than the size of the region where the oscillating sources are located, $\lambda \gg |\vec{r}\,'|$ (an assumption often correct when dealing with long-wavelength radiation such as radio-waves or small emitters such as fluorescent molecules), then:

$$
\vec{A}(\vec{r}, t) = \frac{1}{c} \frac{e^{ikr - i\omega t}}{r} \int \vec{J}(\vec{r}\,') d^3 r'
\tag{3.116}
$$

Integration by parts yield[19] :

$$
\begin{aligned}
\int \vec{J}(\vec{r}\,') d^3 r' &= -\int \vec{r}\,' \vec{\nabla}' \cdot \vec{J} d^3 r' = \int \vec{r}\,' \frac{\partial \rho}{\partial t} d^3 r' \\
&= -i\omega \int \rho \vec{r}\,' d^3 r' = -i\omega \vec{p}
\end{aligned}
\tag{3.117}
$$

---

[19] Using the identity:

$$
\int \vec{\nabla} \cdot (x\vec{J})\, d^3 r = \int x\vec{\nabla} \cdot \vec{J} d^3 r + \int \vec{J} \cdot \vec{\nabla} x d^3 r = \int x\vec{\nabla} \cdot \vec{J} d^3 r + \int J_x d^3 r
$$

The term on the left reduces by the divergence theorem to a surface term that vanishes far from the sources: $\int \vec{\nabla} \cdot (x\vec{J})\, d^3 r = \int x\vec{J} \cdot d\vec{S} \approx 0$. Repeating this argument with y, z replacing x yields: $\int \vec{J} d^3 r = -\int \vec{r}\,\vec{\nabla} \cdot \vec{J} d^3 r$. Using charge conservation $\partial_t \rho + \vec{\nabla} \cdot \vec{J} = 0$, i.e. $\vec{\nabla} \cdot \vec{J} = -\partial_t \rho = i\omega \rho$, yields the result shown in Eq.(3.117).

where $\vec{p} = \int \rho \vec{r}\, d^3r$ is the dipole moment of the oscillating charge distribution (a generalization of the concept introduced for two opposite charges separated by a distance d, see section 3.1.2.2). Therefore[20]:

$$\vec{A}(\vec{r},t) = -\frac{ik\vec{p}}{r} e^{ikr-i\omega t} \tag{3.118}$$

$$\vec{B} = \vec{\nabla} \times \vec{A} \approx \frac{k^2 \hat{r} \times \vec{p}}{r} e^{ikr-i\omega t} \tag{3.119}$$

In the far field region the EM field propagates as a spherical wave: $\hat{k} = \hat{r}$. Since $\vec{\nabla} \times \vec{B} = (1/c)\partial \vec{E}/\partial t$, one obtains $i\vec{k} \times \vec{B} = ik\hat{r} \times \vec{B} \approx -i\omega \vec{E}/c = -ik\vec{E}$ or:

$$\vec{E} = \vec{B} \times \hat{k} = \vec{B} \times \hat{r} \tag{3.120}$$

the time-average power radiated per unit angle is then:

$$\frac{dP}{d\Omega} = \frac{c}{8\pi} Re[r^2 \hat{r} \cdot \vec{E} \times \vec{B}^*] = \frac{c}{8\pi} Re[r^2 \hat{r} \cdot (\vec{B} \times \hat{r}) \times \vec{B}^*] = \frac{c}{8\pi} r^2 |\vec{B}|^2 \tag{3.121}$$

where we used the vector identity, Appendix A.3: $(\vec{B} \times \hat{r}) \times \vec{B}^* = |\vec{B}|^2 \hat{r} - (\vec{B}^* \cdot \hat{r})\vec{B}$ and the fact that $\vec{B} \cdot \hat{r} = 0$ (since $\vec{B}$ being proportional to $\hat{r} \times \vec{p}$ is perpendicular to $\hat{r}$). Hence:

$$\frac{dP}{d\Omega} = \frac{ck^4}{8\pi}(\hat{r} \times \vec{p})^2 = \frac{ck^4 |\vec{p}|^2 \sin^2 \theta}{8\pi} \tag{3.122}$$

Where θ is the angle between the dipole moment and the direction $\hat{r}$ of the propagating wave. The total power radiated by an oscillating dipole is:

$$P_{tot} = \frac{ck^4 |\vec{p}|^2}{8\pi} \int \sin^2 \theta d\Omega = \frac{ck^4 |\vec{p}|^2}{3} = \frac{|\ddot{\vec{p}}|^2}{3c^3} \tag{3.123}$$

This formula is very general and can be used each time the distance r from a source of dimension r' satisfies: $r' \ll \lambda \ll r$.

3.4.2.2 The Dipole Antenna

The simplest antenna consists of a thin conducting rod of length d aligned along the z-axis at $x = y = 0$ fed by a current oscillating at frequency ω, see Fig.3.24. This antenna will loose energy emitted as electromagnetic radiation and will thus appear to possess a resistance R that can be estimated by computing the total power lost (i.e. radiated) for a given current I flowing in the antenna.

---

[20]Using the identity, see Appendix A.3: $\vec{\nabla} \times (\vec{A}\psi) = \psi\vec{\nabla} \times \vec{A} + \vec{\nabla}\psi \times \vec{A}$ one can verify that $\vec{\nabla} \times \vec{A} e^{ikr} \approx ik\hat{r} \times \vec{A} e^{ikr}$.

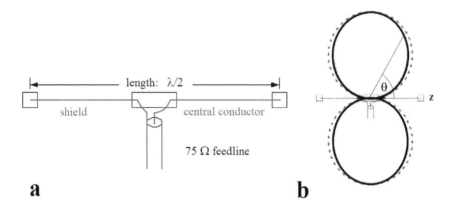

a   b

Figure 3.24 (a) A half-wavelength antenna consists of a conducting rod fed at its center by an alternating current at frequency $f = c/\lambda$. The antenna exhibits a typical resistance (impedance) $R_{\lambda/2} = 75\Omega$ at wavelengths that are twice the antenna's length. (b) The radiation pattern of this antenna (continuous line representing the value of the field as a function of the angle θ) is not much different from that of a radiating dipole (dashed line). Notice that the maximum emission (or sensitivity upon reception) is perpendicular to the axis of the antenna.

Since the current at the antenna's ends is zero and the antenna is thin, the current density can be written as[21]:

$$\vec{J}(z,\omega) = I\hat{z}\sin k(d/2 - |z|)e^{-i\omega t}\delta(x)\delta(y)$$

Plugging that value into the equation for the vector potential $A(\vec{r},t)$, Eq.3.115, yields[22]:

$$\vec{A}(\vec{r},t) = \frac{1}{rc}e^{i(kr-\omega t)}\int_{-d/2}^{d/2} I\hat{z}\sin k(d/2 - |z'|)e^{-ikz'\cos\theta}dz'$$

$$= \frac{2I\hat{z}}{rc}e^{i(kr-\omega t)}\int_{0}^{d/2}\sin k(d/2 - z')\cos(kz'\cos\theta)dz'$$

$$= \frac{2I\hat{z}}{krc}e^{i(kr-\omega t)}\frac{\cos(kd\cos\theta/2) - \cos kd/2}{\sin^2\theta} \tag{3.124}$$

where θ is the angle between the direction of propagation $\hat{r}$ and the antenna's $(z-)$axis, see Fig.3.24. Since $\vec{B} = \vec{\nabla}\times\vec{A} \approx ik\hat{r}\times\vec{A}$ and $\vec{E} = \vec{B}\times\hat{r}$, the energy radiated

---

[21] Notice that since $\int\delta(x)dx = 1$ the dimension of the delta function $\delta(x)$ is m^{-1}. Thus the current density J has units of current/area, as it should.

[22] Using the trigonometric identity: $2\sin\alpha\cos\beta = \sin(\alpha+\beta) + \sin(\alpha-\beta)$ and $\int\sin(ax+b)dx = -\cos(ax+b)/a$ to perform the integration.

per unit angle is:

$$\frac{dP}{d\Omega} = \frac{c}{8\pi}Re[r^2\hat{r}\cdot\vec{E}\times\vec{B}^*] = \frac{c}{8\pi}r^2|\vec{B}|^2$$

$$= \frac{I^2}{2\pi c}[\frac{\cos(kd\cos\theta/2)-\cos kd/2}{\sin\theta}]^2 \qquad (3.125)$$

In the limit of a small antenna (a molecule for example), i.e. when $kd \ll 1$ (using : $\sin k(d/2-|z|) \approx k(d/2-|z|)$), the current density can be written as:

$$\vec{J}(z,\omega) = I_{dip}\hat{z}(1-2|z|/d)e^{-i\omega t}\delta(x)\delta(y)$$

with $I_{dip} = Ikd/2$ and one recovers[23] the power emission of a dipole antenna (equivalent to a rotating dipole with moment: $\vec{p} = qd/2$ and effective current $I_{dip} = q\omega$), Eq.3.122:

$$\frac{dP}{d\Omega} = \frac{I_{dip}^2 k^2 d^2 \sin^2\theta}{32\pi c} = \frac{ck^4|\vec{p}|^2\sin^2\theta}{8\pi}$$

For a half-wavelength antenna ($d = \lambda/2$, i.e. $kd/2 = \pi/2$), the total energy lost through radiation is:

$$P_{tot} = \frac{I^2}{c}\int_0^\pi d\theta\frac{\cos^2(\pi\cos\theta/2)}{\sin\theta} \equiv \frac{I^2 R_{\lambda/2}}{2} \qquad (3.126)$$

Since by numerical integration, the integral is about 1.24, the radiation resistance of the half-wavelength antenna is: $R_{\lambda/2} = 2.48/c = 82.66\ 10^{-12}$ sec/cm $\approx 75\Omega$. In the design of antennas this resistance is very relevant since for the radiated power to be maximal the resistance of the feeding cable has to equal that of the antenna. As shown in Fig.3.24, the radiation pattern of a half-wavelength antenna is not much different from that of a radiating dipole.

In a radio or TV receiver the antenna is not emitting but receiving radiation and generating an oscillating current at the frequencies of the surrounding electromagnetic waves. The antenna's output is fed into a RLC circuit which resonance frequency ω_0 is tuned to the frequency ω of the desired emitting station, see Fig.3.23. Under these conditions, the RLC circuit discussed previously filters out the signals emitted by the unwanted stations, section 3.3.2.4. Its output is then amplified by transistors (or triodes in old receivers - see above) and further processed to extract the audio or video signal from the carrier wave at frequency ω.

---

[23]Using the approximation (valid when $\alpha \ll 1$): $\cos\alpha \approx 1-\alpha^2/2$,

$$[\cos(kd\cos\theta/2)-\cos kd/2]/\sin\theta \approx (k^2 d^2/8)\sin\theta$$

3.4.2.3 The Paradox of Atomic Stability

Dipole radiation presented the late nineteenth century physicists with a conundrum: how to explain the stability of atoms. The scattering of charged particles from matter suggested that atoms consisted of negatively charged electrons (with charge $-e$) orbiting a positively charged nucleus. By Eq.3.123 an electron orbiting the nucleus at a distance r and with a frequency ω would radiate its energy at a rate:

$$\frac{dE}{dt} = -\frac{(\ddot{\vec{p}})^2}{3c^3} = -\frac{(e\ddot{\vec{r}})^2}{3c^3} = -\frac{\omega^4 e^2 r^2}{3c^3} \tag{3.127}$$

The energy of the electron is the sum of its kinetic energy $E_{kin} = m\omega^2 r^2/2$ and its potential energy $E_{pot} = -e^2/r$. The balance of electrostatic attraction and centrifugal force $e^2/r^2 = m\omega^2 r$ therefore implies that $E = E_{kin} + E_{pot} = -e^2/2r$. We thus obtain the following equation for the evolution of the orbiting radius r:

$$\frac{dE}{dt} = \frac{e^2}{2r^2}\frac{dr}{dt} = -\frac{\omega^4 e^2 r^2}{3c^3} = -\frac{e^6}{3m^2 r^4 c^3} \tag{3.128}$$

Starting from a typical radius $r_0 \sim 1\text{Å}$, the electron reaches the nucleus in a finite time $t = (mc^2/V_0)^2(r_0/2c) \approx 0.2nsec$ (where the rest mass of the electron is $mc^2 = 0.5$ MeV and using a typical value for the atomic potential $V_0 = e^2/r_0 \sim 14$ eV). Atoms should therefore be very unstable and quickly radiate away their energy. One way to fix that problem[24] is to assume that Coulomb's law breaks down at atomic scale and that the attraction between the nucleus and the electron is repulsive at very short distances (δ of O(Å)), for example with a potential of the form: $\Phi(r) = e(1 - Ae^{-r/\delta})/r$. When the relative strength of the repulsive potential $A > 1$, the electron settles down at a fixed distance from the nucleus. That assumption was not necessary (Coulomb's law is valid down to sub-atomic scales). The revolutionary proposal which solved that paradox —that atoms do not radiate their energy continuously but in discrete bits or quanta —is at the core of Quantum Mechanics (see Chapter 4 below).

3.4.2.4 The Color of the Sky

One of the many natural phenomena that Maxwell's equations explained was the color of the sky: why is it blue during the day and red at sunset and dawn? The light emitted by the sun consists of electromagnetic waves with a peak of intensity in the so-called visible spectrum (see section 3.5 below) between blue ($\lambda_{blue} = 450$ nm) and red ($\lambda_{red} = 650$ nm) wavelengths. The electric amplitude of the field at a given wavenumber ($k = 2\pi/\lambda$), $\vec{E}_k = \vec{E}_{k,0}\exp i(\vec{k}\cdot\vec{r} - \omega t)$ polarizes the atmosphere creating dipoles of strength: $\vec{p}_k = (\epsilon - 1)\vec{E}_{k,0}/4\pi$, (where ϵ is the dielectric constant of air, see

---

[24] A similar fix was proposed to explain the stability of protons in the nucleus, which should fly away from each other because of their electrostatic repulsion. In that case, the introduction of a short range (δ of O(10^{-15}m)) strong attractive force with potential $V(r) = -Ae^{-r/\delta}/r$, known as the Yukawa potential, explained the stability of nuclei.

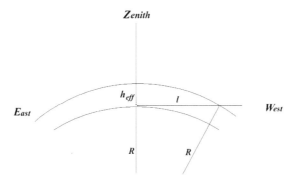

Figure 3.25 The path length of light in the atmosphere l is shorter at noon when the sun is at the Zenith and light-rays are perpendicular to the Earth's surface than at sunrise on the East or sunset on the West when the rays are parallel to it. If R is the Earth radius and h_{eff} the effective thickness of the atmosphere, by Pythagoras' theorem at sunset (or sunrise): $l^2 = (R+h_{eff})^2 - R^2 \sim 2h_{eff}R$.

Eq.3.31). These dipoles oscillate at the frequency ω of the exciting beam and re-emit (scatter) the incident light as dipole radiation thus subtracting from the incident beam a power intensity, Eq.3.123:

$$P_k = \frac{ck^4|\vec{p}_k|^2}{3} \sim \frac{(\epsilon-1)^2k^4}{6\pi}I_k$$

where $I_k = cE_{k,0}^2/8\pi$ is the power intensity of the incident beam at wavenumber k. Notice that the scattered light intensity increases as the fourth power of the wavelength, hence blue light will be scattered four times more efficiently that red light ($\lambda_{red}^4/\lambda_{blue}^4 \sim 4.3$), which gives the sky during day time its light blue hue.

Because of this so-called Rayleigh scattering a light beam passing through an atmospheric thickness dx looses power as $dI_k/dx = -\alpha_k I_k$, with $\alpha_k \sim (\epsilon-1)^2k^4$, i.e. $I_k(x) = I_k^0\exp(-\alpha_k x)$. The penetration length α_k^{-1} increases as the fourth power of the wavelength: it is typically 30km at 410nm (violet), 77km at 520nm (green) and 188km at 650nm (red). Taking into account the decrease of the air density with height, section 5.7.1, the effective thickness h_{eff} of the atmosphere is about 8.2km. When the sun is at the zenith, its radiation is scattered when passing through an atmospheric slab of thickness $l = h_{eff} < \alpha_k^{-1}$ and most of it (about 76% of violet and 96% of red radiation) reaches the Earth. However at dawn and dusk, light passes through a much larger slab of atmosphere: $l = \sqrt{2h_{eff}R} \approx 300$km (where $R = 6300$km is the Earth's radius), see Fig.3.25. In that situation most of the short wavelength radiation is scattered away and little reaches the observer, which sees mostly the red radiation (about 20% of which still reaches him/her). Hence at dawn and dusk the sun and the sky look red.

3.4.2.5 The Doppler-Fizeau Effect and the Expansion of the Universe

If a source of radiation emitting at frequency f_s is moving at velocity v away from an observer, the observed frequency f_o will be different from f_s. The reason is that after each period $T_s = 1/f_s$ the emitted light has to move an extra distance vT_s requiring an extra time vT_s/c. Thus the observed period T_o will be longer: $T_o = T_s + vT_s/c = T_s(1+v/c)$. As pointed out by Christian Doppler in 1842, the observed radiation wavelength will thus be longer $\lambda_o = cT_o = cT_s(1+v/c) = \lambda_s(1+v/c)$ and the observed frequency smaller: $f_o = f_s/(1+v/c)$. The radiation emitted by a source moving *away* is thus said to be *red-shifted* (shifted to the larger wavelengths of the visible spectrum). On the contrary radiation emitted by a source moving *towards* the observer (at velocity $-v$) is said to be *blue-shifted* (the observed wavelength is smaller than the wavelength in the moving frame of the source). As we shall see in the Quantum Mechanical Chapter, atoms can emit (or absorb) radiation at frequency f_s as they transit between different quantum states. Because of the movement of atoms in a gas, this frequency is randomly shifted by the Doppler effect so that the observed frequency f_o is less sharp than the emitted one: it is said to be Doppler broadened. In atomic clock where these transitions are used as the reference for the clock, Doppler broadening often limits the accuracy of the clock, which is improved by cooling the atoms to very low temperatures.

One of the striking observations of the stars, made already in 1848 by Hippolyte Fizeau is that their absorption spectrum is red-shifted with respect to the spectrum of the Sun, see Fig.3.26. We shall see later that these absorption bands are due to absorption by electrons in the hydrogen atoms that consitute most of the stars' outer layer. Fizeau correctly interpreted that red-shift as a motion of the stars away from us. In his and Doppler's honor the effect is known as the Doppler-Fizeau effect. These first observations were systematized in 1929 by Erwin Hubble which found a correlation between the red-shift of the galaxies, i.e. their speed v and their distance D to us: $v = H_0 D$ where H_0, the Hubble constant[25] has the dimensions of 1/time: $H_0 \sim 1/14.4$ billion-years. Hubble's relation is one of the most direct evidence of the expansion of the Universe. In a Universe expanding by a factor $R(t)$ all distances between stars expand by the same factor: $D = R(t)D_0$ and therefore the receding velocity of galaxies obey:

$$v = \frac{dD}{dt} = \frac{dR}{dt}D_0 = \frac{dR/dt}{R}D \equiv H_0 D$$

If the Universe is expanding at constant rate $dR/dt = u_0 = const$ (i.e. $R(t) = u_0 t$), the Hubble time $1/H_0$ is its age t, which is indeed close to the accepted value (13.8 billion-years). We shall see below that the background radiation from the original Bang is an other strong piece of evidence in favor of an expanding Universe.

---

[25] Calling H_0 a constant can be confusing since H_0 varies with time. The quoted value is its value at present.

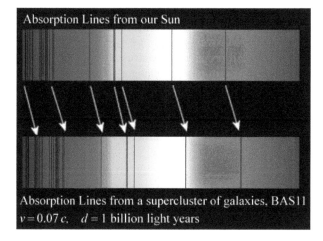

Figure 3.26 Absorption lines in the optical spectrum of a supercluster of distant galaxies (bottom), as compared to absorption lines in the optical spectrum of the Sun (top). The arrows indicate the redshift: the wavelength of the galaxy cluster moving away at a velocity $v = 0.07c$ is shifted to the red (reprinted from ref.[32]).

3.4.2.6 The Darkness of Night

The expanding Universe is solving an other paradox: why is night dark? One might naively say that the Sun is not here to illuminate the night. However there are stars and the light from them could make the night very bright indeed. If the Universe is infinite (or very large) and the density of stars constant then while the radiation of stars a distance r from us decays as $1/r^2$ the number of stars in a shell of radius r from us scale as r^2. As a consequence the light coming from the stars in that shell is independent of r and summing up the contribution from all the shells would yield an intensity which is equal the average intensity emitted by the stars: the night sky would be as bright as the stars! This paradox was first formulated in 1823 by Heinrich Olbers.

However an expanding Universe results in light from the stars to be red shifted and thus invisible to us (like the Cosmic background radiation that we will discuss in the QM chapter). The distance from us where stars emit visible radiation is thus finite and results in a night sky that is much less luminous than it would otherwise be.

It must be said that there are other possible explanations for the dark night-sky: for example a fractal Universe (with matter distributed a bit like in a sponge) where the density of the stars would not be constant but decrease faster that $1/r$ (i.e. a Universe with so-called fractal dimension smaller than 2) would also account for our dark nights. The current observations however seem to rule out a Universe with a fractal dimension below 2.

3.5 OPTICS

The most important result of Maxwell's equations was the prediction that time vary-ing electromagnetic fields will propagate as waves with the velocity of light, thus implying that light is itself a form of electromagnetic radiation. This prediction is verified everyday through our use of radio waves for wireless communication and light for the transfer of information in optical fibers. As summarized in table 3.1 a great variety of radiations (from γ-rays through X-rays to radio-waves) are all elec-tromagnetic waves at different frequencies and wavelengths.

Table 3.1

The nomenclature of electro-magnetic radiation as a function of wavelength and fre-quency.Notice that for short wavelengths radiation, the frequency column is written in energy units (1 eV = 250 THz, in anticipation of Planck's relation, see below: $E = hf$, where h is Planck's constant).

nomenclature	wavelengths	frequencies
radio waves: AM –LW	1 – 2 km	148.5 – 283.5 kHz
AM-MW	176 – 566 m	530 – 1700 kHz
AM-SW	10 – 175 m	1.711 – 30 MHz
FM	2.77 – 3.43 m	87.5 – 108 MHz
microwaves	0.1 – 100 cm	0.3 – 300 GHz
infrared (IR)	0.7 – 300 μm	1 – 430 THz
visible	0.4 – 0.7 μm	430 – 750 THz
Ultraviolet (UV)	10 – 400 nm	3 – 120 eV
X-rays	0.01 – 10 nm	120eV – 120 keV
γ-rays	< 0.01 nm	> 120 keV

Besides communication, electromagnetic waves have found many applications. For example in radar (acronym for RAdio Detection And Ranging), the waves emit-ted as radiation pulses by an antenna are reflected by a conducting material (e.g. an airplane) which distance and speed can be calculated by the time it takes the radia-tion pulses to return. In microwave ovens absorption of the radiation (at $\sim$ 2.45 GHz) inside the cavity by the water, lipids and sugars contained in the food is used to heat it. In infrared cameras, the radiation emitted by warm bodies (see below black-body radiation) can be detected in absence of any source of light that illuminates the scene (i.e. at night), see Fig.3.27.

In the following we shall study the propagation of EM radiation far from their sources, their diffraction by obstacles, their refraction (change of direction) when

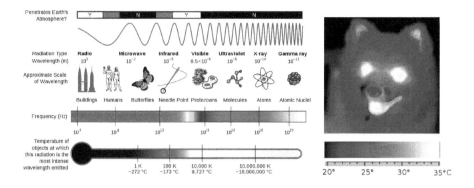

Figure 3.27 Left: The electromagnetic spectrum and the various scales involved (reprinted from ref.[33]). Right: the infrared-light emitted by a warm body such as a live-dog can be used to see it and map its temperature at different locations (the nose is colder than the tongue or the eyes).

passing from one media (say air) into an other (say water) and their use in optics to observe objects as in a microscope or a telescope[34]. For that matter let us consider an important class of solutions of Eq.3.90:

$$\nabla^2 \vec{E} - \frac{1}{c^2} \frac{\partial^2 \vec{E}}{\partial t^2} = 0$$

the so-called plane waves:

$$\vec{E} = \vec{E}_0 e^{i(\vec{k} \cdot \vec{r} - \omega t)}$$
$$\vec{B} = \vec{B}_0 e^{i(\vec{k} \cdot \vec{r} - \omega t)} \tag{3.129}$$
$$\vec{B}_0 = \hat{k} \times \vec{E}_0$$

Away from the radiation sources $\vec{\nabla} \cdot \vec{E} = 0$ and therefore the three vectors $\vec{k}$, $\vec{E}$ and $\vec{B}$ are orthogonal (the physical fields are the real parts of $\vec{E}, \vec{B}$). The electromagnetic wave is thus a transverse wave: the fields oscillate in a plane that is perpendicular to the direction of propagation $\hat{k}$. In that plane one can define two directions of polarization: $\vec{E} \parallel \hat{x}$ (the electric field is parallel to the x−axis and $\vec{B} \parallel \hat{y}$) or $\vec{E} \parallel \hat{y}$ (and $\vec{B} \parallel \hat{x}$). Any plane wave of wave-vector $\vec{k}$ can be described as a superposition of these two polarizations. Plane waves are useful because waves of any shape can be decomposed as a sum (or integral) of plane waves of different frequencies (their Fourier transform, see appendix A.4).

3.5.1 ELECTROMAGNETIC WAVES IN MATTER: REFRACTION

Refraction is a phenomenon associated with the passage of electro-magnetic waves between different media, such as going from air into water or glass. All optical systems use this phenomenon to correct our vision (glasses) or to improve it (microscopes or telescopes). It is used in communication to confine light in optical fibers or to focus it as in solar cells. To study refraction we need to understand how light propagates in transparent (non-absorbing) dielectric media.

The generalization of Maxwell's equation in matter can be deduced from the behavior of the electric and magnetic fields ($\vec{D}$ and $\vec{H}$) which in dielectric materials (e.g. glass, water) satisfy the linear relations: $\vec{D} = \epsilon\vec{E}$ and $\vec{B} = \mu\vec{H}$, where ϵ is the dielectric constant of the material (see Eqs.3.31) and μ its magnetic permeability (see Eq.3.72).

$$\vec{\nabla} \cdot \vec{D} = 4\pi\rho \tag{3.130}$$

$$\vec{\nabla} \times \vec{E} = -\frac{1}{c}\frac{\partial \vec{B}}{\partial t} \tag{3.131}$$

$$\vec{\nabla} \cdot \vec{B} = 0 \tag{3.132}$$

$$\vec{\nabla} \times \vec{H} = \frac{4\pi}{c}\vec{J} + \frac{1}{c}\frac{\partial \vec{D}}{\partial t} \tag{3.133}$$

In absence of sources ($\vec{J} = 0$; $\rho = 0$), as done before one can derive the following wave equation for the electric field:

$$\vec{\nabla} \times \vec{\nabla} \times \vec{E} = -\frac{1}{c}\vec{\nabla} \times \frac{\partial \vec{B}}{\partial t} = -\frac{\mu\epsilon}{c^2}\frac{\partial^2 \vec{E}}{\partial t^2}$$

Using the mathematical identity:

$$\vec{\nabla} \times \vec{\nabla} \times \vec{E} = \vec{\nabla}(\vec{\nabla} \cdot \vec{E}) - \nabla^2\vec{E} = -\nabla^2\vec{E}$$

where the last equality is due to the absence of charges, $\vec{\nabla} \cdot \vec{D} = \epsilon\vec{\nabla} \cdot \vec{E} = 0$, one obtains:

$$\nabla^2\vec{E} - \frac{\mu\epsilon}{c^2}\frac{\partial^2 \vec{E}}{\partial t^2} = \nabla^2\vec{E} - \frac{1}{c'^2}\frac{\partial^2 \vec{E}}{\partial t^2} = 0 \tag{3.134}$$

This is the same equation as derived in vacuum, Eq.3.90, but with a smaller propagation velocity: $c' = c/\sqrt{\epsilon\mu} = c/n$ (where $n \equiv \sqrt{\epsilon\mu}$ is known as the index of refraction of the medium). Similarly the wavelength $\lambda = 2\pi/k$ of plane waves $\vec{E}e^{i(\vec{k}\cdot\vec{r}-\omega t)}$ is shorter in matter than in vacuum, since from Eq.3.134: $k = \omega/c' = k_0 n$, i.e. $\lambda = \lambda_0/n$ (where k_0 and λ_0 are the wavenumber and wavelength in vacuum).

Since the velocity and wavelength of an electromagnetic wave depends on the medium in which it propagates it is interesting to investigate what happens when a plane wave is incident on the interface between two media with refractive indices n_1 and n_2, see Fig.3.28. We know that part of the wave will be reflected and part of the wave will be transmitted. Let $\vec{E}_i$, $\vec{E}_r$ and $\vec{E}_t$ be the incident, reflected and transmitted fields with wavenumbers $\vec{k}_i$, $\vec{k}_r$ and $\vec{k}_t$. For dielectric materials (i.e. most transparent materials) $\mu = 1$ and the continuity of the fields across the interface implies continuity of the magnetic field $\vec{B}$, as well as continuity of the normal component of the

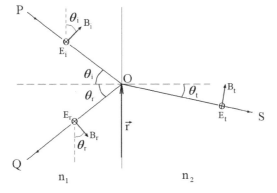

Figure 3.28 Refraction of an electromagnetic wave at the interface between media of refractive indices n_1 and n_2. In this figure the electric field $\vec{E}$ is parallel to the interface.

displacement field $\vec{D} = \epsilon \vec{E}$ (by Gauss's theorem since there are no charges at the interface),

$$D_{1,\perp} = D_{2,\perp}$$

and continuity of the tangential component of $\vec{E}$:

$$E_{1,\|} = E_{2,\|}$$

which can also be written as:

$$\vec{E}_{i,\|} e^{i(\vec{k}_i \cdot \vec{r} - \omega t)} + \vec{E}_{r,\|} e^{i(\vec{k}_r \cdot \vec{r} - \omega t)} = \vec{E}_{t,\|} e^{i(\vec{k}_t \cdot \vec{r} - \omega t)}$$

To be valid at every point $\vec{r}$ on the interface, the terms in the exponentials must be identical, which implies[26]:

$$k_i \sin \theta_i = k_r \sin \theta_r = k_t \sin \theta_t$$

where θ_i, θ_r, θ_t are the angles between the direction of propagation of the waves and the normal to the interface (see Fig. 3.28. Since $k_i = k_r$ one deduces that $\theta_r = \theta_i$: the wave is reflected at an angle equal to the incident angle. Moreover, since $k_i = n_1 k_0$ and $k_t = n_2 k_0$ one deduces Snell's law of refraction:

$$n_1 \sin \theta_i = n_2 \sin \theta_t \qquad (3.135)$$

which is is usually attributed to Willebrord Snellius (who reported his observations in 1621) even though the Arab mathematician Ibn Sahl discovered it as early as

---

[26]The scalar product of two vectors $\vec{A}$ and $\vec{B}$ is: $\vec{A} \cdot \vec{B} = |\vec{A}||\vec{B}| \cos \theta_{AB}$ where θ_{AB} is the angle between the two vectors. In the present case $\theta_{k_i,r} = \theta_i + \pi/2$, hence $\vec{k}_i \cdot \vec{r} = k_i r \cos(\theta_i + \pi/2) = -k_i r \sin \theta_i$.

984. Notice that when light goes from a high index of refraction medium (e.g. glass) to a low index one (e.g. air), i.e. $n_1 > n_2$, there is a maximal angle of incidence $\theta_{i,max}$ (for which $\theta_t = \pi/2$): $n_1 \sin\theta_{i,max} = n_2$ beyond which all light is reflected. This phenomenon, known as total internal reflection, is used to keep light confined in glass fibers for optical communication.

The ratio of the reflected or transmitted EM fields to the incident one (i.e. the reflection or transmission coefficients) can be obtained with a little algebra. If $\vec{E}$ is parallel to the interface, the continuity equation for the electric field $\vec{E}$ implies: $E_i + E_r = E_t$. The continuity equation for the normal component of the magnetic field (e.g. $B_{i,\perp} = B_i \sin\theta_i$, see Fig.3.28) is: $B_{i,\perp} + B_{r,\perp} = B_{t,\perp}$, which by virtue of Faraday's law : $\vec{\nabla} \times \vec{E} = -(1/c)\partial\vec{B}/\partial t$ (e.g. $n_i E_i = B_i$) translates into:

$$n_1 E_i \sin\theta_i + n_1 E_r \sin\theta_r = n_2 E_t \sin\theta_t$$

which by Snell's law is the same as the continuity equation for $\vec{E}$. The continuity equation for the parallel component of the magnetic field (e.g. $B_{i,\parallel} = B_i \cos\theta_i$, see Fig.3.28): $B_{i,\parallel} - B_{r,\parallel} = B_{t,\parallel}$ can be recast as:

$$n_1 E_i \cos\theta_i - n_1 E_r \cos\theta_r = n_2 E_t \cos\theta_t$$

From this equation, the continuity of $\vec{E}$ and Eq.3.135, the reflection coefficient can be obtained as[27]:

$$r \equiv \frac{E_r}{E_i} = \frac{\sin(\theta_i - \theta_t)}{\sin(\theta_i + \theta_t)} \tag{3.136}$$

and the transmission coefficient as:

$$t \equiv \frac{E_t}{E_i} = \frac{2\cos\theta_i \sin\theta_t}{\sin(\theta_i + \theta_t)} \tag{3.137}$$

From these coefficients for the ratio of field amplitudes we can deduce the coefficient of reflection (reflectivity) and transmission (transmittance) for the intensities, i.e. the Poynting vectors, Eq.3.112: $\vec{S}_P = (c/4\pi)\vec{E} \times \vec{B} = (nc/4\pi)E^2\hat{k}$. The amount of energy per unit area incident on the interface is: $I_i = \langle\vec{S}_P\rangle \cdot \hat{n} = (n_1 c/8\pi)E_i^2 \cos\theta_i$, while the amount of energy reflected per unit area is: $I_r = (n_1 c/8\pi)E_r^2 \cos\theta_r$. Since $\theta_i = \theta_r$, and using Eq.3.136, the reflectivity is:

$$\mathcal{R} \equiv \frac{I_r}{I_i} = \frac{E_r^2}{E_i^2} = \frac{\sin^2(\theta_i - \theta_t)}{\sin^2(\theta_i + \theta_t)} \tag{3.138}$$

---

[27]Expressing E_r, E_t as a function of E_i. For example eliminating E_r one gets:

$$n_1 E_i(\sin\theta_i \cos\theta_r + \cos\theta_r \sin\theta_i) = n_2 E_t(\sin\theta_t \cos\theta_r + \cos\theta_r \sin\theta_t) = n_2 E_t \sin(\theta_r + \theta_t)$$

Using the equality $\theta_i = \theta_r$ one obtains: $E_t/E_i = 2n_1 \sin\theta_i \cos\theta_i/[n_2 \sin(\theta_i + \theta_t)]$. Using Snell's law: $n_1/n_2 = \sin\theta_t/\sin\theta_i$ one readily derives Eq.3.137.

Since the amount of energy transmitted per unit area is: $I_t = (n_2 c/8\pi)E_t^2 \cos\theta_t$, from Eq.3.137 the transmittance is:

$$\mathcal{T} \equiv \frac{I_t}{I_i} = \frac{n_2 \cos\theta_t}{n_1 \cos\theta_i} t^2 = \frac{4\sin\theta_i \cos\theta_i \sin\theta_t \cos\theta_t}{\sin^2(\theta_i + \theta_t)} = 1 - \mathcal{R} \qquad (3.139)$$

The last equation simply reflects energy conservation: $I_i = I_r + I_t$.

3.5.2 FERMAT'S PRINCIPLE

Snell's law is a result of Fermat's principle who stated in 1662 that light propagates along the quickest path between two points[15]. If the index of refraction is uniform, the quickest path is also the shortest one: the straight line. However if light encounters a different medium that alters its velocity the straight line is not the quickest path.

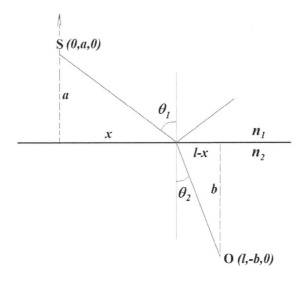

Figure 3.29 Demonstration of Fermat's principle. A ray emitted from point source S at position $(0, a, 0)$ propagates at velocity c/n_1 in a medium with refractive index n_1. It is refracted at position $(x, 0, 0)$ into a medium with refractive index n_2 in which it propagates at velocity c/n_2. The fastest trajectory between points S and O is the one for which $n_1 \sin\theta_1 = n_2 \sin\theta_2$.

Suppose that light propagates from point S=$(0, a, 0)$ to point O=$(l, -b, 0)$ crossing the interface between two dielectric media in the yz-plane at point $(x, 0, 0)$. The velocity of light in the first medium is c/n_1 and the length of its trajectory there is: $\sqrt{a^2 + x^2}$. In the second medium the velocity is c/n_2 and the length of trajectory is: $\sqrt{b^2 + (l - x)^2}$. The propagation time is:

$$t = \frac{\sqrt{a^2 + x^2}}{c/n_1} + \frac{\sqrt{b^2 + (l - x)^2}}{c/n_2}$$

It is minimal when $dt/dx = 0$, i.e. when:

$$\frac{n_1 x}{\sqrt{a^2 + x^2}} - \frac{n_2(l - x)}{\sqrt{b^2 + (l - x)^2}} = 0$$

which by trigonometry (e.g. $\sin\theta_i = x/\sqrt{a^2 + x^2}$) is Snell's law[28]. More generally Fermat's principle states that the phase $\phi = \int k(s)ds = k_0 \int n(s)ds$ (where $n(s)$ is the refractive index along a trajectory parametrized by s and k_0 the wavenumber in vacuum) is minimal along a path s linking the source S to the observation point O.

Fermat's principle encountered strong opposition from the scientific establishment of his time, since it seemed to suggest that Nature had intention and knowledge and that light "chooses" the quickest path. However Fermat's principle is a consequence of light being an electromagnetic wave. Waves emitted from a source S will travel along different paths and arrive at the observation point O with different phases. Most of theses waves will interfere destructively (see Hyugens' principle below, section 3.5.3). The only ones which will interfere constructively are those that have (almost) the same phase, namely those for which the variation of the phase with path-length is null: $\delta\phi/\delta s = 0$ which is the condition for the phase to be minimal (in fact stationary) and for the path to be the quickest. The study of these extremal paths has become the focus of a field known as geometrical optics, which considers light as rays propagating along the quickest path.

In spite of early opposition, Fermat's principle has emerged as one of the most profound observations about physical phenomena. The laws governing the trajectories of physical objects, be they celestial bodies or sub-atomic particles can be derived from a function, known as the action[18], which is minimal along the chosen path. For an electromagnetic wave, the action is simply its phase while for particles the action is the integral along the path of the Lagrangian (the difference between kinetic and potential energies). The derivations of the laws of physics from this simple principle, while illuminating, is beyond the scope of this book.

3.5.2.1 Geometrical Optics

At the time of Maxwell, Geometrical Optics (the propagation of light considered as rays or particles) was a well developed field that could trace its origins back to the Assyrian and Egyptian kingdoms, when lenses and mirrors were used as magnifier glass and when sunlight was focused to ignite a fire. In the middle ages the field of optics blossomed as a result of the translation into latin of Arab works on optics and in particular Ibn al-Haytham's (Alhazen) "Book of Optics" which presented in 1021 an earlier version of Fermat's principle. Alhazen is often viewed as the founder of the scientific method: formulating hypotheses and testing them through experiments. Both he and Ibn Sahl used Snell's law (before it was known as such), Eq.3.135, as a

---

[28]Notice that when $n_1 = n_2$, namely for the reflected beam, the same argument yields $\theta_r = \theta_i$ for the fastest reflected path.

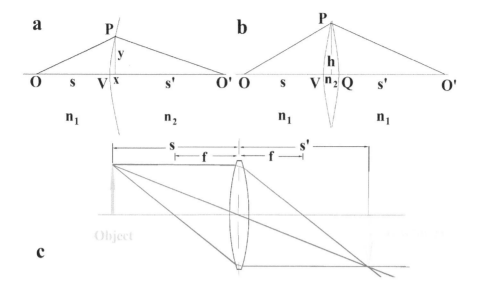

Figure 3.30 (a) Formation of an image of point O $(s,0,0)$ at point O' $(s',0,0)$ by refraction through an interface between two media of refractive indices n_1 and n_2. For the image to form the time traveled by a light ray along path OPO' must equal to the time traveled along path OVO'. (b) formation of an image of point O $(s,0,0)$ at point O' $(s',0,0)$ by refraction through a spherical lens of refractive index n_2 and radii of curvature r_1 and r_2. The maximal thickness of the lens is given by a simple geometric calculation: $\overline{VQ} = h^2/2r_1 + h^2/2r_2$. (c) real image of an object through a thin lens. The magnification M is the ratio of the image to the object size which by geometry (intercept or Thales' theorem) is: $M = s'/s$.

foundation for the design of spherical lenses (the common lenses which are defined by surfaces of spheres) and of more sophisticated aspherical lenses that are free of spherical aberrations!

To demonstrate the power of Fermat's principle, we shall use it to derive some basic results in Geometrical Optics[34]. Consider an interface separating two media of indices of refraction n_1 and n_2, see Fig.3.30(a), such as the interface between air and glass. To interfere constructively, light rays that are emitted from source O (at position $(-s,0,0)$ and focused at image point O' (at position $(s',0,0)$) have to take the same time t going along path OPO' as along path OVO' (for which $ct = n_1 s + n_2 s'$), namely they must satisfy:

$$ct = n_1 \sqrt{(s+x)^2 + y^2} + n_2 \sqrt{(s'-x)^2 + y^2} = n_1 s + n_2 s'$$

For an arbitrary shaped refractive interface the position of the image, i.e. the value of s' will usually depend on x,y, i.e. the point source at O will be spread out, an unwanted feature known as an aberration. For ideal lenses the goal is to minimize this aberration and to ensure that the position of the image s' is only a function of

s and does not depend on x, y. For simplicity, we shall here assume that the angle $\angle$ POV is small, so that $x, y \ll s, s'$. In that case we can expand the square-root in the previous equation to get:

$$n_1 s(1 + \frac{x}{s} - \frac{x^2}{2s^2} + \frac{x^2 + y^2}{2s^2} + ...) + n_2 s'(1 - \frac{x}{s'} - \frac{x^2}{2s'^2} + \frac{x^2 + y^2}{2s'^2} + ...) = n_1 s + n_2 s'$$

which yields:

$$\frac{n_1 y^2}{2s} + \frac{n_2 y^2}{2s'} \simeq (n_2 - n_1)x$$

The position of the image s' will be independent of x and y, if $x = y^2/2r$, namely if the interface between the media is the surface of a parabola or to a good approximation for thin lenses ($x \ll r$) a sphere of radius r. In that case the positions of source s and image s' obey:

$$\frac{n_1}{s} + \frac{n_2}{s'} = \frac{n_2 - n_1}{r}$$

Notice that light coming from infinity in medium 1 ($s = \infty$) will focus at point: $s' = f_2 = n_2 r/(n_2 - n_1)$ and light coming from infinity in medium 2 will focus at $s = f_1 = n_1 r/(n_2 - n_1)$. The property that $f_1/n_1 = f_2/n_2$ is general and valid for any optical system, not just the one considered here. Spherical surfaces are thus capable of creating images of objects with minimal aberrations (i.e. the position of the image s' does not depend on where the light passed through the interface (i.e. the coordinate x, y)). However this is an approximation valid when $\angle$ POV $\ll 1$, i.e. when $x \ll r$. For larger angles, spherical interfaces give rise to spherical aberrations which can be corrected by the aspherical (e.g. parabolic) interfaces first invented by Ibn Sahl[29]

An interesting application of the previous exercise is the case where one observes (in air $n_2 = 1$) a fish in a water pond ($n_1 \simeq 4/3$; $r = \infty$). The fish appears to be at a shorter distance $s' = -s/n_1 \simeq -3s/4$ from the surface of the pond (the minus sign indicates that the image of the fish is on the water side). Conclusion: when trying to catch a fish aim at a deeper distance than it appears to be!

We may now use Fermat's principle to derive the equation of a spherical lens. Consider the lens shown in Fig.3.30(b) consisting of two spherical surfaces of radius r_1 and r_2. The extra time going along the path OPO', $n_1(h^2/2s + h^2/2s')$ must be balanced by the time going through the lens from point V to Q: $(n_2 - n_1)(h^2/2r_1 + h^2/2r_2)$ from which one gets:

$$\frac{1}{s} + \frac{1}{s'} = (\frac{n_2}{n_1} - 1)(\frac{1}{r_1} + \frac{1}{r_2}) \equiv \frac{1}{f} \tag{3.140}$$

where f is the focal distance of the lens. This equation is known as the thin lens formula. Light coming from infinity ($s = \infty$, e.g. the sun) is concentrated at the focal

---

[29]A chromatic aberration may still be observed when using light sources with multiple wavelengths (colors). This aberration is due to the frequency dependence of the refractive index, see section 3.5.4, so that different wavelengths are focused at different positions. As a result a point source of white light is spread out into a multicolored extended spot. These aberrations can be corrected by a proper choice of media (glasses) to create what are known as achromatic lenses.

point a distance f from the lens. The magnification of an object by such lens is given by $M \equiv l'/l = s'/s$, see Fig.3.30(c).

3.5.3 DIFFRACTION

Diffraction is a phenomenon arising from the interference between waves. It is evidenced by the modulation of the intensity of a wave after it encounters certain objects (e.g. an aperture or an obstacle). Wave interference is a widely encountered phenomena: musician use it to tune their instruments by listening to the beat frequency, while some music fans use active headphones that eliminate external noise by generating sound waves that interfere destructively with it.

In diffraction the interference is between waves scattered by different parts of the object. The effect is stronger the smaller the size d of the object or the larger the distance L from it. As we shall see below, diffraction effects are noticeable when $d^2/L < \lambda$. Diffraction sets a limit on the resolving power of our eyes, our microscopes and telescopes, namely on how detailed can our observations be, see Appendix A.7.2.1. In Quantum Mechanics the observation of diffraction patterns in the scattering of particles is at the heart of the wave-particle duality and the Heisenberg uncertainty principle (see section 4.4.1).

As noted by Huygens as early as 1678, the propagation of a plane wave can be viewed as resulting from the emission of a multitude of spherical waves e^{ikr}/r originating from the points on the plane wave at a previous time (or position), see Fig.3.31. To demonstrate that point let us consider a plane wave $\Psi(x,y,z)$ propagating along the z−axis such that: $\Psi(x,y,0) = \Psi_0$. Since Ψ is a plane wave the field at position (x,y,z) is the same as at position $(0,0,z)$. Therefore to prove Huygens intuition, we need to show that the plane wave field $\Psi(0,0,z) = \Psi_0 e^{ikz}$ can be obtained as a sum of spherical waves originating from point sources at position (ρ,ϕ) (in 2D angular coordinates) on the plane wave $\Psi(x,y,0) \equiv \Psi(\rho,\phi,0)$. This is done using the following identity (with $k_\epsilon = k + i\epsilon$):

$$
\frac{1}{i\lambda} \lim_{\epsilon \to 0} \int_0^\infty \int_0^{2\pi} \rho d\rho d\phi \Psi_0 \frac{e^{ik_\epsilon \sqrt{\rho^2+z^2}}}{\sqrt{\rho^2+z^2}} =
$$
$$
-\Psi_0 \lim_{\epsilon \to 0} \int_0^\infty d\rho \frac{d}{d\rho}[e^{ik_\epsilon \sqrt{\rho^2+z^2}}] = \tag{3.141}
$$
$$
= \Psi_0 e^{ikz} = \Psi(0,0,z)
$$

Huygens' principle is very convenient for computing the field diffracted by an orifice (in the plane $z = 0$) onto which a plane wave $\vec{E}_i e^{ikz}$ is incident as in Fig.3.31. Using Huygens' representation of a plane wave, Eq.3.141, the field $\vec{E}_{diff}$ diffracted by the orifice at position (x,y,z) is obtained as an integration over point sources at positions (x',y') within the aperture only (since the plane wave is partially blocked

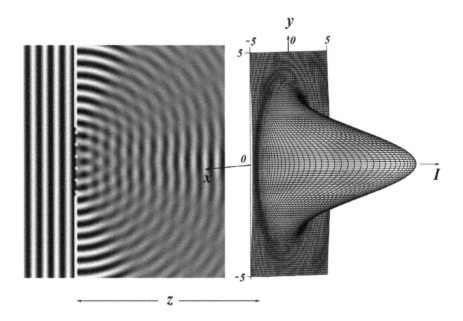

Figure 3.31 Demonstration of the Huygens principle for a plane wave diffracted from a hole in a screen and its far-field distribution of intensity ($\propto |J_1(r)/r|^2$, which is plotted here). The yellow dots represent some of the multitude of spherical waves emitted from point sources in the aperture (adapted from ref.[35]).

by the orifice):

$$
\vec{E}_{diff}(x,y,z) = \frac{1}{i\lambda} \oint dx'\,dy'\,\vec{E}_i \frac{e^{ik\sqrt{(x-x')^2+(y-y')^2+z^2}}}{\sqrt{(x-x')^2+(y-y')^2+z^2}}
$$

$$
\approx \frac{\vec{E}_i e^{ikz}}{i\lambda z} \oint dx'\,dy'\,e^{ik((x-x')^2+(y-y')^2)/2z} \tag{3.142}
$$

The integration ($\oint dx'\,dy'$) is performed over the aperture, assuming that we are in the Fresnel limit: $z^2 \gg (x-x')^2 + (y-y')^2$ in order to expand the square-root[30]. This equation is known as the Fresnel diffraction formula. By developing the square in the exponent one obtains:

$$
\vec{E}_{diff}(x,y,z) = \frac{\vec{E}_i e^{ik(z+(x^2+y^2)/2z)}}{i\lambda z} \oint dx'\,dy'\,e^{-i(k_x x'+k_y y')} e^{2\pi i(x'^2+y'^2)/2\lambda z}
$$

---

[30] Notice that: $\sqrt{(x-x')^2+(y-y')^2+z^2} \approx z + [(x-x')^2+(y-y')^2]/2z$

where $k_x = kx/z$ and $k_y = ky/z$. The so-called far field or Fraunhofer limit: $x'^2 + y'^2 \ll \lambda z$ corresponds to a situation where the orifice size $d \ll \sqrt{\lambda z}$. In that limit the field diffracted by the aperture is:

$$\vec{E}_{diff}(x,y,z) = \frac{\vec{E}_i e^{ik(z+(x^2+y^2)/2z)}}{i\lambda z} \oint dx'dy' e^{-i(k_x x' + k_y y')} \tag{3.143}$$

And the diffraction intensity $I_{diff} = \frac{c}{8\pi}|\vec{E}_{diff}(x,y,z)|^2$ is:

$$I_{diff}(x,y,z) = \frac{I_i}{(\lambda z)^2}\left|\oint dx'dy' e^{-i(k_x x' + k_y y')}\right|^2 \tag{3.144}$$

Hence, the intensity of Fraunhofer diffraction is proportional to the power of the Fourier transform of the diffracting aperture.

Examples. For a circular aperture of radius a, the integration variables x', y' can be written in polar coordinates:

$$x' = \rho' \cos\phi' ; \qquad\qquad y' = \rho' \sin\phi'$$

Similarly in the Fresnel limit ($\theta \ll 1$), it is advantageous to express the coordinates (x,y,z) of the observation point in terms of spherical coordinates (r,θ,ϕ):

$$z = r\cos\theta \approx r ; \quad x = r\sin\theta\cos\phi \approx z\sin\theta\cos\phi ; \quad y = r\sin\theta\sin\phi \approx z\sin\theta\sin\phi$$

With $k_x = k\sin\theta\cos\phi$ and $k_y = k\sin\theta\sin\phi$, the intensity in the far-field, i.e. at a distance $z \gg a^2/\lambda$, is then:

$$\begin{aligned}I_{diff}(z,\theta,\phi) &= \frac{I_i}{(\lambda z)^2}\left|\oint \rho'd\rho'd\phi' e^{-i(k\rho' \sin\theta\cos(\phi'-\phi))}\right|^2 \\ &= \frac{4\pi^2 I_i}{(\lambda z)^2}\left|\int_0^a \rho'd\rho' J_0(k\rho' \sin\theta)\right|^2\end{aligned} \tag{3.145}$$

where we used the mathematical identity[31]: $J_0(x) = (1/\pi)\int_0^\pi \cos(x\cos\phi)d\phi$. Since one of the properties of Bessel functions is: $x^n J_n(x) = \int_0^x d\xi \xi^n J_{n-1}(\xi)$, the integration of Eq.3.145 (upon change of variable: $\xi = k\rho' \sin\theta$) yields:

$$I_{diff}(z,\theta,\phi) = \frac{k^2 a^4}{z^2}I_i\left|\frac{J_1(ka\sin\theta)}{ka\sin\theta}\right|^2 \tag{3.146}$$

The intensity is maximal at the center ($\theta = 0$) and vanishes at the zeros of $J_1(x)$, the first of which occurs when $ka\sin\theta = 3.83$, i.e. when $\sin\theta = 1.22\lambda/d$ (where $d = 2a$ is the diameter of the hole). In the plane of observation the radial distance of the

---

[31] Changing the integration variable $\phi' - \phi = \psi$ and using the identity $e^{i\alpha} = \cos\alpha + i\sin\alpha$. By symmetry the integral $\int \sin(x\cos\psi)d\psi = 0$ since $\sin(x\cos(\psi + \pi)) = -\sin(x\cos\psi)$.

first minima to the central intensity peak is thus: $\rho = z \sin\theta = 1.22\lambda z/d$. Notice that the diffraction pattern is larger (i.e. ρ is larger) the smaller the hole diameter d and vice-versa.

In section 3.5.2.1, we have studied the propagation of light as if it was a particle propagating along the quickest path between two points. Studying the propagation of light as a wave in a system of lenses, such as a microscope or a telescope, is the subject of a field known as physical optics[34] and sketched in Appendix A.7.2. Light diffraction by the finite aperture of a lens sets a limit on the resolution of a microscope or a telescope, i.e. on the identification of two separate point sources of light. As a result of diffraction the image of each source is spread over a disk of radius: $\rho = 1.22\lambda s'/D$ (where D is the diameter of the lens and s' the distance of the image to the lens, see Fig.3.30), i.e. it is spread over an angle: $\delta\theta = \rho/s' = 1.22\lambda/D$.

The overlap of the images of two point sources prevents their separate identification: they appear as one blob instead of two. A criterion, established in the 19th century by Lord Rayleigh and Ernst Abbe, limits the angular separation between two point sources (e.g. two stars) that can be resolved by a telescope to : $\delta\theta_{12} \geq 1.22\lambda/D$. For the Hubble telescope ($D = 2.4$m) this corresponds to $\delta\theta_{12} \geq 2.5\ 10^{-7}$ rad. A planet orbiting Proxima Centauri (the closest star to the Sun at a distance $L = 4\ 10^{13}$ km) could be resolved if it orbits the star at a distance $R \geq 10^7$ km (about the size of the orbit of Mercury) such that $R/L = \delta\theta_{12} \geq 2.5\ 10^{-7}$ and if its reflects enough light!.

In a microscope Abbe's criterion imply that for two point sources to be resolved, their distance $\delta_{12} = s\delta\theta_{12} > 1.22\lambda f/D = 0.61\lambda/N.A \approx 0.25\mu m$, where $N.A = D/2f$ is the numerical aperture ($N.A. \approx 1.4$ for the best microscope objectives). Hence a microscope cannot resolve features in a cell or a tissue that are smaller than ≈ 250nm, but see section 4.12.3.1 for recent approaches that bypass that limit.

3.5.4   DISPERSION OF LIGHT AND THE COLORS OF THE RAINBOW

In the previous discussion on refraction we assumed that the refractive index was the same for all wavelengths or frequencies. This is often not the case with the concomitant result that the optical properties of materials may depend on the wavelength (color) of the light beam. To understand how the refractive index depends on the frequency ω of the electromagnetic field, we have to understand how the dielectric constant ϵ depends on ω.

We have seen in section 3.1.6 that: $\vec{D} \equiv \epsilon\vec{E} = \vec{E} + 4\pi\vec{P}$, where $\vec{P}$ is the induced dipole moment of the material due to its interaction with the electric field. Electrons in a dielectric medium such as air or glass (but not in a conducting medium such as a metal) are not free to move: they are bound to the molecules by electrostatic attraction. The description of their binding and distribution around atoms and molecules occupies a large part of Quantum Mechanics(QM) and is at the foundation of the chemical bond. We will discussed those issues extensively in the chapters devoted to QM. For the moment we shall follow H.A.Lorentz and assume that an electron is bound by a potential $V(x)$. At its equilibrium position x_0, the force on the electron is by definition null: $F(x_0) = -dV/dx|_{x_0} = 0$. Near x_0, the potential can thus be expanded as the following Taylor series in $\delta x = x - x_0$:

$$V(x) = V(x_0) + k\delta x^2/2 + \ldots \text{ with } k \equiv d^2V/dx^2 \big|_{x_0}$$

which is similar to the potential energy of a spring of stiffness k. This result is generic for any system at equilibrium: a deviation from equilibrium by δx leads to an elastic restoring force: $F = -dV/dx = -k\delta x$. If an electron of charge q and mass m is subjected to an oscillating electric field $E(t) = E_0 e^{-i\omega t}$, by Newton's law its motion obeys the ordinary differential equation of a forced damped harmonic oscillator:

$$m\frac{d^2}{dt^2}\delta x + m\gamma\frac{d}{dt}\delta x + k\delta x = qE(t) = qE_0 e^{-i\omega t} \tag{3.147}$$

Where γ (typically $\sim 10^{11} - 10^{13}$ Hz) represents damping of the electron motion (through radiation, shocks, etc.). The displacement of the electron from equilibrium is thus:

$$\delta x(t) = \delta x_0 e^{-i\omega t}$$

$$\text{with :} \quad \delta x_0 = \frac{qE_0/m}{\omega_0^2 - \omega^2 - i\gamma\omega}$$

where $\omega_0 = \sqrt{k/m}$ is the natural frequency of oscillations of a bound electron. This frequency vary wildly depending on the type of motion involved: rotational, vibrational or electronic. In the QM chapters discussed below we shall see that rotational motions are associated with frequencies in the microwave range, whereas vibrational modes are excited at infrared frequencies. Finally excitation of the system to higher electronic energy states is associated with visible or UV frequencies. Thus, in a material containing N electrons per unit volume of which a fraction ϕ_n (such that $\sum \phi_n = 1$) are oscillating at frequency $\omega_n = \sqrt{k_n/m}$ with damping coefficient γ_n, the dipole moment P induced by the electric field is:

$$P = N\sum_n \phi_n q\delta x_n = \frac{Nq^2}{m}E_0 \sum_n \frac{\phi_n}{\omega_n^2 - \omega^2 - i\gamma_n\omega}$$

For gases N is roughly equal to the gas density ($\rho = N_A/V_A \sim 3 \ 10^{19}$ cm^{-3}, see section 3.1.2.1). In a solid (assuming a typical distance of ~ 2Å between atoms), $N \sim 1/(2\text{Å})^3 \sim 10^{23}$ cm^{-3}. As shown earlier, section 3.5.1, the refractive index of a medium n_0 obeys: $n_0^2 = \epsilon\mu = 1 + 4\pi P/E_0$ (assuming a magnetic permeability $\mu = 1$, as in most transparent dielectrics). The frequency dependence of the index of refraction thus satisfies:

$$n_0^2(\omega) = 1 + \frac{4\pi Nq^2}{m}\sum_n \frac{\phi_n}{\omega_n^2 - \omega^2 + i\gamma_n\omega)} = 1 + \omega_p^2\sum_n \frac{\phi_n}{(\omega_n^2 - \omega^2 - i\gamma_n\omega)} \tag{3.148}$$

Where $\omega_p = \sqrt{4\pi Nq^2/m}$ is known as the plasma frequency. In gases in standard conditions, see section 5.5, $\omega_p \sim 3 \ 10^{13}$ rad/sec whereas in solid dielectrics (e.g. glass): $\omega_p \sim 3 \ 10^{15}$ rad/sec.

Since the major molecular components of air N_2 and O_2 have no dipole moment, their rotational motion does not couple to the EM-field. Their lowest resonant frequency is associated to vibrational motion at infrared frequencies ($\omega_n = 2\pi f_n \sim 10^{15}$ rad/sec). Therefore at low frequencies ($\omega \ll \omega_n$ the refractive index $n_0 \simeq \sqrt{1 + \sum_n \omega_p^2 \phi_n / \omega_n^2}$ is in air (and many gases) very close to one, whereas it can be significantly larger than one in solid dielectrics.

As the frequency of the electro-magnetic wave increases and approaches one of the resonant frequencies ω_n, the dielectric constant increases and acquires a large imaginary part: $\epsilon(\omega_n) \simeq i\omega_p^2 \phi_n / \gamma_n \omega_n$. As a result the refractive index becomes complex:

$$n_0(\omega_n) \simeq \sqrt{\omega_p^2 \phi_n / \gamma_n \omega_n} \frac{1+i}{2}$$

The amplitude of electromagnetic radiation at frequency $\omega \approx \omega_n$ propagating in such a medium decreases as $\exp[-Im(n_0)\omega x/c]$, i.e. radiation is adsorbed by the medium[32]. Thus, for a medium to be transparent in a given frequency range, a necessary requirement is that there are no resonant frequency modes in the range. Water which possesses many rotational and vibrational modes in the microwave and infrared range is thus opaque to radiation in that range (which is why microwave radiation —which is absorbed by water —is used to heat water-containing food). Water and glasses are however transparent to visible radiation having no resonant modes in this range. The combination of both properties is the cause of the greenhouse effect: visible radiation penetrates through the atmosphere is adsorbed by the Earth and re-emitted as infrared (heat) radiation which is adsorbed by the water in the atmosphere thereby heating it.

Both water and glasses have resonant modes in the UV range and thus absorb UV light, an effect that protects us from damaging (ionizing) UV irradiation. Moreover as the frequency of visible light increases and approaches the UV, the refractive index $n_0(\omega)$ increases, resulting in stronger refraction (bending) of light with shorter wavelenghts, as observed in a prism and in a rainbow.

The rainbow has fascinated humanity since time immemorial. In the Bible it is interpreted as a sign of God's promise to Noah that life would never again be destroyed by a flood. In various mythologies, the rainbow is an attribute of the gods, like thunder and lightning. However a scientific explanation of the phenomena was only given in the thirteen century AD by the Persian physicist Kamal al Farisi who proposed an essentially correct model of the rainbow where "the ray of light from the sun was refracted twice by a water droplet, one or more reflections occurring between the two refractions". He further performed experiments on spherical vessels filled with water to test his hypothesis.

Indeed as shown in Fig.3.32 during a rainbow, incident light from the sun is refracted into a water droplet, reflected from the back surface and refracted out towards the observer. Since as argued above the index of refraction increases with the

---

[32]$Im(n)$ stands for the imaginary part of the complex index of refraction $n(\omega)$

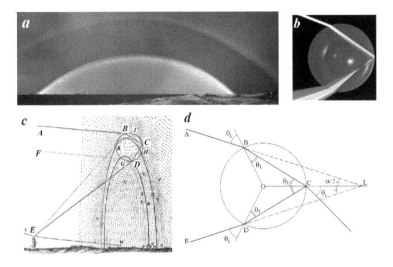

Figure 3.32   (a) Double rainbow on a rural landscape (reprinted from ref.[36]). (b) The refraction of a beam of white light (composed of all visible wavelengths) in a water droplet (reprinted from ref.[37]): notice the larger bending of blue rays as compared to red ones which is at the origin of the colors of the rainbow. (c) Descartes's original explanation of the rainbow. A beam (continuous line) coming from point A is refracted at point B into a water droplet. It is reflected from point C and refracted again at point D where it exits the drop to reach the observer at point E. Descartes also explained the faint second rainbow sometimes observed above the main one as a result of the refraction at point G of a beam parallel to the first one (dotted line), twice reflected (at points H and I) and exiting the droplet at point K. (d) Schematics of a light beam incident on the droplet at B with angle θ_i and refracted into the droplet with angle θ_t. Since the triangles OBC and OCD are isosceles, the incident angle at the reflection point C and exit point D is also θ_t. The angle α between the incoming and outgoing beams can be determined by the sum of the angles in the triangle CBL: $\theta_t = \alpha/2 + \theta_i - \theta_t$.

frequency of visible light, blue light gets deflected more strongly than red light: the water droplets in the cloud act like a prism.

A simple geometric calculation, see Fig.3.32 shows that the angle α between the incoming and outgoing beams satisfies $\alpha = 4\theta_t - 2\theta_i$. This angle is extremal (namely least sensitive to variation in the angle of incidence) at a value $\theta_{i,e}$ of the incidence angle such that:

$$d\alpha/d\theta_i = 4d\theta_t/d\theta_i - 2 = 0$$

Since by Snell's law: $\sin\theta_i = n_w \sin\theta_t$ (where $n_w \simeq 1.33$ is the index of refraction of water) we find that α is extremal when: $d\theta_t/d\theta_i = \cos\theta_{i,e}/n_w \cos\theta_{t,e} = 1/2$, i.e. when the angle of incidence $\theta_{i,e}$ satisfies:

$$2\cos\theta_{i,e} = n_w \cos\theta_{t,e} = n_w \sqrt{1 - \sin^2\theta_{i,e}/n_w^2}$$

from which we find that $\sin^2 \theta_{i,e} = (4 - n_w^2)/3$, i.e. $\theta_{i,e} = 59.6°$ and $\alpha_e \sim 42.5°$. Since the angle α is much less sensitive to variation in the incident angle when $\theta_i \sim \theta_{i,e}$ (namely when $\alpha = \alpha_e$, the intensity of the refracted light is brighter (less diffuse) when it makes an angle α_e with the incoming light. This simple explanation proposed first by Descartes accounts for the position of the rainbow in the sky (the angle between the sun, the rainbow and the observer has to be $\sim 42°$). Because of its dependence on the index of refraction, α_e is slightly smaller for blue light than for red light which explains why the inner ring of a rainbow is blue while its outer ring is red.

4 Quantum Mechanics

Quantum mechanics (QM) is a theory of matter and its interactions with force fields (here we will only care about electromagnetic fields). While classical mechanics and electromagnetism are intuitive (one has a direct experience of gravitation, light, electricity, magnetism, etc.) quantum mechanics is not. As we shall see below, the description of matter that arises from the QM formalism is totally at odds with our daily experience: particles can pass through walls, can be at two different places and in different states at the same time, can behave as waves and interfere with each other. Worse, QM is a non-deterministic description of reality: it only predicts the probability of observing events. This aspect deeply disturbed Einstein who could not accept that QM was the correct final description of reality (as he famously quipped: "God does not play dice"). He and many others came up with alternative descriptions of QM introducing hidden variables (unknowable to the observer) to account for its non-deterministic aspects. But in 1964 John Bell showed that if these hidden variables existed then some measurements would satisfy certain inequalities. The experiments performed by Alain Aspect and his collaborators in the 1970's showed that Bell's inequalities were violated as predicted by QM, but not by the hidden variable theories thereby falsifying them[38].

Yet, for all its technical prowess Aspect's experiment was only addressing a philosophical issue concerning the foundations and interpretation of QM. The theory itself had been amply vindicated earlier by its enormous predictive power: QM explains the stability of atoms, their spectra, the origin of a star's energy and of the elements and their properties, the nature of the chemical bond, the origin of magnetism, conductivity, superconductivity and superfluidity, the behavior of semiconductors and lasers, etc., etc.. All of today's micro-electronic industry is derived from applications of QM (transistors, diodes, integrated circuits, etc.). The development of the chemical industry is also a result of the QM understanding of the chemical bond and of course the nuclear industry would have been impossible without an understanding of the nucleus and the nuclear forces that QM provided.

So, for all its weirdness Quantum Mechanics is the most successful explanation of the World ever proposed by Mankind. It beats Platonicism, the Uppanishads, Kabbalah, Scholasticism, etc., yet it is non-intuitive and cannot be understood except by following its mathematical formalism to its logical conclusions. "The great book of Nature is written in the language of mathematics", Galileo's quip is truer for QM than for any other scientific theory. Facing this fact, one of the founder's of QM, Eugene Wigner, wrote in an article entitled "the unreasonable effectiveness of mathematics in the natural sciences", that "the miracle of the appropriateness of the language of mathematics for the formulation of the laws of physics is a wonderful gift which we neither understand nor deserve"[1].

It is with this mind set that I would like you to approach the study of QM[39, 40, 41, 42, 43, 44, 45]. Like an apprentice sorcerer learning the tricks of his master

DOI: 10.1201/9781003218999-4

without fully understanding them, yet always at awe confronting their power. As we have done with electromagnetism, we will approach QM by following as far as we can the historical narrative. We will see why the radiation of a black-body was such a puzzle that it prompted Max Planck to introduce the idea that energy was quantized; why the stability of atoms and their spectra prompted Bohr, Sommerfeld and others to suggest that the energy levels of atoms were also quantized; how the idea that particles could also have wavelike behavior was first suggested by de Broglie and brought to fruition by Schrödinger, Heisenberg and Dirac; and how from then on, QM revolutionized the understanding of matter, the chemical bond, magnetism, conduction, etc.

4.1 ◆ THE PUZZLES OF MATTER AND RADIATION

At the end of the 19th century, scientists had at their disposal a very successful and revolutionary framework, both scientific and technological, that could explain many of the problems known at that time. Newtonian mechanics was amazingly successful in predicting the motion of celestial bodies. Its most striking success was the prediction by Le Verrier in 1846 of the existence of the planet Neptune. Analyzing some anomalies in the motion of Uranus, he predicted Neptune's precise location in the sky, a prediction which was immediately confirmed by German astronomers. In 1861, Maxwell unified electricity, magnetism and optics opening the area of electrical appliances and wireless communication: Edison invented the light bulb in 1879 and founded "General Electric" in 1892 while Marconi established the "Marconi Wireless Telegraph Company" in 1897. Finally, thermodynamics was assisting the advance of the industrial revolution as thermal engines were driving industry and railways. In spite of terrible social inequalities (as described by C.Dickens, E.Zola and others) this was a time of peace, prosperity and optimism, illustrated by the nascent Impressionist movement.

Yet, many fundamental scientific questions remained unsolved and paradoxical. The chemical properties of the various elements were not understood. The periodicity of these properties as a function of the mass of the elements as determined in 1869 by Mendeleev in his famous Periodic Table of the Elements was a mystery. Nonetheless on the basis of his ad-hoc classification Mendeleev predicted the existence of two new elements, Gallium and Germanium, which were duly discovered in 1875 and 1886 and are essential in today's semiconductor industry! The existence of atoms (indivisible particles of matter characteristic of each element) postulated by Dalton to explain the properties of molecules was not generally accepted. Because of the successful applications of continuum mechanics (in the design of bridges, buildings such as the Eiffel Tower, etc.) and fluid dynamics (in explaining the tides, water waves, etc.), matter was generally believed to be some sort of continuum akin to a gel, not a swarm of particles. It was Einstein who in 1905 finally managed to convince the scientific world of the existence of atoms and molecules by showing that the erratic motion exhibited by dust particles on the surface of water (first observed by the botanist R.Brown in 1827) was due to the shocks caused by the water molecules. The continuum pre-conception also sustained the interpretation of electromagnetic waves. Since all known waves at the times were observed to propagate in a continuous medium (such as water, air, etc.) at a velocity $v = \sqrt{1/\kappa\rho}$ (where κ is the compressibility and ρ the density of the medium), the electromagnetic waves predicted by Maxwell and discovered by Hertz were assumed to propagate in some continuous medium, the ether, the properties of which determined their velocity. However all attempts to detect the "wind" of the ether resulting from the motion of the Earth in it proved negative. This prompted Einstein to formulate his theory of relativity which postulated the constancy of the speed of light and got rid of any notion of ether, see appendix A.6.

Then there were questions related to the emission and absorption spectra of elements that exhibited discrete lines rather than an undifferentiated continuum of absorption or emission, as was the case for sound and water waves and as we have seen for scattered and refracted light. Not only did the elements exhibit specific absorption/emission lines but those differed from element to element, see Fig. 4.1. These observations did not fit with the prevailing conception then of matter as a continuum.

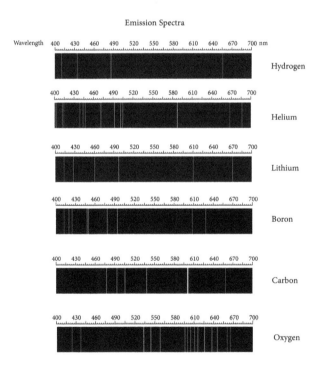

Figure 4.1   The emission spectrum in the visible range for a few elements. Notice the fine spectral lines and the different spectral characteristics for the different elements. This was one of the puzzles that QM solved (reprinted from ref.[46]).

4.1.1 ◆ BLACK-BODY RADIATION

Finally there was the problem of the radiation from a black-body, a material (such as the tungsten filament in a light bulb) which emits radiation uniformly at all frequencies (and also absorbs it uniformly). Notice however that many bodies (e.g. the elements shown in Fig.4.1) are not black-bodies as they adsorb/emit only at certain frequencies. At a given temperature, the radiation inside a black-body cavity is at thermal equilibrium with the walls of the cavity that absorb and re-emit it. When computing the electromagnetic radiation energy emitted by a black-body at a given temperature, one found its energy to diverge because the number of modes at high frequencies diverged. This was not only absurd but, more importantly, in contradiction with the experiments which studied the energy distribution inside the cavity (an oven) by measuring the energy leaking out of it (for example through a small hole) as a function of frequency.

To see how this comes about, consider a cubic oven of side a at temperature T. As we have seen, a body at a given temperature emits electromagnetic radiation, see Fig.3.27. Imagine that the walls of the oven are made of small thermally excited oscillators emitting radiation at angular frequency ω (like the oscillators we considered when studying the frequency dependence of the refractive index, see section 3.5.4). Stationary waves of the form $\sin k_x x \sin k_y y \sin k_z z \cos \omega t$ will be present in the cavity if its walls are reflecting (though for the energy of the EM field to equilibrate with the walls, those cannot be 100% reflective). To satisfy the boundary conditions on the walls we shall require that the wavenumbers: $k_x = \pi l/a$, $k_y = \pi m/a$, $k_z = \pi n/a$ (where n, m, l are real numbers greater than zero, so that $\sin k_x a = \sin \pi l = 0$, etc.). Hence: $k^2 = (l^2 + m^2 + n^2)(\pi/a)^2 \equiv (\pi s/a)^2$ (where the distance s is defined as $s^2 \equiv l^2 + m^2 + n^2$). The number of modes dN_{lnm} with wavenumber k, i.e. within a shell of radius s and thickness ds in the positive octant $l, m, n \geq 0$, is:

$$dN_{lnm} = 2\frac{4\pi s^2}{8}ds = \pi(\frac{ka}{\pi})^2\frac{adk}{\pi} = 8\pi a^3(\frac{k}{2\pi})^2\frac{dk}{2\pi}$$

The factor 2 results from the two possible polarizations of the fields. According to the equipartition theorem of statistical mechanics (see section 5.6) the average energy of each oscillatory mode at frequency v is[1]: $\langle E(v)\rangle = k_B T$ (where $k_B = 1.38 \ 10^{-23}$ J/°K is known as Boltzmann constant). Using the relations: $k = \omega/c \equiv 2\pi v/c$, the energy density of the emitted radiation $du = dN_{lnm}\langle E(v)\rangle/a^3$ becomes:

$$du = \frac{8\pi k_B T}{c^3}v^2 dv \qquad (4.1)$$

hence the total energy, the integral of the energy density over all frequencies, diverges as v^3. This divergence became known as the Jeans' (or ultra-violet) catastrophe. While the data agreed with that formula at low frequencies, it markedly differed at high frequencies (i.e. small wavelengths, in the ultra-violet).

---

[1] The electric mean energy contributes $k_B T/2$ and the magnetic mean energy an other $k_B T/2$

Rather than questioning the equipartition theorem which was verified in other contexts or the possibility of atoms to emit light of arbitrarily high frequencies, Planck suggested in 1900 that light was emitted by the cavity walls in very small discrete quantities, quanta of energy: $\epsilon = h\nu$, where h, the Planck constant is:

$$h = 6.626 \ 10^{-27} \text{ erg sec} = 4.135 \ 10^{-15} \text{ eV sec}$$

so that light of energy E_n is made up of n quanta: $E_n = nh\nu$. In that case the average energy $E(\nu)$ emitted at frequency ν is the sum over all possible energies E_n, weighted by their Boltzmann probability (see below section 5.9.1):

$$P(E_n) = \frac{e^{-E_n/k_B T}}{\sum_n e^{-E_n/k_B T}}$$

Defining $\beta \equiv 1/k_B T$, the average energy is[2]:

$$\langle E(\nu) \rangle = \sum_n E_n P(E_n) = \frac{\sum_n nh\nu e^{-n\beta h\nu}}{\sum_n e^{-n\beta h\nu}} = -\frac{d}{d\beta} \log \sum_n e^{-n\beta h\nu}$$

$$= \frac{d}{d\beta} \log\left(1 - e^{-\beta h\nu}\right) = \frac{h\nu}{e^{h\nu/k_B T} - 1}$$

At low frequencies, $h\nu \ll k_B T$, upon expanding $e^{h\nu/k_B T} \approx 1 + h\nu/k_B T$ one recovers the previous result: $\langle E(\nu) \rangle = k_B T$. However at large emission frequencies the average energy decays as $\langle E(\nu) \rangle \sim h\nu \exp(-h\nu/k_B T)$. Planck therefore suggested to modify the previous result, Eq.4.1 to yield, see Fig.4.2:

$$\rho(\nu) \equiv du/d\nu = \frac{8\pi h\nu^3}{c^3} \frac{1}{e^{h\nu/k_B T} - 1} \tag{4.2}$$

Where $\rho(\nu)$ is known as the spectral density of radiation. Identifying the smallest quanta of energy with a "particle" of light (a photon of energy $h\nu$), Eq.4.2 states that the spectral density of photons (the number of photons per unit volume and unit frequency[3])) in a black-body is:

$$\frac{dN_p}{d\nu} \equiv \eta(\nu) = \frac{\rho(\nu)}{h\nu} = \frac{8\pi\nu^2}{c^3} \frac{1}{e^{h\nu/k_B T} - 1} \tag{4.3}$$

Notice that the total energy density in the cavity is now finite:

$$U_{tot} = \int_0^\infty \rho(\nu)d\nu = \frac{8\pi(k_B T)^4}{h^3 c^3} \int_0^\infty dx \frac{x^3}{e^x - 1} = \frac{8\pi^5 k_B^4}{15 h^3 c^3} T^4 \tag{4.4}$$

Where we changed variables to $x = h\nu/k_B T$ and used the equality: $\int_0^\infty dx \ x^3/(e^x - 1) = \pi^4/15$. The black-body brightness is defined as the power radiated per unit area and

---

[2]Using the formula for the sum of a geometric series: $\sum_n q^n = 1/(1-q)$, with $q = e^{-\beta h\nu}$

[3]The units of $\eta(\nu)$ are sec cm^{-3}

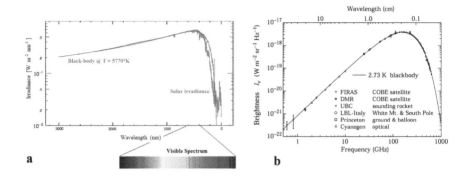

Figure 4.2 The emission spectra of the Sun (a) and the universe (b). The Sun irradiance ($I_{irr}(\lambda)$, data courtesy of Dr.Meftah, LATMOS-CNRS) is pretty well fit by the spectrum of a black-body at 5770°K, however notice the existence of some specific absorption bands in the visible and UV spectrum. The universe on the other hand presents a spectrum that is perfectly matched by a black-body at 2.726°K, figure courtesy of NASA/ARCADE [47].

angle at a given frequency: $I_\nu = c\rho(\nu)/4\pi$, while its irradiance is the total power radiated per unit area at a given frequency through a small hole in the cavity:

$$I_{irr}(\nu) = I_\nu \int \hat{k}\cdot\hat{n}d\Omega = 2\pi I_\nu \int_0^1 \cos\theta d(\cos\theta) = \pi I_\nu = c\rho(\nu)/4$$

The total power radiated per unit area by a black-body is obtained by integrating the irradiance over all frequencies:

$$I_{rad} = \int_0^\infty I_{irr}(\nu)d\nu = \frac{c}{4}\int_0^\infty \rho(\nu)d\nu = \frac{c}{4}U_{tot} = \sigma_{SB}T^4 \tag{4.5}$$

which is known as Stefan's law and where the Stefan-Boltzmann constant:

$$\sigma_{SB} = \frac{2\pi^5 k_B^4}{15h^3c^2} \equiv \frac{\pi^2 k_B^4}{60\hbar^3 c^2} = 5.67\ 10^{-5}\ \mathrm{erg\ sec^{-1}cm^{-2}\ °K^{-4}} = 5.67\ 10^{-8}\ \mathrm{W\ m^{-2}\ °K^{-4}}$$

Therefore by measuring the total intensity of the radiation leaking out from an oven one can measure the temperature of that oven. One can test the validity of Planck's law (actually how close to a black-body the oven really is) by measuring the dependence of the intensity on the radiation wavelength. From the wavelength λ_{max} at which the intensity is maximal an other estimate of the temperature can be deduced: $k_B T \simeq hc/5\lambda_{max}$. For example at 300°K (which corresponds to a thermal energy $k_B T \simeq 25$ meV), the maximum of emission is in the far-infrared at $\lambda_{max} \sim 10\mu$m. The thermal cameras that visualize humans and warm animals (see Fig.3.27) must therefore be sensitive to far-infrared light.

4.1.1.1 The Sun and Earth Temperatures

The Sun is to a very good approximation a black body, see Fig.4.2. The radiation emitted by the fusion reactions occurring at its core (at temperature of $3 \ 10^7 \ °\text{K}$) is at thermal equilibrium with the reacting nuclear particles and diffuse out to the Sun's surface which is much cooler. By fitting the spectrum of sunlight to Planck's formula one can determine the Sun's surface temperature: $T_S = 5770°\text{K}$. The total power generated at the Sun's core and emitted at its surface is:

$$P_S = 4\pi R_S^2 \sigma_{SB} T_S^4 = 3.85 \ 10^{26} \ \text{W}$$

where $R_S = 6.96 \ 10^8$ m is the Sun's radius. Since the radius of the Sun's core is estimated to be $\sim R_S/5$ the volume of the core is: $V_{core} = 1.13 \ 10^{25} \text{m}^3$ and the average power per unit volume generated in the Sun's core is: $P_S/V_{core} = 34 \ \text{W/m}^3$. This is much less than the power generated by our body to keep warm!!

There are a few ways to verify this: let us assume an average daily calory intake of $\simeq 3000\text{kcal} \simeq 1.2 \ 10^7 \text{J}$, which comes to a power consumption of $1.2 \ 10^7 \text{J/day} \simeq 150W$ (1day = 86400sec) of which about half is spent on keeping our body warm. Assume an average human weight of 75kg, i.e. a typical volume of $0.075 \ \text{m}^3$. The power spent on heating our body is thus $\simeq 1000 \ \text{W/m}^3$, about 30 times larger than the power density in the Sun's core!

Alternatively one can use Stefan's law to estimate the losses between a body at 37°C ($T_b = 310°\text{K}$) and an environment at 27°C ($T_e = 300°\text{K}$) (this is a crude estimate since other effects such as perspiration regulate our temperature): $\Delta I = \sigma_{SB}(T_b^4 - T_e^4) \simeq 64\text{W/m}^2$. At steady state, these losses must be compensated by internally generated heat. Modeling a typical human as a cylinder of height 1.7m and radius 12cm (to match the aforementioned typical volume), the power per unit volume generated by a typical human $\simeq 1090\text{W/m}^3$. From these consistent estimates we deduce that our power consumption per unit volume is much larger than the Sun's! What makes the Sun so bright and hot is its huge mass, not its rather inefficient thermonuclear reactions[4].

Let us now estimate the temperature of the Earth T_E resulting from the balance between its absorption of the Sun's radiation and the losses due to its own radiation at temperature T_E. The sunlight impinging on the Earth at a distance from the Sun $R_{SE} = 1.496 \ 10^{11}$m has an intensity:

$$I_E = P_S/4\pi R_{SE}^2 = (R_S/R_{SE})^2 \sigma_{SB} T_S^4 = 1.37 \ \text{kW/m}^2$$

Of that radiation a fraction (known as the Earth's albedo) $\alpha \sim 30\%$ is reflected, mostly by the clouds, snow and ice-caps. The Sun radiation power arriving at the surface of

---

[4]The typical time for two Hydrogen nuclei (protons) in the Sun's core to fuse in a thermonuclear reaction to form Deuterium (with emission of a positron) is about 1 billion years! So in spite of the high density of protons in the Sun's core (about 10^{32} protons per m³) there are only about $1.5 \ 10^{13}$ Hydrogen fusion reactions per m³ per second releasing about 1.3 MeV ($2 \ 10^{-13}$J) per reaction or a total of $3 \ \text{W/m}^3$, about 10% of the total energy (the rest being contributed by faster deuterium/proton and He^3/He^3 fusion reactions).

the Earth is thus about 1 kW/m^2. This is an important number to remember when designing solar energy plants: it sets the maximal power per unit area available from the Sun. Notice that by measuring the light intensity impinging on Earth and the angle subtended by the Sun: $\theta_S = (2R_S/R_{SE})$ one can also get an estimate of the Sun's temperature: $T_S = (4I_E/\sigma_{SB}\theta_S^2)^{1/4}$. The energy absorbed by the Earth heats it and is reradiated (to a good approximation) like a black-body at temperature T_E. We can compute the temperature of the Earth by a simple energy balance. At steady-state the energy radiated is equal to the energy absorbed:

$$\sigma_{SB}T_E^4\,4\pi R_E^2 = (1-\alpha)I_E\,\pi R_E^2 \quad \text{or} \quad T_E^4 = (1-\alpha)(\theta_S/4)^2 T_S^4$$

From which we get $T_E = ((1-\alpha)I_E/4\sigma_{SB})^{1/4} = 255°K = -18°C$. The Earth is actually slightly warmer because of the greenhouse effect that results in the atmosphere reflecting a large part (about 40%) of the emitted energy back to Earth. Notice that alternatively, knowing the temperature on Earth we can get an estimate of the Sun's surface temperature: $T_S = 2(1-\alpha)^{-1/4}\theta_S^{-1/2}T_E$.

4.1.1.2   The Universe as a Perfect Black-Body

While it is difficult to design a perfect black-body, since as we shall see below bound electrons adsorb or emit only specific frequencies (as is partially the case in the outer layers of the Sun), the Universe as a whole turned out to be the best known example of a black-body, see Fig.4.2. The Universe is bathed in a uniform radiation field of very low frequency whose spectral distribution is perfectly matched by a black-body at 2.726°K. This phenomena was predicted by George Gamow in 1948 and observed serendipitously by Arno Penzias and Robert Wilson in 1964 when measuring the noise of a microwave antenna they had built. This noise was larger than they had expected as they were actually detecting the $\approx 3°K$ radiation of the Universe and not the (smaller) noise of the antenna. This background radiation is the most striking evidence for the existence of the Big-Bang. According to this scenario, the Universe began as a big explosion of matter and radiation. At the beginning light and ionized matter interacted continuously and were in thermal equilibrium (as they are in the Sun's core). But then as the Universe expanded it cooled. After about 380000 years when it reached a temperature of $T_{dec} \sim 3000°K$ (and a size a) Hydrogen atoms started to form that could not absorb non-resonant light (see section 4.1.4): radiation decoupled from matter. At that point the radiation spectrum was that of a black body at temperature T_{dec}. It is the relics of that original radiation that we are observing today as an isotropic cosmic background radiation. Let us see why it exhibits a black-body spectrum at a temperature of 2.726°K.

Since once hydrogen atoms formed radiation largely stopped to interact with matter, the number of photons at initial frequency v (see Eq.4.3): a^3dN_p remained constant. But as the Universe expanded to a size $a' > a$, the radiation wavelength similarly increased (recall that in a box $k = 2\pi/\lambda$ is a multiple of π/a), i.e. the frequency v' of the radiation decreased by the expansion factor $\alpha_e = a'/a$: $v' = v/\alpha_e$. Thus, the

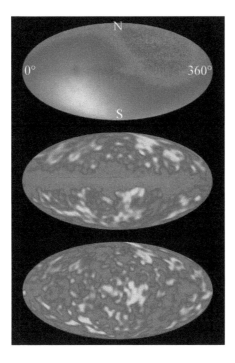

Figure 4.3 The temperature of the cosmic microwave background measured across the sky by the COsmic Background Explorer (COBE) satellite. The top image is the raw data which is red/blue shifted due to the movement of our galaxy through the universe at ~ 370km/s. Correcting for this Doppler shift yields the middle image which is still "polluted" at the equator by the light emitted from the stars in our galaxy. Subtracting that measurable emission yields the bottom image where the temperature fluctuations of the microwave background across the Universe are as small as $10\mu°K$. These small fluctuations nonetheless served as nucleation points for the formation of the stars and galaxies shortly after decoupling. Reprinted from Physics Today (June 1992), with permission of the American Institute of Physics [48].

energy density du' of the present background photons at frequency v' obeys:

$$(a')^3 du' = hv' a^3 dN_p = hv' a^3 \frac{8\pi v^2}{c^3} \frac{1}{e^{hv/k_B T} - 1} dv$$

$$= (a')^3 \alpha_e^{-3} hv' \alpha_e^3 \frac{8\pi v'^2}{c^3} \frac{1}{e^{hv'/k_B(T/\alpha_e)} - 1} dv' \qquad (4.6)$$

which is the energy density of a black-body at a temperature T', smaller than the temperature at decoupling $T = T_{dec}$ by the expansion factor α_e: $T' = (T_{dec}/\alpha_e)$:

$$du' = \frac{8\pi hv'^3}{c^3} \frac{1}{e^{hv'/k_B T'} - 1} dv'$$

Because the Universe expanded by a factor $\alpha_e \sim 1100$ since decoupling[5], one obtains a current temperature for the background radiation of $T' = 2.72°$K. The precise agreement on the value of that temperature is not very important as is the observation that the cosmic background radiation is the best black-body ever observed. It is also highly isotropic in the rest frame of the Universe. As the combined velocity of the Sun and our galaxy the Milky Way is ≈ 370 km/sec with respect to the cosmic background, the Doppler effect (see section 3.4.2.5) red-shifts the radiation in one direction and blue-shifts it in the opposite one. This effect can be subtracted from the measured distribution of radiation intensities. One also needs to subtract the residual contribution from the stars in the galaxy (which fortunately emit at much higher frequencies, in the visible mostly). The measured variations in the temperature of the Universe at different angular positions are then smaller than 10^{-5} °K, yet these small fluctuations served as the nucleation points for the galaxies and can account for their observed distribution, see Fig.4.3. As E.Wigner wrote it is a "miracle ... that we neither understand nor deserve" that a theory devised to explain (approximately) the radiation of hot bodies has turned out to provide such an amazingly precise and powerful description of the Universe!

4.1.2   THE PHOTO-ELECTRIC EFFECT

Besides the emission spectrum of atoms and the black body radiation, another experiment stood in apparent contradiction with Maxwell's electromagnetic theory: the photo-electric effect revealed that electrons were emitted from a conducting material with an energy that depended on the color (the frequency) of the radiation not on its intensity. This was at odds with Maxwell's electromagnetic theory that asserted that the energy of radiation was related to its intensity not its frequency (see Eq.3.112)! Einstein knew that to solve the puzzle of the radiation from a black-body Planck had to assume that radiation was emitted by the walls of the cavity in small quanta. In 1905 Einstein went further and assumed that all light actually comes in bunches, photons, the energy of which is proportional to their frequency: $E = h\nu$. When such a photon is absorbed by an electron its energy is used to tear the electron from the binding potential Φ of the material and move it at velocity v. By energy conservation then:

$$h\nu = \frac{1}{2}mv^2 + e\Phi \qquad (4.7)$$

Hence an electron current can only be observed if light of high enough frequency is used to remove it from the material. The kinetic energy of the electron increases then linearly with the illumination frequency. The current emitted is however proportional to the number of adsorbed photons, i.e. to the light intensity. At the time

---

[5]Estimated from the largest observed red-shifts, section 3.4.2.5, which - from Einstein theory of gravitation - measure the ratio of the Universe size between now and when the radiation was emitted.

Einstein proposal was revolutionary since it assumed that energy came in discrete packets that could not be infinitely divided and it appeared to contradict Maxwell's equations. It took 16 years and confirming experiments to establish the validity of his model, for which he got the Nobel prize in 1921 (and not for his more profound and revolutionary theories of relativity and gravitation). Incidentally from Einstein's relation between energy and momentum: $E^2 = (mc^2)^2 + (pc)^2$ (see appendix A.6) one deduces that if the energy of the photon (of mass $m = 0$) is quantized so must be its momentum: $p = E/c = h/\lambda \equiv \hbar k$ ($\hbar \equiv h/2\pi$).

4.1.2.1 Digital Cameras and Solar Cells

Einstein's understanding of the photo-electric effect has had an enormous techno-logical impact. All digital cameras are based upon it. These CCD (Charge Coupled Device) cameras consist of an array of small capacitors (a few micron in size) each defining a pixel (= picture element). When light (with frequency in the infrared or higher) impinges on a given pixel it kicks off an electron from one side of the ca-pacitor to the other and loads it with a charge proportional to the light intensity (an analogy often made is with a bucket partially filled with water). The charges in a given row of capacitors are then read out by transferring them from one capacitor to the next along the line as in a "water bucket brigade". Thus is the image read row after row and stored. By covering the array of pixels with a mask-array that fil-ters different colors (Red, Green or Blue) the device can be transformed into a color camera where adjacent pixels respond to different colors.

Similarly all of today's solar cells are based on the photo-electric effect using light to generate a current by transferring electrons in a semiconductor material from the so-called valence band into the conduction band (leaving a positively charged "hole" behind, on which more in section 4.3.4 below).

4.1.3 BOHR'S ATOM AND THE SPECTRUM OF HYDROGEN

Following on the footsteps of Planck and Einstein who proposed that energy and momentum were quantized: $E = n\hbar\omega$ and $p = n\hbar k$, Niels Bohr suggested in 1913 that the angular momentum of electrons in an atom was similarly quantized: $L \equiv \vec{r} \times \vec{p} = mvr = n\hbar$ (where n is an integer that specifies the number of quanta of angular mo-mentum) thereby explaining their paradoxical stability (see section 3.4.2.3). Indeed given the balance of electrostatic and centrifugal forces: $e^2/r^2 = mv^2/r = n^2\hbar^2/mr^3$ and the assumed quantization of angular momentum one can deduce that, in the Hy-drogen atom, the orbits of the electron (also known as atomic orbitals) are quantized with a radius: $r = n^2\hbar^2/me^2 \equiv n^2 r_0$ ($r_0 \equiv \hbar^2/me^2 = 0.53$Å is known as the Bohr radius of Hydrogen). Similarly the electron's velocity is $v = e^2/n\hbar$ and its energy:

$$E_n = \frac{1}{2}mv^2 - \frac{e^2}{r} = -\frac{me^4}{2\hbar^2 n^2} = -\frac{e^2}{2r_0}\frac{1}{n^2} = -\frac{13.6\text{eV}}{n^2} \tag{4.8}$$

Thus the energy to ionize a hydrogen atom, i.e. kick off its electron from its ground state at $n = 1$ is 13.6 eV. Because of energy quantization an electron orbiting the

nucleus will not radiate continuously, but emit (or absorb) radiation in quanta of
energy:

$$h\nu_{nl} = E_l - E_n = 13.6\text{eV}(\frac{1}{n^2} - \frac{1}{l^2}) \tag{4.9}$$

when transiting between energy levels characterized by quantum numbers n, l. Hence
the emission (or absorption) spectra of atoms consists of discrete lines corresponding to electronic transitions between states, formally designated by Dirac[39, 44] as
$|n\rangle$, $|l\rangle$, with different quantum numbers n, l (see section 4.2). For the hydrogen atom
the visible emission lines (the so-called Balmer series) correspond to transitions from
states $|l\rangle$ ($l > 2$) to state $|2\rangle$ with wavelengths in the red: 656.3nm ($l = 3$); in the blue
486.1nm ($l = 4$) and in the UV range: 434.1nm ($l = 5$) and 410.2nm ($l = 6$). Transi-
tions to state $|1\rangle$ are in the UV range, while transitions to state $|3\rangle$ (and higher) are in
the infra-red, see Table 4.1 and Fig.4.1. The explanation of the stability of atoms and
the excellent agreement with the emission lines of hydrogen was a major success of
the Bohr model for which he was awarded the Nobel prize in 1922.

Table 4.1

The major emission lines in the hydrogen atom. The wavelength (in nm) is tabulated
for various values of the initial (l) and final state($n < l$) [49].

$l =$	2	3	4	5	6	7	8
Lyman series (n=1)	121.567	102.572	97.254	94.974	93.780	93.074	92.622
Balmer series (n=2)	-	656.464	486.270	434.169	410.290	397.120	389.012
Pashen series (n=3)	-	-	1875.61	1282.16	1094.11	1005.21	954.83

4.1.3.1 Particle-Wave Duality

In 1924, Louis de Broglie (in his Ph.D thesis!) generalizing Einstein's ideas about the
wave-particle duality of light suggested that the same duality also described particles
of matter. So, just as one can assign an EM wave $E(x) = E_0 e^{ikx}$ to a photon possessing
a momentum $p = \hbar k = h/\lambda$, one can also assign a plane wave $\Psi(x) = e^{ikx}$ to an electron
with momentum p (with a similar relation $p = \hbar k$). Hence, Bohr's quantization rule
could be explained as the constructive interference of electronic waves orbiting at a
distance r from the atomic nucleus: $2\pi r/\lambda = pr/\hbar = n$, i.e. $pr = mvr = n\hbar$. It took a
few more years to show that electrons could be diffracted from crystals just as X-
rays are and their pattern of interference could be used to determine the crystalline
structure. For this fruitful insight de Broglie was awarded the Nobel prize in Physics
in 1929.

4.1.3.2 The Bohr-Sommerfeld Quantization Rule

De Broglie's insight also explained a generalization of Bohr's ideas proposed earlier by Arnold Sommerfeld who suggested that for a bound particle (moving along a closed path such as in an atom, an oscillator, etc.) a quantity known in classical mechanics as the action, see section 3.5.2, was also quantized:

$$\oint \vec{p} \cdot d\vec{q} = nh \tag{4.10}$$

where $\vec{p}, \vec{q}$ are the momentum and coordinate of the particle. Again with the identification $p = h/\lambda$, this quantization rule is equivalent to the requirement that the particle-wave interferes constructively along its path to form a standing wave. The Bohr (or Wilson)-Sommerfeld quantization rule, Eq.4.10, reduces to Bohr's quantization of the angular momentum in the case of the hydrogen atom since we have $\oint pdq = 2\pi mvr$.

The same rule also explains why the oscillators of frequency ω assumed by Planck to exist in the walls of a black-body would emit radiation in quanta of energy $\hbar\omega$. For a harmonic oscillator (spring) with frequency $\omega = \sqrt{k/m}$, the energy is:

$$E_{osc} = \frac{p^2}{2m} + \frac{kq^2}{2} = \frac{p^2}{2m} + \frac{m\omega^2 q^2}{2} = \frac{m\omega^2 q_m^2}{2}$$

where q_m is the maximal extension of the spring, from which one derives[6]:

$$nh = \oint pdq = 2 \int_{-q_m}^{q_m} dq \sqrt{2mE_{osc} - (m\omega q)^2} = 2m\omega \int_{-q_m}^{q_m} \sqrt{q_m^2 - q^2}\,dq = \pi m\omega q_m^2 = 2\pi E_{osc}/\omega$$

Hence: $E_{osc} = n\hbar\omega$, which was the assumption made by Planck.

The Bohr-Sommerfeld quantization rule can also be used to find the energy level of a quantum rotor rotating at frequency ω about one of its major axes with moment of inertia I. In that case the angular moment is $I\omega = l\hbar$ ($l = 0, 1, 2, ...$) and the angular energy is $E_l = I\omega^2/2 = \hbar^2 l^2/2I$, quite close to the exact result to be derived later, section 4.10.2.

4.1.4 ABSORPTION AND STIMULATED EMISSION

Based on the Bohr-Sommerfeld model, Einstein proposed in 1917 an intuitive theory of light-matter interaction[50] which provided a microscopic explanation of Planck's formula, set the foundation for the understanding of light absorption and fluorescent emission and would be (40 years later) the basis for the invention of the laser (see section 4.12.3). Einstein considered a system in thermal equilibrium with the EM

---

[6]The integration is performed by making the change of variables: $q = q_m \sin\theta$ to obtain: $\int_{-\pi/2}^{\pi/2} \cos^2\theta d\theta = \pi/2$.

radiation field. At thermal equilibrium the transition rate between any state $|i\rangle$ of energy E_i and any state $|f\rangle$ (of energy $E_f > E_i$) is balanced by the transition rate from $|f\rangle$ to $|i\rangle$.

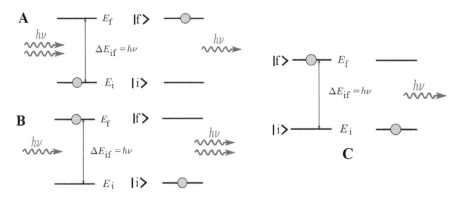

Figure 4.4 (A) The absorption of radiation by an atom in initial state $|i\rangle$. (B) the stimulated emission of a photon in presence of radiation by an atom in a higher energy state $|f\rangle$: notice that this process is the time reversal of absorption. (C) The spontaneous emission of a photon (in absence of radiation) by an atom from its higher energy state $|f\rangle$.

The transition $|i\rangle \rightarrow |f\rangle$ is induced by absorption of a photon of energy: $h\nu_{fi} = E_f - E_i > 0$, see Fig.4.4(A). The associated transition rate k_{fi} (in sec^{-1}) depends on the density of photons available at the transition frequency, namely on the spectral density of photons, Eq.4.3: $\eta(\nu) = \rho(\nu)/h\nu$ (the number per unit volume of photons in a frequency bandwidth $d\nu$ around ν_{fi}):

$$k_{fi} = B_{fi}\eta \qquad (4.11)$$

where B_{fi} is known as the Einstein coefficient for absorption (its units are cm^3/sec^2). Einstein pointed out that state $|f\rangle$ could decay back to state $|i\rangle$ by a process (known ever since as stimulated emission) which is the time reversal of absorption: the **light-driven emission** of a photon of energy $h\nu$ from high energy state $|f\rangle$ to lower energy state $|i\rangle$ as a result of the interaction of the atom/molecule with photons of density η, see Fig.4.4(B).

$$k_{if} = B_{if}\eta \qquad (4.12)$$

B_{if} is known as the Einstein coefficient for stimulated emission. A large part of QM is devoted to the calculation of these coefficients, see sections 4.3.2 and 4.12. Since at the atomic level a process and its time reversal are equivalent: $B_{fi} = B_{if}$ (at our scale this doesn't hold: glass breaks but the pieces never spontaneously reassemble, see Chapter 5).

The laser (acronym for Light Amplification by Stimulated Emission of Radiation) is a direct outcome of Einstein's insight, see section 4.12.3.2. It implements the amplification of an EM wave by stimulated emission in an out-of-equilibrium system, where there is a higher probability P_f of observing atoms/molecules in the excited state than in the ground state, P_i. In that situation, the probability of stimulated emission is larger than the probability of absorption. Consequently the intensity of an EM wave at frequency ν_{fi} passing through such a medium will be amplified.

Einstein finally pointed out that high energy states are inherently unstable and may decay spontaneously to a lower energy state within a mean time τ_s (i.e. at rate $k_s = 1/\tau_s$). That decay (which occurs even in absence of surrounding radiation) is accompanied by emission of a photon of appropriate energy, namely by fluorescent emission.

At steady state (and equilibrium) the rate of absorption by atoms in state $|i\rangle$ must equal the rate of emission (stimulated and spontaneous) of atoms in state $|f\rangle$:

$$B_{fi}\eta P_i = B_{if}\eta P_f + k_s P_f \qquad (4.13)$$

from which one derives[7]:

$$\frac{P_f}{P_i} = \frac{\eta}{\eta + k_s/B_{fi}} < 1 \qquad (4.14)$$

Einstein was concerned with a microscopic explanation of Planck's black-body radiation formula, Eq.4.3. He thus considered the case of a thermodynamic equilibrium between the radiation field and the absorbing/emitting atoms, in which case: $P_f/P_i = e^{-h\nu_{fi}/k_BT}$, see section 5.5.1 . Hence from Eq.4.14, Planck's black body formula is obtained:

$$\eta(\nu_{fi}) = \frac{k_s}{B_{fi}}\frac{1}{e^{h\nu_{fi}/k_BT}-1} = \frac{8\pi\nu_{fi}^2}{c^3}\frac{1}{e^{h\nu_{fi}/k_BT}-1} \qquad (4.15)$$

$$= \frac{8\pi}{c\lambda_{fi}^2}\frac{1}{e^{h\nu_{fi}/k_BT}-1} \qquad (4.16)$$

with the identification:

$$B_{fi}\tau_s = c\lambda_{fi}^2/8\pi \qquad (4.17)$$

This equation is fundamental: it relates the coefficient of absorption B_{fi} to the intrinsic (radiative) lifetime τ_s of the excited state. The longer τ_s the smaller the absorption coefficient. Einstein intuitive, phenomenological understanding of light matter interactions has had a profound impact on our understanding of light emission (e.g. fluorescence) and the development of the laser, see section 4.12.3.

---

[7]Notice that the condition $P_f > P_i$ (a situation known as population inversion) cannot be achieved simply by illuminating the sample. However some lasers, such as the Ruby laser, use illumination at one wavelength to induce a population inversion for amplification at an other (longer) wavelength.

4.2 ◆ QUANTUM MECHANICAL FORMALISM

The early 20th century investigations by Planck, Einstein, Bohr, Sommerfeld, de Broglie, etc. revealed a picture of matter that was different from the infinitely divisible continuum of energy and momentum that prevailed until then. Many of the properties of matter could be explained by assuming that these quantities were discrete rather than continuous. This "old"-quantum theory had many successes: it predicted the radiation spectrum of a black body, explained the photo-electric effect and the discrete emission and absorption spectra of single-electron atoms, the rotational and vibrational spectra of molecules, the stability of atoms and molecules, etc.. These early efforts suggested that matter and radiation shared similar properties: light came as photons, particles of zero mass but possessing definite energy and momentum. Similarly, as suggested by de Broglie, electrons had wave-like properties and could interfere with themselves, as was shown by their diffraction from crystals.

However, it was clear that something was missing as this early theory failed to tackle the spectra of more complicated atoms and could not calculate the intensities of the observed spectral lines. As we shall see below de Broglie's insight opened the way for the formal development of quantum mechanics from its analogy with optics: classical mechanics becoming to quantum mechanics what geometrical optics is to physical optics. Thus a particle with momentum p can be described as a wave-packet $\Psi(x) = A(x)e^{ikx}$ with mean wavenumber $k = p/\hbar$, see Fig.4.5(a,b). This wavepacket is spread over a distance Δx and following the properties of Fourier transforms (see Appendix A.4) a concomitant spread in wavenumber $\Delta k \approx 1/\Delta x$ (and hence $\Delta x \cdot \Delta p \approx \hbar$ which is Heisenberg uncertainty principle, see below). What was not clear then (and is still not clear today) is the physical nature of that wave. What is it? What does the wave amplitude Ψ represent? Schrödinger, Dirac, Heisenberg, Bohr and many others considered that wave to be a pure mathematical construction where the probability of observing a particle with certain properties (position, momentum, energy, etc.) is given by $|\Psi|^2$. Thus they interpreted the diffraction by a crystal of electrons of momentum $p = \hbar k$, see Fig.4.5(c) as a sum of electron plane waves

$$\Psi(x, p) \sim e^{ikx} + e^{ikx+\phi} + e^{ikx+2\phi} + \ldots$$

that interfere constructively only when $\phi = 2kd \sin\theta = 2\pi n$. The probability $P(\theta) = |\Psi(x, p)|^2$ of observing an electron diffracted at an angle θ is thus non-zero only when $2pd \sin\theta = nh$.

This point of view was anathema to Einstein and de Broglie who believed that this probabilistic interpretation of the particle-wave duality reflected a lack of knowledge (i.e. the existence of hidden (unknown) variables that controlled the interaction between the wave and the particle) rather than some fundamental aspect of Nature. Thus de Broglie believed that Ψ represented the amplitude of a pilot-wave that guides the particle (by pushing it perpendicular to surfaces of constant phase). He believed that the reflection of these pilot-waves from the crystal guided the particle to the positions of their constructive interference. Recent hydrodynamics experiments of a water droplet on top of a vibrating water surface exhibit some of the features of the de Broglie picture. Thus the droplet interacts with the water waves (vibrations) like

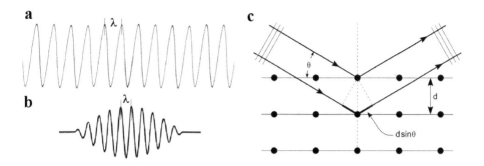

Figure 4.5 a) An infinite plane wave $\Psi(x) = e^{ikx}$ which real part is a sinusoidal wave of wavelength $\lambda = 2\pi/k$. According to Bohr's interpretation of Quantum Mechanics, it describes a particle in a pure state of momentum $p = \hbar k$, i.e. the particle has probability $P(p) = 1$ of having a momentum p. However it is completely delocalised, it has probability $P(x) = |\Psi(x)|^2 = 1$ of being anywhere along the x-axis. (b) A wavepacket $\Psi(x) = A(x)e^{ikx}$ which real part is a sinusoidal wave confined in a region of size Δx, resulting in a spread of wavenumbers $\Delta k = 1/\Delta x$. According to Bohr's interpretation it describes a particle with mean momentum $p = \hbar k$ found at position x with probability $P(x) = |A(x)|^2$. c) The plane waves reflected by the successive crystalline planes interfere constructively when the phase difference between successive reflections $2kd\sin\theta = 2\pi n$. This well known property of X-rays is also observed with particles (e.g. electrons or neutrons) and provided the first definite proof of the wave-particle duality inherent to Quantum Mechanics. It is nowadays used to determine the structure of crystals.

a particle with its pilot-wave[51]. However, QM experiments (see below) stand in contradiction with these hidden variable theories and support the unintuitive formal interpretation of Quantum Mechanics.

Even if it was accepted in the 1920's that particles had wave-like properties, it was not clear what type of wave equation they satisfy. The breakthrough came with the works of Werner Heisenberg and Max Born in 1925 and Erwin Schrödinger in 1926. The latter in particular wrote an equation for the probability of finding a particle at a given position whose solutions share many similarities with EM waves. Hence, as we shall see later, the eigenvalues of the famed Schrödinger equation yield the energy levels of a bound system, much as one determines the resonant modes (stationary waves) of electromagnetic radiation in a cavity: in both cases one solves a Helmholtz equation, see appendix A.5.

Heisenberg proposed a matrix formulation of Quantum Mechanics that was later shown by Schrödinger to be equivalent to his own formulation. Heisenberg's approach however inspired the mathematically rigorous and clear formulation of Quantum Mechanics presented in 1930 by Paul A.M.Dirac in his landmark book ("the Principles of Quantum Mechanics"). In the following we shall follow Dirac's lead[39].

The Bohr atom picture suggested that an electron orbiting the nucleus could be observed in discrete (quantized) states denoted by Dirac as $|n\rangle$, ($|n\rangle$ stands for the electronic wavefunction in the n'th orbital sometimes represented by their real-space amplitude: $\phi_n(\vec{r})$). These states are (complex) functions of the position, angular-momentum, etc. with energy ϵ_n. An electron prepared in a state $|n\rangle$ is always observed with energy ϵ_n, i.e. the probability of measuring an energy $E = \epsilon_n$: $P(\epsilon_n) = 1$. States such as $|n\rangle$ are pure states (or eigenstates) of the energy. Pure states are orthogonal: a system prepared in an eigenstate $|n\rangle$ has probability zero of being in eigenstate $|m\rangle$ (with $\epsilon_m \neq \epsilon_n$).

An eigenstate may not be discrete or necessarily a pure state of the energy. For example the state $|x_0\rangle \sim \delta(x - x_0)$ is an eigenstate of the position operator: if a particle is measured to be precisely at position x_0, it is at that position with probability $P(x) = 1$. However it is not a pure (or eigen-) state of the energy: a free electron in state $|x_0\rangle$ can have any momentum or energy.

A plane wave $e^{ipx/\hbar}$ on the other hand is an eigenstate $|p\rangle$ of the momentum operator: if a particle is observed with a momentum that is exactly p, the pure state associated with it is a plane wave of wavelength $\lambda = h/p$. That pure state is also an eigenstate of the energy for a free electron, i.e. the energy associated to state $|p\rangle$ is $E = p^2/2m$ with probability one. That state is however not a pure state of the position operator: a particle in state $|p\rangle$ can be anywhere in space.

Prior to an observation however the fundamental and totally unintuitive assumption of Quantum Mechanics is that a particle can exist in a mixed state $|\Psi\rangle$ which is a superposition of these pure states (just as any electromagnetic wave can be described as a superposition of plane waves):

$$|\Psi\rangle = \sum_n \alpha_n |n\rangle \qquad (4.18)$$

Notice that since $|\Psi\rangle$ is in general a complex function its conjugate[8] is written as: $\langle\Psi| = \sum_n \alpha_n^* \langle n|$, so that $\langle\Psi|\Psi\rangle = \sum_{n,m} \alpha_m^* \alpha_n \langle m|n\rangle = \sum_n |\alpha_n|^2 \langle n|n\rangle$ (the last equation is a result of the orthogonality of the pure states). For example, an electron in an atom can be prepared in a state which is a superposition of different pure states $|n\rangle$ of energy ϵ_n. However, when the energy of the electron in that mixed state is measured, only discrete values $E = \epsilon_n$ are observed. The electron wavefunction is said to "collapse" into one of the associated energy eigenstates, though no one understands what this "collapse" means! The probability of observing an electron in state $|n\rangle$, i.e. measuring an energy value ϵ_n is postulated to be:

$$P(\epsilon_n) \equiv |\alpha_n|^2 \qquad (4.19)$$

Hence the complex numbers α_n are referred to as probability amplitudes. To satisfy the requirement that $\sum_n P(\epsilon_n) = \sum_n |\alpha_n|^2 = 1$, pure states are normalized such that:

---

[8]The state $|\Psi\rangle$ is sometimes called a ket-state; its complex conjugate $\langle\Psi|$ is called a bra-state, and the projection of one onto the other is called a braket.

$\langle m|n \rangle = \delta_{mn}$ and thus $\langle \Psi|\Psi \rangle = \sum_n |\alpha_n|^2 = 1$. The mean value of the electron energy in state $|\Psi\rangle$ is:

$$\langle E \rangle = \sum_n \epsilon_n P(\epsilon_n) = \sum_n \epsilon_n |\alpha_n|^2 \tag{4.20}$$

The "collapse" of the wavefunction into one of the pure states of the measured variable (or operator, i.e. energy, momentum, position, angular momentum, etc.) is one of the mysteries of QM, which gave rise to an epic controversy between Einstein and Bohr. In particular QM implied that if two particles interacted such as to end-up in a state of zero total angular momentum, measurement of the angular momentum of one particle would cause the two (no matter how far apart) to immediately collapse into a joint state of zero angular momentum. Even if the angular momentum of one particle was randomly distributed, the angular momentum of its far-away partner would be perfectly anti-correlated. Einstein was outraged by what he dubbed "spooky" action at a distance. Recent experiments have however shown that Bohr was right: these "spooky" actions are real.

Dirac drew an analogy between the picture of a physical system emerging from QM and linear algebra. In that analogy, the pure (eigen-) states $|n\rangle$ of an observable $\hat{O}$ (such as the energy, momentum, etc.[9]) form an orthonormal basis (i.e. a set of orthogonal unit-vectors, since $\langle m|n \rangle = \delta_{mn}$) that define a coordinate system in the space of complex continuous functions, known as the Hilbert space. A wavefunction $|\Psi\rangle$ in that space is a unit vector (since $\langle \Psi|\Psi \rangle = 1$) which can be described as a linear combination, Eq.4.18, of these basis-vectors (pure states), see Fig.4.6(a). The coefficients α_n of that superposition are given by the value of the projection of the function (i.e. vector) $|\Psi\rangle$ onto the eigen-vectors $|n\rangle$ of $\hat{O}$. That projection is a scalar product (a complex number) defined as: $\alpha_n = \langle n|\Psi \rangle$. If the states are represented by functions of the spatial coordinates ($|n\rangle \sim \psi_n(x)$ and $|\Psi\rangle \sim \Psi(x)$) then the scalar product is simply:

$$\alpha_n = \langle n|\Psi \rangle = \int dx \psi_n^*(x) \Psi(x)$$

The probability of observing the system in state $|n\rangle$ with eigenvalue O_n is:

$$P_n = |\langle n|\Psi \rangle|^2 = |\alpha_n|^2$$

In Dirac's notation when a variable $\hat{O}$ is measured, the system is perturbed and collapses with probability P_n into an eigenstate $|n\rangle$ of $\hat{O}$ with eigenvalue O_n. That is

---

[9]We shall denote by $\hat{O}$ an operator (not to confuse with a unit vector such as $\hat{r}$) and by O_{mn} its matrix representation, a representation required when actually attempting to compute something, see section 4.2.1. This matrix can be computed in **any orthonormal basis** $\{|\phi_n\rangle\}$ used to describe a physical state with the appropriate boundary conditions, since $\hat{O}|\phi_n\rangle = \sum_l O_{ln} |\phi_l\rangle$, with O_{mn} some complex numbers given by:

$$\langle \phi_m|\hat{O}|\phi_n \rangle = \sum_l O_{ln} \langle \phi_m|\phi_l \rangle = O_{mn}$$

where we used the orthonormality of the set: $\langle \phi_m|\phi_l \rangle = \delta_{ml}$. The eigenstates of $\hat{O}$ define a particular orthonormal basis in which the matrix representation of $\hat{O}$ is diagonal.

formally written as:

$$\hat{O}|\Psi\rangle = \sum_n \alpha_n \hat{O}|n\rangle = \sum_n \alpha_n O_n |n\rangle \quad \text{with} \quad \hat{O}|n\rangle = O_n|n\rangle \qquad (4.21)$$

In other words as discussed above, if $|n\rangle$ is a pure state of $\hat{O}$ it has no projection on other pure (eigen-)states:

$$O_{mn} \equiv \langle m|\hat{O}|n\rangle = O_n \delta_{mn}$$

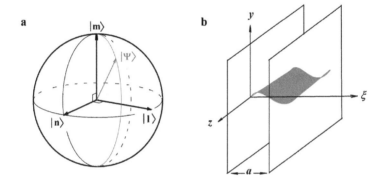

Figure 4.6 (a): Illustration of Dirac's picture of Quantum Mechanics. The orthonormal eigenstates (labeled here as $|n\rangle$, $|m\rangle$ and $|l\rangle$ and sketched as unit vectors) of an operator $\hat{O}$ (such as the energy or angular-momentum of a given system) span a multi-dimensional space know as a Hilbert space. An arbitrary state $|\Psi\rangle$ of the system (e.g. Hydogen atom) is a unit vector in that space characterized by its projections along the different eigenstates (or as here axes). The norm of that projection (its absolute value squared) defines the probability of finding the system in the corresponding eigenstate. While the states are here illustrated as vectors in 3D space they are actually complex functions (of space, momentum, angular momentum, etc.). (b): The state or wavefunction $|\Psi\rangle$ of a particle confined between two planes a distance a apart, such as an electron in a conductor sandwiched between two insulating materials.

In linear algebra the eigenstates $|n\rangle$ of the observable (operator) $\hat{O}$ are the states that diagonalize the matrix O_{mn}. These eigenstates are often unknown, however the matrix O_{mn} can always be evaluated (even if only numerically) in any orthonormal basis $|\phi_n\rangle$. It can then be diagonalized and its eigenstates[10] expressed in this basis (see example below).

For a state: $|\Psi\rangle = \sum_n \alpha_n |n\rangle$ the mean value of the observable (or operator) $\hat{O}$ in

---

[10] See footnote 9

that state becomes:

$$\langle \Psi | \hat{O} | \Psi \rangle = \sum_{n,m} (\langle n | \alpha^*_n) \hat{O}(\alpha_m | m \rangle)$$

$$= \sum_{n,m} O_m \, \alpha^*_n \alpha_m \, \langle n | m \rangle \tag{4.22}$$

$$= \sum_n |\alpha_n|^2 O_n = \sum_n P_n O_n = \langle O \rangle \tag{4.23}$$

The last equation is the usual definition of the average of a stochastic variable. Quantum mechanics in Dirac's formulation is thus reduced to linear algebra: physical states are represented as complex unit vectors in the space of complex functions and the physical observables are complex matrices with real eigenvalues (the results of the measurement), i.e. Hermitian matrices satisfying $A_{mn} = A^*_{nm}$ (also written as $A^\dagger = A$). If the eigenstates of one observable $\hat{A}$ are also the eigenstates of an other observable $\hat{B}$ then $\hat{A}$ and $\hat{B}$ commute:

$$\hat{A}\hat{B}|n\rangle = \hat{A}b_n |n\rangle = a_n b_n |n\rangle = b_n a_n |n\rangle = \hat{B}a_n |n\rangle = \hat{B}\hat{A}|n\rangle$$

If the observable (operators) commute they can be both measured simultaneously: they are diagonalized by (i.e they share) the same eigenstates and thus measurement of one variable drives the system in a pure-state of the other. If on the other hand the eigenstates of $\hat{A}$ and $\hat{B}$ are not identical, their simultaneous measurement is not possible: measurement of one variable drives the system into a state that is a superposition of the eigenstates of the other. Let $\{|n\rangle\}$ be the eigenstates of $\hat{A}$, then

$$\hat{B}\hat{A}|n\rangle = \hat{B}a_n |n\rangle = a_n\hat{B}|n\rangle = \sum_m a_n |m\rangle\langle m|\hat{B}|n\rangle = \sum_m a_n B_{mn} |m\rangle$$

The last two equations are a representation of the state $\hat{B}|n\rangle$ in the basis of the eigenstates of $\hat{A}$: $\hat{B}|n\rangle = \sum_m \langle m|\hat{B}|n\rangle |m\rangle = \sum_m B_{mn} |m\rangle$. On the other hand:

$$\hat{A}\hat{B}|n\rangle = \sum_m \hat{A}|m\rangle\langle m|\hat{B}|n\rangle = \sum_m a_m B_{mn} |m\rangle \neq \sum_m a_n B_{mn} |m\rangle = \hat{B}\hat{A}|n\rangle$$

hence the operators do not commute[11]:$[\hat{A}, \hat{B}] \equiv \hat{A}\hat{B} - \hat{B}\hat{A} \neq 0$. We shall see later that the non-commutability of operators (which is quite common with matrix operations) is at the core of Heisenberg uncertainty principle which states that the position and momentum of a particle cannot be simultaneously determined.

The energy being an important observable in physics, the energy operator (or Hamiltonian, $\hat{H}$) plays a central role in Quantum Mechanics. Its eigenvalues are the measured energies of the system (which can be discrete or continuous) and its eigenstates are like the resonant modes of an oscillator or the specific orbits of the electron in Bohr's atom:

$$\hat{H}|n\rangle = \epsilon_n |n\rangle$$

---

[11] the operator $\hat{C} = [\hat{A}, \hat{B}]$ is known as the commutator of $\hat{A}$ and $\hat{B}$. $\hat{C} \equiv 0$ when the operators commute.

From Planck and Einstein, we know that there is a relation between energy and frequency: $\epsilon_n = \hbar\omega_n$, so that an eigenstate with given frequency ω_n evolves in time as:

$$|n(t)\rangle = \exp(-i\omega_n t)|n\rangle = \exp(-i\epsilon_n t/\hbar)|n\rangle$$

As expected from energy conservation, if a system has been prepared in an eigenstate of the energy $|n\rangle$, it remains in that state with probability $P = |\exp(-i\omega_n t)|^2 = 1$. If however the initial state is a superposition of energy eigenstates: $|\Psi(0)\rangle = \sum_n \alpha_n |n\rangle$ it evolves as:

$$|\Psi(t)\rangle = \sum_n \alpha_n e^{-i\epsilon_n t/\hbar} |n\rangle$$

From which one derives Schrödinger's equation:

$$\frac{\partial}{\partial t}|\Psi(t)\rangle = \frac{1}{i\hbar}\sum_n \alpha_n \epsilon_n e^{-i\epsilon_n t/\hbar}|n\rangle = \frac{1}{i\hbar}\sum_n \alpha_n e^{-i\epsilon_n t/\hbar}\hat{H}|n\rangle = \frac{1}{i\hbar}\hat{H}|\Psi(t)\rangle$$

or in the more common notation:

$$i\hbar\frac{\partial}{\partial t}|\Psi(t)\rangle = \hat{H}|\Psi(t)\rangle \tag{4.24}$$

This is essentially all of Quantum Mechanics: a definition of physical states as vectors in a complex space, measurements as matrix operations on these vector states, the outcome of the measurements as eigenvalues of those matrices and a description of the time evolution of the physical states by Schrödinger's equation. The rest is application of this linear algebra formalism!

4.2.1 ◆ EXAMPLE: PARTICLE IN A BOX

Consider for example a particle of mass μ confined in a one-dimensional box: $0 \leq \xi \leq a$. This problem once considered a thought (gedanken) experiment, is nowadays commonly realized when creating a structure consisting of a thin conducting plane sandwiched between two insulators. This configuration, known as a quantum well, confines the electrons moving freely in the conduction zone to a one-dimensional box between the two insulators, see Fig.4.6(b). Quantum wells are used in solar cells and are at the core of the diode lasers used in DVDs, laser pointers or as sources for fiber-optics communication. These devices are based on the absorption/stimulated emission of light via the electronic transitions between energy levels in a quantum well, see sections 4.3.4 and 4.11.2.

For simplicity we shall work with the dimensionless variable $x = \xi/a$ (i.e. $0 \leq x \leq 1$). Since the probability of finding the particle outside the box is zero, continuity of the wavefunction (and the associated probability) implies that it be zero at the boundaries. To describe the states $|\Psi\rangle$ of the particle in the box we may thus choose **any complete set** of states $|\phi_n\rangle$, with real space representation $\phi_n(x)$ satisfying the boundary conditions: $\phi_n(0) = \phi_n(1) = 0$. For example the set of functions:

$$\phi_n(x) = A_n x(1-x)P_n(x)$$

will do. Here A_n are normalization constants and $P_n(x)$ are polynomials chosen to ensure the orthogonality of the functions in the set: $\langle m|n \rangle = \int_0^1 \phi_m^*(x)\phi_n(x)dx = \delta_{nm}$. The first functions are[12]:

$$\phi_1(x) = \sqrt{30}x(1-x)$$

$$\phi_2(x) = \sqrt{210}x(1-x)(1-2x)$$

$$\phi_3(x) = \sqrt{810}x(1-x)(1 - \frac{14x}{3} - \frac{14x^2}{3})$$

$$\vdots$$

Any state $|\Psi\rangle$ describing a particle in a box can be described as a superposition of functions from the set $\{|\phi_n\rangle\}$.

$$|\Psi\rangle = \sum_n b_n |\phi_n\rangle$$

A recurring problem in QM is to find the eigenvalues E_n and eigenstates ψ_n of the energy operator $\hat{H}$. We shall later see (section 4.4) that the energy operator $\hat{H} = -(\hbar^2/2\mu)\partial_\xi^2$ or in the dimensionless variable x: $\hat{H} = -(\hbar^2/2a^2\mu)\partial_x^2$. In the set $\{|\phi_n\rangle\}$, the energy operator can thus be written as a matrix:

$$H_{mn} = \langle\phi_m|\hat{H}|\phi_n\rangle$$

$$\text{e.g.:} \quad H_{31} = \langle\phi_3|\hat{H}|\phi_1\rangle = -\frac{\hbar^2}{2a^2\mu}\int_0^1 dx\phi_3(x)\frac{d^2\phi_1(x)}{dx^2} = 2\sqrt{3}\frac{\hbar^2}{2a^2\mu}$$

Notice that since $H_{31} \neq 0$, the energy matrix is not diagonal and the states $\{|\phi_n\rangle\}$ are not eigenstates of the Hamiltonian. As is often the case with real problems (such as in quantum chemistry), while finding the eigenstates of a given observable (e.g. the energy) is often difficult, the energy matrix can always be computed in some appropriate set of basis vectors[13] which span the space of functions with the required boundary conditions, here the set $\{|\phi_n\rangle\}$. The energy matrix can always be truncated to any desired size and its eigenvalues and eigenstates evaluated (usually numerically). In the example studied here the result turns out to be (upon truncation at $n = 3$):

$$H_{mn}^{(3)} = \frac{\hbar^2}{2a^2\mu} \begin{pmatrix} 10 & 0 & 2\sqrt{3} \\ 0 & 42 & 0 \\ 2\sqrt{3} & 0 & 102 \end{pmatrix}$$

The lowest eigenvalue of this truncated energy matrix is

$$E_1^{(3)} = (56 - 4\sqrt{133})(\hbar^2/2a^2\mu) = 9.869749...(\hbar^2/2a^2\mu)$$

---

[12]The constants A_n are set by the condition $\langle\phi_n|\phi_n\rangle = 1$. With $P_1 = 1$, the $n-1$ coefficients a_{nm} of the polynomial $P_n(x) = 1 + \sum_1^{n-1} a_{nm}x^m$ are determined recursively from the $n-1$ orthogonality conditions: $\langle\phi_l|\phi_n\rangle = 0$ for $l < n$.

[13]See footnote 9

Since the problem of a particle in a box is exactly soluble, we may compare that approximate result with the exact one. Indeed the equation for the eigenvalues and eigenstates of the Hamiltonian $\hat{H}$, see Eq.4.21:

$$\hat{H}\psi_n(x) = -(\hbar^2/2a^2\mu)\partial_x^2\psi_n(x) = E_n\psi_n(x)$$

has for eigenstates: $\psi_n(x) = \sqrt{2}\sin n\pi x$ with eigenvalues: $E_n = \hbar^2 n^2\pi^2/2a^2\mu$. The approximate ground state energy $E_1^{(3)}$ is thus very close to the exact value $E_1 = \hbar^2\pi^2/2a^2\mu = 9.869604...(\hbar^2/2a^2\mu)$. Similarly the state $\phi_1(x)$ is very close to the ground state $\psi_1(x)$. If the system is prepared in state $|\phi_1\rangle$, that state can be written as superposition of energy eigenstates: $|\phi_1\rangle = \sum_n a_n |\psi_n\rangle$ with coefficients a_n given by:

$$a_n = \langle\psi_n|\phi_1\rangle = \sqrt{60}\int_0^1 dx\sin n\pi x\, x(1-x) = \frac{2[1-(-1)^n]\sqrt{60}}{n^3\pi^3}$$

The coefficient for even n is zero. The leading coefficient is $a_1 = 4\sqrt{60}/\pi^3 = 0.9992$. The probability of observing the system in its lowest energy state $|\psi_1\rangle$ if it is prepared in state $|\phi_1\rangle$ is $P(E_1) = |a_1|^2 = 0.9984 \approx 1$. Thus, even though $|\phi_1\rangle$ is not an eigenstate of $\hat{H}$ (the matrix H_{mn} is not diagonal in the set $\{|\phi_n\rangle\}$) it is a pretty good approximation of its ground state $|\psi_1\rangle$.

The ability to tune the energy levels (and thus the light being absorbed or emitted) using boxes of different sizes a is what makes quantum wells so attractive for many applications. For electrons (of mass $\mu = 9.1\ 10^{-28}$g) in a well of size $a = 10$nm, the energies are given by $E_n = E_1 n^2$ with $E_1 = 0.6\ 10^{-14}$erg ≈ 4meV, with corresponding absorption lines in the infrared.

4.3   SIMPLE QM SYSTEMS

4.3.1   THE CHIRAL AMMONIA MOLECULE

Our first application of the QM formalism described above will be the ammonia molecule and the ammonia maser (the microwave equivalent of the laser). We will consider here the chiral ammonia molecule NHDT (D and T stand for the isotopes of Hydrogen: Deuterium and Tritium), rather than the achiral NH_3 considered by Feynman (in Vol.III of his "Lectures on Physics"[40]). This choice exemplifies the queer nature of QM better than the achiral molecule[52]. Notice that by working at low enough temperatures the molecule can be assumed to be non-rotating (i.e. it is in the rotation ground state, with $l = 0$, see section 4.1.3.2 for the energy levels of the quantum rotor in the Bohr-Sommerfeld approximation). NHDT is a tetrahedron with the four atoms sitting at the four apexes. The molecule possesses distinct enantiomers, i.e. distinct states: $|L\rangle$ (with left-handed chirality) and $|R\rangle$ (with right-handed chirality) that are mirror images of each other depending on whether the HDT plane is on the right or the left of Nitrogen atom, see fig.4.7. These states are NOT eigenstates of the Hamiltonian as the molecule in state $|L\rangle$ can end up in state $|R\rangle$ by tunneling of the nitrogen atom through the HDT plane, like a left-handed glove can be transformed into a right-handed one by turning it inside out. Even though this energetically costly transition is classically impossible, there is in QM always a small probability for such a process to happen (this tunneling through an energetically forbidden zone (a wall, see section 4.5.4) is one of the oddities of QM).

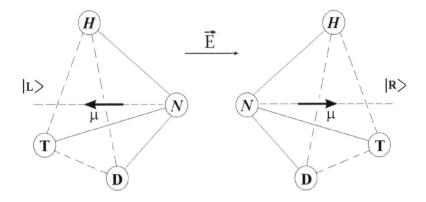

Figure 4.7 The chiral ammonia molecule, NHDT consist of a nitrogen atom bound to the different isotopes of hydrogen(H): deuterium(D) and tritium(T). This molecule exist with different chirality: left-handed $|L\rangle$ or right-handed $|R\rangle$ which are mirror images of each other. Since nitrogen is slightly more electrophilic (negatively charged) than the hydrogen isotopes the molecule possesses a small electric dipole moment[15] μ

Due to the symmetry of states $|L\rangle$ and $|R\rangle$, the Hamiltonian of this two-state system can thus be written[16] as:

$$H_0 = \begin{pmatrix} E_0 & -A \\ -A & E_0 \end{pmatrix}$$

Its eigenvalues (eigen-energies) are: $E^0_{I,II} = E_0 \pm A$ and its eigenvectors (eigenstates) are:

$$|I\rangle = \frac{|L\rangle - |R\rangle}{\sqrt{2}} \equiv \frac{1}{\sqrt{2}} \begin{pmatrix} 1 \\ -1 \end{pmatrix} \qquad |II> = \frac{|L\rangle + |R\rangle}{\sqrt{2}} \equiv \frac{1}{\sqrt{2}} \begin{pmatrix} 1 \\ 1 \end{pmatrix}$$

You can check that $H_0|I\rangle = E^0_I|I\rangle$ and $H_0|II\rangle = E^0_{II}|II\rangle$. In the eigenstate basis the Hamiltonian is diagonal:

$$H_0^D = \begin{pmatrix} E_0 + A & 0 \\ 0 & E_0 - A \end{pmatrix}$$

---

[15]Notice that the direction of the electric dipole in physics is from the more negatively charged site, i.e. Nitrogen to the more positively charged one, i.e Hydrogen, so that the dipole tends to align along the electric field. In chemistry the convention is often opposite: it is from the less electrophilic ("electron-loving") atom (Hydrogen or its isotopes) to the more electrophilic one it is bonding with, i.e. Nitrogen.

[16]E_0 is known as the on-site energy and A as the energy overlap between states $|L\rangle$ and $|R\rangle$, which are NOT eigenstates of the Hamiltonian and for which therefore the matrix representation of H is NOT diagonal. A justification for this Hamiltonian can be found in section 4.10.1.1.

It is a general result from linear algebra that the matrix of eigenvectors:

$$\Lambda = \frac{1}{\sqrt{2}} \begin{pmatrix} 1 & 1 \\ -1 & 1 \end{pmatrix}$$

diagonalizes the original matrix: $H_0^D = \Lambda^T H_0 \Lambda$, see Appendix A.2.

4.3.1.1 Schrödinger's Cat

Notice that the energy eigenstates for the chiral ammonia molecule consist of a coherent superposition of a left- and a right-handed state with probability 1/2! This is a classically absurd situation: how can a system be simultaneously both left and right-handed?! It is similar to the famous Schrödinger's cat paradox, see Fig.4.8. In this gedanken experiment Schrödinger proposed to couple a cat enclosed in a box to a two-state QM system: the cat is dead if the system is in state $|L\rangle$ and alive if in state $|R\rangle$. According to QM, before one looks into the box the cat exists as a superposition of the two states: dead and alive; just like our chiral ammonia molecule is described as a superposition of left- and right-handed states. However once a measurement is made the cat is either dead or alive; just as the chiral ammonia molecule is —when observed —either left- or right-handed.

Figure 4.8 The gedanken (thought) experiment that Schrödinger proposed to test the validity of QM. In a closed box isolated from the external world there is a cat and a radioactive source of say β−particles (electrons). If a particle is emitted and detected by a Geiger counter a poison vial is broken that kills the cat. Otherwise the cat is alive. As the state of the particle is a superposition of bound and emitted particle, one must consider the cat to be in a superposition of a live and dead animal (reprinted from ref.[53]). For Einstein that thought experiment demonstrated that QM was incomplete since it gave rise to absurd assumptions. Yet, when isolated from the external world (a tall order for macroscopic systems) all experiments so far are consistent with this "absurd" superposition of states.

So did Einstein ask: how was the cat "really" before we looked into the box? He claimed that the superposition is absurd and that the cat is either dead or alive. He argued that some unknown variables (hidden variables) of the microscopic two-state system that set the cat's state result in the observed QM probabilistic predictions. However as mentioned earlier, the experimental violation of Bell's inequalities suggest that Einstein was wrong and that such "Schrödinger cats" exist as a superposition of dead and alive states, even if we have no clue what that means! We shall discuss again this point more quantitatively below.

A crucial ingredient enters into the QM picture which is the measurement process. For a system to exhibit the interference effects resulting from QM state superpositions it must not interact with the environment. These interactions are like independent measurements of the system and they destroy its coherence (e.g. the two-state superposition) by implicating (entangling them with) many external uncontrolled states. Now it is very easy for a macroscopic system like a cat (even if isolated) to interact with the world outside the box (for example through the radiation it emits or adsorbs). Hence, quantum experiments with large objects are notoriously difficult to perform. One of the largest molecule for which quantum interference effects have been demonstrated is the buckyball: C_{60} (see below).

Since the chiral states are given by:

$$|L\rangle = (\,|I\rangle + |II\rangle)/\sqrt{2}$$
$$|R\rangle = (-|I\rangle + |II\rangle)/\sqrt{2} \tag{4.25}$$

if the molecule has left-handed chirality to begin with: $|\Psi(0)\rangle = |L\rangle$, it will evolve as:

$$|\Psi(t)\rangle = \frac{e^{-iE_I^0 t/\hbar}|I\rangle + e^{-iE_{II}^0 t/\hbar}|II\rangle}{\sqrt{2}}$$

The probability $P_L(t)$ of finding the molecule with a left-handed chirality (state $|L\rangle$) after a time t is set by the component of $|\Psi(t)\rangle$ along $|L\rangle$, i.e. its projection on $|L\rangle$:

$$P_L(t) = |\langle L|\Psi(t)\rangle|^2 = \left|\frac{(\langle I| + \langle II|)}{\sqrt{2}} \frac{(e^{-iE_I^0 t/\hbar}|I\rangle + e^{-iE_{II}^0 t/\hbar}|II\rangle)}{\sqrt{2}}\right|^2$$

$$= \frac{|e^{-iE_I^0 t/\hbar} + e^{-iE_{II}^0 t/\hbar}|^2}{4} = \cos^2(At/\hbar) \tag{4.26}$$

and the probability of finding the system with right-handed chirality is $P_R(t) = \sin^2(At/\hbar)$. When a given molecule is measured its chirality is well defined (either left or right), but measurements over many molecules yield the oscillating probability distribution just computed. Since $\cos^2(At/\hbar) = (1 + \cos 2At/\hbar)/2$ the oscillation frequency is related to the difference between the energy levels of the eigenstates $\omega_0 = 2A/\hbar$. For NH_3 that frequency $\nu_0 = \omega_0/2\pi = 24$ GHz is in the microwave range. For NHDT it is slighly lower due to the higher mass of Deuterium and Tritium. The observation of this QM oscillation between enantiomers requires an extremely fast measurement of the molecules' chirality at very low temperatures (below $0.3°$K to ensure a negligible occupation of the higher energy state via thermal fluctuation), which hasn't been done so far[52].

4.3.1.2 Bell's Inequalities

The essence of Bell's inequalities may be grasped from this simple example. Imagine that a NHDT molecule prepared in a definite chiral state ($|L\rangle$ or $|R\rangle$) is observed a time δt later so that the probability of finding it in the same state is 99%. If it is measured again a time δt later it will be observed in the same state as previously with probability 99%. One may now ask: what is the probability of observing the system in its initial state if we look at it a time not δt but $2\delta t$ later? If, as Einstein believed, the system is at δt in a definite state that is only once in a hundred times different from the initial state, then at $2\delta t$ the state of the system would be at worst twice in a hundred times different from the initial state, namely the probability of observing the system in its initial state would be at worst 98%. However the QM prediction is: $P(2\delta t) = \cos^2(2A\delta t/\hbar) \approx 1 - 4(A\delta t/\hbar)^2 = 0.96$ (since we assumed that $P(\delta t) = 0.99$, i.e. $A\delta t/\hbar \approx 0.1$). The QM prediction (0.96) violates the lower bound (0.98) set by "realistic" theories which assume (as Einstein did) that the system is, at intermediate times, in a definite state which we have simply no way of determining[17]. As mentioned earlier the experimental results (measured not on chiral ammonia molecules but some other two-state systems) vindicate the QM prediction which assumes that at intermediate times the system is in a "meaningless" superposition of left and right-handed molecules. These experiments rule out the "realistic" theories.

4.3.2 THE AMMONIA MOLECULE IN AN ELECTRIC FIELD

Since nitrogen is more electrophilic than hydrogen, it tends to be slightly more negatively charged than the hydrogen isotopes and the molecule ends up with a permanent electric dipole moment $\vec{\mu}$, as shown in Fig.4.7. In presence of an electric field along the x-axis as in Fig.4.7, $\vec{E} = E\hat{x}$ the energy of a dipole is $W = -\vec{\mu} \cdot \vec{E} = -\mu_x E$, (Eq.3.13). We thus expect state $|L\rangle$ which dipole moment is opposite to the field direction to have higher energy than state $|R\rangle$, which dipole is oriented along the field. The Hamiltonian of the molecule in an external electric field is thus:

$$H = \begin{pmatrix} E_0 + W & -A \\ -A & E_0 - W \end{pmatrix}$$

---

[17]This argument has been formalized by Leggett and Garg[54] who considered a two-level system prepared at $t = 0$ in state Q_0 ($= \pm 1$), observed (or not) at time δt later in state Q_1 and observed a further time δt later in state Q_2. If the probability of remaining in the same state during a time δt is p ($1 - p$ of switching states), then within "realistic" theories the correlation $C_{01} \equiv \langle Q_0 Q_1 \rangle = C_{12} \equiv \langle Q_1 Q_2 \rangle = p - (1 - p) = 2p - 1$, while the correlation $C_{02} \equiv \langle Q_0 Q_2 \rangle = (2p - 1)^2$. The Leggett-Garg (LG) inequality $C_{01} + C_{12} + C_{02} + 1 = 4p^2 \geq 0$ holds. On the other hand if QM is correct: $C_{01} = C_{12} = \cos^2 \theta - \sin^2 \theta = \cos 2\theta$ and $C_{02} = \cos 4\theta$ where the dephasing of the wavefunction within a time δt is θ ($= At/\hbar$ for the Ammonia molecule considered here). Thus $C_{01} + C_{12} + C_{02} + 1 = 2\cos 2\theta + \cos 4\theta + 1$ which can be less than zero (e.g. for $\theta = \pi/3$): the LG inequality is violated. The experimental observation[55] confirms the QM prediction: the LG inequality is violated.

Notice that in the eigenstate representation (where the unperturbed Hamiltonian H_0 is diagonal), the perturbed Hamiltonian can be written as:

$$H' = \Lambda^T H \Lambda = H_0^D + \delta H = \begin{pmatrix} E_I^0 & 0 \\ 0 & E_{II}^0 \end{pmatrix} + \begin{pmatrix} 0 & W \\ W & 0 \end{pmatrix} \quad (4.27)$$

with $W = \langle I|\mu_x E|II\rangle \equiv \mu_{12}E$ (μ_{12} is known as the dipole overlap between the energy eigenstates $|I\rangle$ and $|II\rangle$). The eigenvalues (and eigenvectors) of H' (and H) can be obtained exactly: $E_{I,II} = E_0 \pm \sqrt{A^2 + W^2}$. It is however more instructive to look at the case where the perturbation W is small with respect to the difference in energy between the two eigenstates. The equation for the eigen-energies is:

$$(E_I^0 - E)(E_{II}^0 - E) - W^2 = 0$$

Since W is small: $E_I \approx E_I^0$ and $E_{II} \approx E_{II}^0$ and therefore to order W^2:

$$E_I = E_I^0 + \frac{W^2}{E_I^0 - E_{II}^0} = E_I^0 + \frac{\delta H_{12}\delta H_{21}}{E_I^0 - E_{II}^0} > E_I^0 \quad (4.28)$$

$$E_{II} = E_{II}^0 + \frac{W^2}{E_{II}^0 - E_I^0} = E_{II}^0 + \frac{\delta H_{21}\delta H_{12}}{E_{II}^0 - E_I^0} < E_{II}^0$$

As we shall prove in section 4.11, this is a general result when the diagonal Hamiltonian is perturbed by an amount δH. Notice that in presence of an electric field the energy of the higher energy state $|I\rangle$ increases, while that of the lower energy state $|II\rangle$ decreases. These results suggest a way to separate the eigenstates by passing a beam of ammonia molecules through a strong electric field gradient. The field gradient generates a force on the molecules:

$$\vec{F}_{I,II} = -\vec{\nabla}E_{I,II} = \mp\frac{\mu_{12}^2}{2A}\vec{\nabla}E^2$$

which separates them: state $|I\rangle$ is deflected to regions of small electric fields (which minimizes the increase in its energy), while state $|II\rangle$ is deflected to regions of high electric fields (which maximizes the decrease in its energy). A population inversion can thus be generated where low energy ammonia molecules are separated from higher energy ones. This high energy population can be used to amplify microwave radiation by stimulated emission. The resulting device, known as a maser, was the first implementation of a stimulated radiation amplification device and served as the first atomic clock.

4.3.2.1 The Ammonia Maser.

In an ammonia maser, Fig.4.9, the high energy state $|I\rangle$ of the ammonia molecule selected as described above, enters a resonant cavity (tuned to a frequency $\nu_0 = (E_I^0 - E_{II}^0)/h = 24$ GHz) creating a population inversion situation. A small electric field oscillating at frequency $\omega \approx \omega_0$ is thus amplified by stimulated emission, with the consequent generation of a highly coherent microwave beam[40].

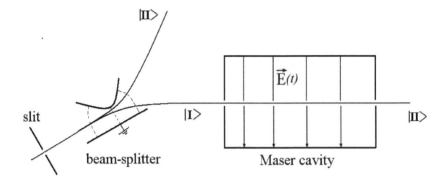

Figure 4.9 Principle of operation of a maser. An ammonia beam is sent trough a slit into a beam splitter that consists of a strong nonuniform electric field which splits the eigenstates of the energy: $|I\rangle$, $|II\rangle$. The high energy eigenstate, $|I\rangle$, is sent into a cavity where it is stimulated to decay into the low energy state $|II\rangle$ by emitting radiation at a frequency $\omega_0 = (E_I - E_{II})/\hbar$

To analyze the operation of a maser, let us assume that the ammonia molecules enter a cavity in which they experience a time varying electric field directed along the x-axis of amplitude $2\mathcal{E}$ (and intensity $I = c\mathcal{E}^2/2\pi$, see Eq.3.113):

$$\vec{E} = 2\mathcal{E}\hat{x}\cos\omega t = (\mathcal{E}e^{-i\omega t}\hat{x} + \mathcal{E}e^{i\omega t}\hat{x})$$

As for the static electric field considered above, this field couples with the dipole moment $\vec{\mu}$ of the molecules to modulate their coupling by: $\delta H_{12}(t) = W(t) = -\mu_{12}\mathcal{E}(e^{-i\omega t} + e^{i\omega t})$[18]. The perturbed Hamitonian in the eigenbasis ($|I\rangle$ and $|II\rangle$) is then, see Eq.4.27:

$$H' = H_0^D + \delta H(t) = \begin{pmatrix} E_I^0 & 0 \\ 0 & E_{II}^0 \end{pmatrix} + \begin{pmatrix} 0 & W(t) \\ W^*(t) & 0 \end{pmatrix} \qquad (4.29)$$

We seek a general solution $|\Psi(t)\rangle$ of Schrödinger's equation: $i\hbar\partial\Psi/\partial t = H'\Psi$ as a time varying superposition of the eigenstates $|I\rangle$ and $|II\rangle$ with time dependent probability amplitudes, $C_{I,II}(t)$, see Eq.4.18:

$$|\Psi(t)\rangle = C_I(t)|I\rangle + C_{II}(t)|II\rangle$$

and initial conditions: $C_I(0) = 1$ and $C_{II}(0) = 0$.

Eq.4.24 yields : $i\hbar\partial_t C_I = E_I^0 C_I + W(t)C_{II}$

$i\hbar\partial_t C_{II} = W^*(t)C_I + E_{II}^0 C_{II}$

---

[18] In the following we will only consider the first term, $(-\mu_x\mathcal{E}e^{-i\omega t}$, the contribution of the second term is obtained by setting $\omega \to -\omega$). Each term contributes either to absorption or to stimulated emission.

Looking for a solution $C_{I,II} = \alpha_{I,II}(t)\exp\left(-iE_{I,II}^0 t/\hbar\right)$ yields the following equation for $\alpha_{I,II}$:

$$i\hbar\partial_t\alpha_I = W(t)e^{-i(E_{II}^0-E_I^0)t/\hbar}\alpha_{II} = -\mu_{12}\mathcal{E}e^{-i(\omega-\omega_0)t}\alpha_{II}$$

$$i\hbar\partial_t\alpha_{II} = W^*(t)e^{-i(E_I^0-E_{II}^0)t/\hbar}\alpha_I = -\mu_{12}\mathcal{E}e^{i(\omega-\omega_0)t}\alpha_I \qquad (4.30)$$

While these equations can be solved exactly to yield $\alpha_I(t)$ and $\alpha_{II}(t)$, it is instructive to consider their short time behavior when $\alpha_I(t) \approx 1$ and $\alpha_{II}(t) \ll 1$:

$$\partial_t\alpha_{II} \approx i\frac{\mu_{12}\mathcal{E}}{\hbar}e^{i(\omega-\omega_0)t}$$

which can be integrated to yield:

$$\alpha_{II}(t) = i\frac{\mu_{12}\mathcal{E}}{\hbar}te^{i(\omega-\omega_0)t/2}\frac{\sin(\omega-\omega_0)t/2}{(\omega-\omega_0)t/2} \equiv i\frac{\mu_{12}\mathcal{E}}{\hbar}te^{i(\omega-\omega_0)t/2}\text{sinc}(\omega-\omega_0)t/2$$

The probability of observing the molecule in its ground state $|II\rangle$ at time t, i.e. the transition probability from state $|I\rangle$ to state $|II\rangle$, is thus:

$$P_{I\to II}(t) = |\alpha_{II}|^2 = \frac{\mu_{12}^2\mathcal{E}^2}{\hbar^2}t^2\text{sinc}^2(\omega-\omega_0)t/2$$

Since, the function sinc $x \equiv \sin x/x$ decays rapidly for values of $|x| > \pi$, the transition probability is significant only for frequencies which are very close to resonance: $|\omega - \omega_0| < 2\pi/t$. Hence the longer the residence time t in the cavity, the smaller the allowed detuning $\Delta\omega = \omega - \omega_0$ for stimulated amplification. This explains the interest in Penning traps (section 3.2.3.3) for precise spectroscopic studies on ionized molecules at low temperatures. For example if the molecule remains in the cavity for 1 sec, the relative possible detuning : $|\nu/\nu_0 - 1| = 1/\nu_0 t \sim 4\ 10^{-11}$. Only frequencies within that very narrow range will be amplified by stimulated emission. A maser is thus a very high Q resonator, Eq.3.84. It has been used as a very precise frequency generator and clock.

Because of the sharpness of the function $\text{sinc}^2(\omega - \omega_0)t/2$ one often rewrites the previous equation using the approximation[19]

$$\text{sinc}^2(\Delta\omega t/2) \approx \frac{2\pi}{t}\delta(\Delta\omega) \qquad (4.31)$$

The stimulated emission rate $k_{I,II}$ from high energy state state $|I\rangle$ to lower energy state $|II\rangle$ is then:

$$k_{I,II} = \frac{dP_{I\to II}(t)}{dt} = 2\pi\frac{\mu_{12}^2\mathcal{E}^2}{\hbar^2}\delta(\Delta\omega) = \frac{2\pi|\delta H_{12}|^2}{\hbar}\delta(E_I - E_{II} - \hbar\omega) \qquad (4.32)$$

---

[19]Using the identity: $\int_{-\infty}^{\infty} dx\ \text{sinc}^2 x = \pi$, see Appendix A.4, for any function $g(\omega)$ which doesn't vary much over the range $\Delta\omega > \pi/t$, the following is valid: $\int d\omega g(\omega)\text{sinc}^2(\omega-\omega_0)t/2 \approx 2\pi g(\omega_0)/t = (2\pi/t)\int g(\omega)\delta(\omega-\omega_0)d\omega$.

This relation between the rate of stimulated emission (or absorption) and the square of the coupling matrix element $|\delta H_{12}|^2$ is very general and known as Fermi's Golden rule, see section 4.12.

Notice that had we assumed the ammonia beam to enter the cavity in state $|II\rangle$, i.e. with $\alpha_{II}(0) = 1$, $\alpha_I = 0$, then the solution for $\alpha_I(t)$ would yield the transition rate for absorption from state $|II\rangle$ to $|I\rangle$:

$$k_{II,I} = \frac{dP_{II \to I}(t)}{dt} = \frac{2\pi |\delta H_{21}|^2}{\hbar}\delta(E_{II} - E_I + \hbar\omega) = k_{I,II}$$

which was the intuitive assumption Einstein made: the rate of stimulated emission $B_{I,II}$ is equal to the rate of absorption $B_{II,I}$. The transition rate $k_{I,II}$ can be expressed in terms of the intensity of the stimulated radiation $I = c\mathcal{E}^2/2\pi$ and is related to the Einstein coefficients via Eq.4.12:

$$k_{I,II} = \frac{2\pi\mu_{12}^2 I}{c\hbar^2}\delta(\nu - \nu_0) = B_{I,II}\frac{I}{h\nu c}\delta(\nu - \nu_0)$$

from which we derive:

$$B_{I,II} = \frac{4\pi^2\nu_0}{\hbar}\mu_{12}^2 \tag{4.33}$$

This result is general: the only thing that varies from system to system is the estimate of the dipole coupling element μ_{12}, see section 4.12.1.

The ammonia maser was the first atomic (or molecular) clock. The present generation of atomic clocks uses a microwave transition in Cesium (Cs) as the reference frequency for the clock. The atoms are cooled to very low temperatures ($\mu°$K) to reduce the Doppler broadening of their emission lines and to keep them as long as possible in the cavity, usually implemented as a Penning trap (see Fig.3.14). As a result the record for the frequency precision of atomic clock is: $|\nu/\nu_0 - 1| \sim 10^{-16}$ (an error of about 3 sec per billion years!).

4.3.3   THE ENERGY SPECTRUM OF AROMATIC MOLECULES

A simple generalization of the two-state system considered above is the n-state system consisting for example of a circular chain of n identical atoms such as benzene C_6H_6 ($n = 6$) around which electrons can hop[20]. If an electron has energy E_0 when on a particular orbital associated with the lth atom, a state designated as $|l\rangle$, and if it can hop only between nearest neighbors the Hamiltonian for this system is:

$$H = \begin{pmatrix} E_0 & -A & 0 & \cdots & -A \\ -A & E_0 & -A & \cdots & 0 \\ \vdots & & & & \vdots \\ -A & \cdots & 0 & -A & E_0 \end{pmatrix} \tag{4.34}$$

---

[20]We shall see later, section 4.10, that chemical bonds result from the sharing (overlapping or hoping) of a pair of electrons between neighboring atoms. E_0 is the on-site energy (the energy associated to the atomic orbital) and A is the overlap energy (measuring the overlap between the atomic orbitals of neighboring atoms). The more external the orbital is the larger the overlap.

The eigenstates of that system obey: $H|\Psi_m\rangle = E_m|\Psi_m\rangle$. In the basis $\{|l\rangle\}$ of the orbitals at position l along the chain, the eigenstates can be written as: $|\Psi_m\rangle = \sum_{l=1}^{n} C_l^m |l\rangle$ and where $m = 1, ..., n$, from which we derive the following equations for the probability amplitudes C_l^m:

$$\frac{E - E_0}{A} C_1^m + C_2^m + C_n^m = 0$$

$$C_1^m + \frac{E - E_0}{A} C_2^m + C_3^m = 0$$

$$\vdots$$

$$C_{l-1}^m + \frac{E - E_0}{A} C_l^m + C_{l+1}^m = 0 \qquad (4.35)$$

$$\vdots$$

$$C_1^m + C_{n-1}^m + \frac{E - E_0}{A} C_n^m = 0$$

Looking for a properly normalized solution: $C_l^m = e^{i2\pi lm/n}/\sqrt{n}$ we get:

$$\frac{E_m - E_0}{A} = -(e^{i2\pi m/n} + e^{-i2\pi m/n}) = -2\cos 2\pi m/n$$

Thus the eigen-energies of the system are:

$$E_m = E_0 - 2A\cos 2\pi m/n \qquad (4.36)$$

For benzene ($n = 6$) we have: $E_6 = E_0 - 2A$; $E_{1,5} = E_0 - A$; $E_3 = E_0 + 2A$ and $E_{2,4} = E_0 + A$. Energies $E_{1,5,6}$ which are smaller that E_0 are associated to so-called molecular bonding orbitals or wave-functions, whereas energies $E_{2,3,4} > E_0$ are associated to molecular anti-bonding orbitals, on which we shall have more to say later. Notice that the eigenstate associated to the lowest energy (maximally bonding orbital, $m = 6$) is fully symmetric (a general result for ground state orbitals): $|\Psi_6\rangle = (1, 1, 1, 1, 1, 1)/\sqrt{6}$, while that associated to the maximally anti-bonding orbital ($m = 3$) is antisymmetric to the permutation of nearest neighbors: $|\Psi_3\rangle = (1, -1, 1, -1, 1, -1)/\sqrt{6}$.

4.3.4 CONDUCTION BANDS IN SOLIDS

An interesting generalization of the previous analysis is the case of a long chain of n atoms a distance a apart. In that case we have: $C_l^m = \exp[i(2\pi m/na)la]$. Since the position of atom l is $x = la$ we may write the probability amplitudes as: $C_k(x) = \exp ikx$ with $k = 2\pi m/na$. Like the oscillation modes on a string of length na, the electron's eigenstates are 1D transverse waves of wavelength $\lambda = 2\pi/k = na/m$. The energy of such a mode is:

$$E(k) = E_0 - 2A\cos ka \qquad (4.37)$$

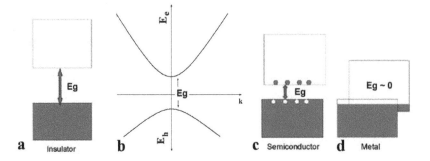

Figure 4.10 Band-gap theory of material. (a) If the gap between the valence and conduction band is large (a few electron-Volts) the material is an insulator. (b) Variation of the band-gap energy with wave-vector k. Top curve: energy of electrons E_e; bottom curve: energy of holes E_h (increases towards the bottom). (c) A semiconductor with some electrons in the conduction band and some holes in the valence band due to thermal excitation or illumination. (d) a conductor is a material for which the valence and conduction band overlap or equivalently for which an energy band is not filled.

The energy of the electron is bounded: $E_0 - 2A < E(k) < E_0 + 2A$[21]. If each atom contributes two electrons, these $2n$ electrons will occupy all the n-energy states (we shall see later that each energy eigenstate can accommodate at most 2 electrons) and electron hoping in this energy-band will be impossible[22] . If however the on-site energy E_0 can possess discrete values (E_l as the orbitals in Bohr's atom do), the coupling of the n−atoms will open around each energy value E_l an energy band of size $4A_l$. This forms the basis of the band-theory of conduction in materials, see Fig.4.10 and section 5.9.2.2: a material will conduct if there are empty states into which electrons can hop by thermal excitation. If the low energy (valence) band is filled with electrons and the next (conduction) band is empty but many electron-volts (eV) above it, then the electrons have no states in which to thermally hop and the material is an insulator. If the conduction band overlaps with the valence band then the electrons have empty states to hop to and current can flow in the material which is therefore an electrical conductor. The interesting and technologically important case is the situation where the band-gap E_g between the conduction and valence bands is small: $E_g \lesssim 1\text{eV}$ (the gap for Silicon (Si) is: $E_g = 1.1\text{eV}$; for Germanium (Ge): $E_g = 0.72\text{eV}$). In that case electrons can be transferred from the valence to the conduction band by thermal agitation or via the photo-electric effect resulting in a

---

[21] In a 3D cubic crytals of size a: $E(k) = E_0 - 2A \cos k_x a - 2A \cos k_y a - 2A \cos k_z a$, with $-\pi/a < k_x, k_y, k_z < \pi/a$

[22] If each atom contributes one electron, the n electrons will fill the lowest (bonding) $n/2$ molecular orbitals. Hoping into the unoccupied higher orbitals will be possible if the gap between the highest occupied molecular orbital (HOMO) and the lowest unoccupied molecular orbital (LUMO) isn't too large, see section 4.10.1.1

material that can behave either as a metal or an insulator depending on the external conditions (temperature, voltage, illumination wavelength and intensity, etc.).

In particular the introduction of atomic impurities (doping) that donate electrons to (or accept electrons from) a semiconductor lattice creates a situation where few electrons occupy an almost empty conduction band (or few holes (electron vacancies) occupy an almost full valence band), see section 5.9.2.2 below. As we shall see in section 4.9 the electrons in an atom occupy certain orbitals or shells around the nucleus. Atoms (such as Phosphate or Arsenic) that have more electrons in their outer shell than the bulk semiconductor (typically Silicon) will usually donate an electron to the lattice, whereas atoms that have less electrons in their outer shell (such as Boron or Aluminium) will accept an electron from the lattice. The doping creates so called n- and p-type semiconductors that have electrons in their conduction-band (n-type) or holes in their valence-band (p-type), see Fig.4.11 and section 5.9.2.2.

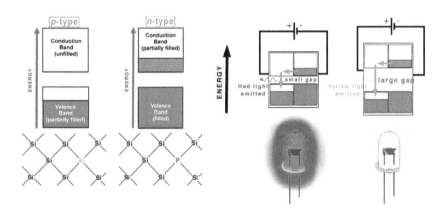

Figure 4.11 A p-type semiconductor consists of a material (usually Silicon, Si) doped with an element (such as Aluminium) which, having less electrons in its outer shell than Si, see section 4.9.1, tend to accept electrons from the lattice leaving an electron vacancy instead: a hole. This depletes the valence band of electrons allowing conduction through motion of holes. A n-type semiconductor consists of a material doped with an electron donor, an element (such as Phosphate) which has more electrons in its outer shell than Si. Electrons in such elements hop into the conduction band allowing the material to conduct electricity. A pn junction is formed when p-type and n-type semiconductors are brought in close contact, see section 5.9.2.2. Such a junction can be used as a Light Emitting Diode (LED) when electron from the n-side of the junction recombine with holes from the p-side. The wavelength of emitted light is determined by the energy-gap between the conduction and valence bands. Reprinted with permission from ref.[56].

The energy of the electrons moving near the bottom of the conduction band ($E_{min} = E_{0,c} - 2A_c$, i.e. when $ka << 1$ in Eq.4.37) is (after expanding $2A_c \cos ka \approx$

$2A_c - A_c a^2 k^2$) :

$$E_e(k) = E_{min} + A_c a^2 k^2 = E_{min} + \hbar^2 k^2 / 2m_c \qquad (4.38)$$

If, as proposed by de Broglie, we identify the momentum of the electron as $p = \hbar k$, then Eq.4.38 represents the kinetic energy of an electron with effective mass $m_c = \hbar^2 / 2a^2 A_c = \hbar^2 / \partial_k^2 E(k)\big|_{k=0}$ at the bottom of the conduction band, which can be different from the mass of a free electron. Similarly near the top of the valence band at $E_{max} = E_{0,v} + 2A_v$ the energy is:

$$E_h(k) = E_{max} - \hbar^2 k^2 / 2m_v \qquad (4.39)$$

The movement of holes in the valence band is similar to that of air bubbles in water: just as the motion of water displaces the bubble up, the electrons moving in the opposite direction to the hole minimize their energy by displacing it to the top of the valence band. As a result holes with effective mass m_v have minimal energy at the top of the valence band ($k = 0$) and their energy increases as $\hbar^2 k^2 / 2m_v$.

Doped semiconductors are the basic ingredients of all the semiconductor industry (transistors, diodes, integrated circuits, etc.). For example the coupling of p-type and n-type semiconductors, generate a pn junction which acts like a diode (current flows in only one direction), see section 5.9.2.3. It can also be used as a powerful and efficient light source. Electrons in the n-type part of the junction can recombine with holes in the p-type (i.e. transit to the valence band) with emission of light (as in the atomic or molecular transitions discussed earlier in the context of stimulated emission). The advantage of these Light Emitting Diodes (LED) is that by appropriate tuning of the energy gap (appropriate choice of the semi-conducting material) one can tune the wavelength at which the LED will emit light. For example in the red and infrared part of the spectrum Gallium-Arsenide (GaAs) is the material of choice for LEDs. The intensity of light is controlled by the current flowing through the junction. These LED are used in all kinds of electronic and TV displays, in high efficiency spot-lamps and traffic lights, in the remote control of various electronic devices, in the laser diodes of DVD players, etc.

4.4 ◆ MOMENTUM, SPACE AND ENERGY OPERATORS

In the previous examples we considered the case of a QM system that could occupy only discrete states $|n\rangle$. It is easy to generalize this approach to free particles that can be found at continuous positions $|x\rangle$. In such case the general wave-function in position (i.e. real) space can be formally written as we did for an electron along a 2D chain, see section 4.3.4 above:

$$|\Psi\rangle = \sum_x \Psi(x)|x\rangle$$

Where $\Psi(x)$ is the probability amplitude of finding the particle at position $|x\rangle$, so that:

$$P(x) = \Psi^*(x)\Psi(x) \tag{4.40}$$

Evidently, the probability of finding the particle somewhere is one, so that the wavefunction $\Psi(x)$ is normalized:

$$\int dx \Psi^*(x)\Psi(x) = 1 \tag{4.41}$$

For example, the wavefunction of a Gaussian wavepacket is:

$$\Psi_G(x) = \frac{1}{(2\pi\sigma^2)^{1/4}} e^{-(x-x_0)^2/4\sigma^2}$$

with probability distribution : $\quad P_G(x) = \frac{1}{\sqrt{2\pi}\sigma} e^{-(x-x_0)^2/2\sigma^2}$

$P_G(x)$ is a Gaussian probability distribution with mean position $\langle x \rangle = x_0$ and standard deviation $\sigma_x \equiv \sqrt{\langle (x-x_0)^2 \rangle} = \sigma$. In general the mean position of a particle is[23] :

$$\langle x \rangle = \int dx P(x)x = \int dx \Psi^*(x)x\Psi(x) = \int dxdx' \Psi^*(x')x\delta(x'-x)\Psi(x)$$

$$\equiv \int dxdx' \Psi^*(x')\hat{x}\Psi(x)$$

In real space the position operator $\hat{x} = x\delta(x'-x)$ is diagonal, as it should since real space is the eigenspace of position: $\hat{x}|x\rangle = x|x\rangle$. Notice that we can express the state $|\Psi\rangle$ of a free particle in any basis spreading the space of complex functions of the coordinates. For example, as noticed for EM fields, section 3.5, we can write any function $\Psi(x)$ as a sum of plane waves e^{ikx} of amplitude $\Psi(k)$ (i.e. the Fourier transform of $\Psi(x)$, appendix A.4) :

$$\Psi(x) = \frac{1}{\sqrt{2\pi}} \int dk \Psi(k)e^{ikx} = \frac{1}{\sqrt{2\pi\hbar}} \int dp \Psi(p)e^{ipx/\hbar} \tag{4.42}$$

---

[23] Recall that the average of an operator $\hat{O}$ is given by Eq.4.23 $\langle \Psi|\hat{O}|\Psi \rangle = \int dx \Psi^*(x)\hat{O}\Psi(x) = \int dp \Psi^*(p)\hat{O}\Psi(p)$ whether $\hat{O}$ is expressed in real space x or momentum space p.

where we used de Broglie's relations: $p = \hbar k$. Conversely:

$$\Psi(p) = \frac{1}{\sqrt{2\pi\hbar}} \int dx \Psi(x) e^{-ipx/\hbar} \tag{4.43}$$

$\Psi(p)$ is the wavefunction in the momentum (or Fourier) space[24]: $|\Psi\rangle = \sum_p \Psi(p)|p\rangle$. For the Gaussian wavepacket introduced above the momentum wavefunction is:

$$\Psi_G(p) = \frac{e^{-ipx_0/\hbar}}{\sqrt{2\pi\hbar}(2\pi\sigma^2)^{1/4}} \int_{-\infty}^{\infty} dx\, e^{-(x-x_0)^2/4\sigma^2} e^{-ip(x-x_0)/\hbar}$$

$$= \frac{e^{-ipx_0/\hbar} e^{-p^2\sigma^2/\hbar^2}}{\sqrt{2\pi\hbar}(2\pi\sigma^2)^{1/4}} \int_{-\infty}^{\infty} dx\, e^{-[x-x_0+2ip\sigma^2/\hbar]^2/4\sigma^2}$$

$$= \frac{\sqrt{2\sigma/\hbar}}{(2\pi)^{1/4}} e^{-ipx_0/\hbar} e^{-p^2\sigma^2/\hbar^2}$$

with probability : $P_G(p) = \Psi_G^*(p)\Psi_G(p) = \dfrac{2\sigma}{\sqrt{2\pi\hbar}} e^{-2p^2\sigma^2/\hbar^2}$

$P_G(p)$ is a Gaussian distribution centered on $\langle p \rangle = 0$ with standard deviation $\sigma_p = \sqrt{\langle p^2 \rangle} = \hbar/2\sigma \equiv \hbar/2\sigma_x$. Therefore the more localized the wavepacket is in space (the smaller σ_x), the greater its spread in momentum (the larger σ_p). As we shall see in section 4.4.1 below, this is the limiting case of Heisenberg uncertainty principle: $\sigma_x\sigma_p \geq \hbar/2$ (the equality being achieved by Gaussian wavepackets).

Being the Fourier transform of $\Psi(x)$, $\Psi(p)$ satisfies the normalization condition:

$$1 = \int dx \Psi^*(x)\Psi(x) = \int dp dp' \frac{\Psi^*(p')\Psi(p)}{2\pi\hbar} \int dx e^{i(p-p')x/\hbar}$$

$$= \int dp \Psi^*(p)\Psi(p) = \int dp P(p) \tag{4.44}$$

where we used the identity (see appendix A.4):

$$\int dx e^{\pm ipx/\hbar} = 2\pi\hbar\delta(p) \tag{4.45}$$

Eq.4.44 is known in the theory of Fourier transforms as Parseval's theorem, see Appendix A.4. The mean value of the momentum satisfies:

$$\langle p \rangle = \int dp \Psi^*(p)p\Psi(p) = \int dp' dp \Psi^*(p')p\delta(p' - p)\Psi(p) \equiv \int dp' dp \Psi^*(p')\hat{p}\Psi(p)$$

In momentum space the momentum operator $\hat{p}$ is a diagonal matrix, $\hat{p} = p\delta(p' - p)$ as it should since that space is the eigenspace of momentum: $\hat{p}|p\rangle = p|p\rangle$.

---

[24]The representation of a physical state in a real space basis $|x\rangle$ or momentum space basis $|p\rangle$ is equivalent and arbitrary, just as for the particle in a box, we could express a state using the complete set $|\phi_n\rangle$ or $|\psi_n\rangle$, see section 4.2.1.

We shall now find the representation of the momentum operator $\hat{p}$ in real space. Using the identity Eq.4.45 and Eq.4.42 the mean momentum $\langle p \rangle$ can be expressed in terms of the real space wavefunction $\Psi(x)$:

$$\langle p \rangle = \int dx \Psi^*(x)\hat{p}\Psi(x) = \frac{1}{2\pi\hbar}\int dx dp' dp \Psi^*(p')e^{-i(p'-p)x/\hbar}p\Psi(p)$$

$$= \frac{1}{2\pi\hbar}\int dx \int dp' \Psi^*(p')e^{-ip'x/\hbar}\int dp \frac{\hbar}{i}\frac{\partial}{\partial x}\Psi(p)e^{ipx/\hbar}$$

$$= \int dx \Psi^*(x)\frac{\hbar}{i}\frac{\partial}{\partial x}\Psi(x)$$

Hence we identify the momentum operator representation in real space as:

$$\hat{p} = \frac{\hbar}{i}\frac{\partial}{\partial x} \equiv -i\hbar\partial_x \qquad (4.46)$$

The kinetic energy $\hat{E}_k = \hat{p}^2/2m$ can then be written in real space as the following operator:

$$\hat{E}_k = -\frac{\hbar^2}{2m}[\frac{\partial^2}{\partial x^2} + \frac{\partial^2}{\partial y^2} + \frac{\partial^2}{\partial z^2}] = -\frac{\hbar^2}{2m}\nabla^2 \qquad (4.47)$$

The Hamiltonian operator which is the sum of kinetic energy E_k and potential energy $V(\vec{x})$ is thus:

$$\hat{H} = \frac{\hat{p}^2}{2m} + V(\vec{x}) = -\frac{\hbar^2}{2m}\nabla^2 + V(\vec{x}) \qquad (4.48)$$

By writing the mean position $\langle x \rangle$ in momentum space we can similarly identify the representation of the position operator $\hat{x}$ in momentum space as:

$$\hat{x} = i\hbar\frac{\partial}{\partial p} \equiv i\hbar\partial_p \qquad (4.49)$$

Notice that momentum and space-operators do not commute. In real space:

$$\langle x|[\hat{x},\hat{p}]|x\rangle = \int dx \Psi^*(x)(x\hat{p}-\hat{p}x)\Psi(x) = \frac{\hbar}{i}\int dx \Psi^*(x)[x\partial_x\Psi - \partial_x(x\Psi)] = i\hbar \quad (4.50)$$

We would have obtained the same result by computing the commutator in momentum space: $\langle p|[\hat{x},\hat{p}]|p\rangle = \int dp \Psi^*(p)(\hat{x}p - p\hat{x})\Psi(p) = i\hbar$.

As discussed earlier, in order to simultaneously measure two observables, they have to share the same eigenstates, namely they have to commute. Since the position and momentum operator of a particle (along a given axis) do not commute, they cannot be simultaneously determined with absolute precision.

From Schrödinger's equation $i\hbar\partial_t\Psi = \hat{H}\Psi$, one can show that a similar result holds for the commutator of energy and time:

$$\langle x|[\hat{H},t]|x\rangle = \langle x|\hat{H}t - t\hat{H}|x\rangle = \int dx \Psi^*(x)[i\hbar\partial_t(t\Psi) - ti\hbar\partial_t\Psi] = i\hbar \qquad (4.51)$$

4.4.1 ◆ HEISENBERG UNCERTAINTY PRINCIPLE

We shall now derive Heisenberg uncertainty principle from the non-commutability of momentum and space, Eq.4.50. Consider the action on state $|\psi\rangle$ of the operator $\hat{x}_o + i\lambda\hat{p}_o$ where λ is a real number and the operators $\hat{x}_o$ and $\hat{p}_o$ are the deviation of the position and momentum operators from their mean: $\hat{x}_o = \hat{x} - \langle x \rangle$ and $\hat{p}_o = \hat{p} - \langle p \rangle$:

$$|\phi\rangle = (\hat{x}_o + i\lambda\hat{p}_o)|\psi\rangle$$

Since $[\hat{x}_o, \hat{p}_o] = [\hat{x}, \hat{p}] = i\hbar$ the positiveness of the probability implies that for any real λ:

$$0 \leq \langle\phi|\phi\rangle = \langle\psi|(\hat{x}_o - i\lambda\hat{p}_o)(\hat{x}_o + i\lambda\hat{p}_o)|\psi\rangle$$
$$= \langle\psi|\hat{x}_o^2 + i\lambda[\hat{x}_o, \hat{p}_o] + \lambda^2\hat{p}_o^2|\psi\rangle$$
$$= \langle\psi|\hat{x}_o^2|\psi\rangle - \hbar\lambda + \langle\psi|\hat{p}_o^2|\psi\rangle\lambda^2 \equiv P_2(\lambda)$$

For the quadratic polynomial $P_2(\lambda)$ to be non-negative for any real λ, the equation $P_2(\lambda) = 0$ must not have a real root namely its determinant has to satisfy:

$$\hbar^2 - 4\langle\psi|\hat{x}_o^2|\psi\rangle\langle\psi|\hat{p}_o^2|\psi\rangle \leq 0$$

Or in terms of the position and momentum standard deviations: $\langle\Delta x^2\rangle \equiv \langle\psi|\hat{x}_o^2|\psi\rangle$, $\langle\Delta p^2\rangle \equiv \langle\psi|\hat{p}_o^2|\psi\rangle$:

$$\langle\Delta x^2\rangle\langle\Delta p^2\rangle \geq \frac{\hbar^2}{4} \qquad (4.52)$$

This is Heisenberg's principle: it sets a limit on the precision with which one can measure both the position $\delta x = \sqrt{\langle\Delta x^2\rangle}$ and the momentum $\delta p = \sqrt{\langle\Delta p^2\rangle}$ of a physical system. The smallest uncertainty (the equality in Eq.4.52) is obtained for a Gaussian probability distribution as shown in the previous section. Since the Hamiltonian and time operators do not commute a similar relation can be obtained for the energy and time uncertainties:

$$\langle\Delta E^2\rangle\langle\Delta t^2\rangle \geq \frac{\hbar^2}{4} \qquad (4.53)$$

Notice that Heisenberg principle is a direct mathematical consequence of the QM description of physical systems by a complex wave-function $\Psi(x)$ (Eq.4.19) and of de Broglie's relation between wavelength and momentum. Heisenberg uncertainty principle is a tautology: a consequence of the definition of Fourier transforms (see appendix A.4). In the context of optics it has been known for a long time that the diffraction pattern from a hole is larger the smaller the hole is. It is also well known in the context of communication where a very short time signal is spread over a very large range in frequency.

4.5 ◆ SCHRÖDINGER'S EQUATION

We have determined the real space representation of the Hamiltonian operator $\hat{H}$, Eq.4.48. Schrödinger's equation, Eq.4.24 then becomes:

$$i\hbar \frac{\partial}{\partial t} \Psi(\vec{x}, t) = -\frac{\hbar^2}{2m} \nabla^2 \Psi(\vec{x}, t) + V(\vec{x}) \Psi(\vec{x}, t) \tag{4.54}$$

Since in QM the probability of finding a particle at point x: $P(x) = |\Psi(x)|^2$, it is crucial to check that Schrödinger's equation conserves probability (just as it was to verify that Maxwell's equations conserve charge). Multiplying Eq.4.54 by Ψ^* and its complex conjugate by Ψ and subtracting the two yields:

$$i\hbar \partial_t P = -\frac{\hbar^2}{2m} (\Psi^* \nabla^2 \Psi - \Psi \nabla^2 \Psi^*)$$

$$= -\frac{\hbar^2}{2m} \vec{\nabla} \cdot (\Psi^* \vec{\nabla} \Psi - \Psi \vec{\nabla} \Psi^*)$$

which can be recast as a conservation law for the probability distribution $P(x)$ (compare with Eq.3.43 for charge conservation in EM):

$$\partial_t P + \vec{\nabla} \cdot \vec{J} = 0 \quad \text{with} \quad \vec{J} = \Psi^* \frac{\hbar}{2im} \vec{\nabla} \Psi + c.c. = \frac{\Psi^* \hat{v} \Psi + c.c.}{2} \tag{4.55}$$

where $c.c.$ stands for the complex conjugate and $\hat{v} = \hbar \vec{\nabla}/im = \hat{p}/m$ is the velocity operator. Eq.4.55 is a crucial self-consistency check of Quantum Mechanics, since for $P(x)$ to be interpreted as a probability distribution it must satisfy a conservation law. This equation expresses the intuitive expectation that the change in the probability of finding a particle in a given volume is equal to the gradient in the particle flux (the difference between the particles entering and leaving the given volume).

4.5.1 PARTICLE-WAVE DUALITY: DIFFRACTION

If the potential is null $V(x) = 0$ then the eigenstates of Schrödinger's equation, Eq.4.54:

$$i\hbar \frac{\partial \Psi}{\partial t} = -\frac{\hbar^2}{2m} \nabla^2 \Psi \tag{4.56}$$

are plane waves:

$$\Psi(\vec{x}, t) = e^{i\vec{k} \cdot \vec{x} - iE_k t/\hbar}$$

with $\vec{p} = \hbar \vec{k}$ and $E_k = \hbar^2 k^2/2m$. These plane-waves are also eigenstates of the momentum operator:

$$\hat{p} \Psi(\vec{x}, t) = -i\hbar \vec{\nabla} \Psi(\vec{x}, t) = \vec{p}\, \Psi(\vec{x}, t)$$

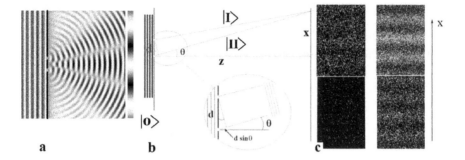

Figure 4.12 The double-slit or Young's experiment. (a) A wave passing through two slits in a screen generates two wave-sources which interference creates on a far-away screen a pattern of interference consisting of alternating minima and maxima of intensity(adapted from ref.[57]), see section 3.5.3. (b) A particle in state $|O\rangle$ impinging on a double slit generates two states $|I\rangle$ and $|II\rangle$ corresponding to its passage through slit 1 or 2. The phase of these states evolves as $\exp ikl$. If their coherence is maintained they can interfere on a screen a large distance z from the slits, generating an oscillating pattern related to their phase difference: $\phi_{int} = k(l_2 - l_1) \simeq kd\sin\theta$. (c) Observation of the interference pattern of electrons passing through a double slit and impinging on a camera. Each electron is observed as a particle (white dot) with a well defined position on the camera. The QM interference pattern is only visible when a sufficiently large number of particles has been observed. Reprinted with permission from ref.[58] copyright (2005) by the National Academy of Science, USA.

One of the most striking confirmations of the QM mechanics picture is the observation of a diffraction pattern like the one seen with electro-magnetic radiation (see section 3.5.3) when a free particle is passed through one or a few slits, see Fig.4.12(a). Let the particle be in a plane-wave eigenstate of momentum $p = \hbar k$ as it impinges on an obstacle set in the xy-plane (at $z = 0$):

$$|O\rangle = \Psi_{in}(\vec{x}, t) = e^{ikz - iE_k t/\hbar}$$

The obstacle is absorbing except for two slits of width a, a distance d apart. In the far-field, i.e. at distances $z \gg d^2/\lambda$ the wave amplitude is given by Huygens's principle, Eq.3.143:

$$\Psi_{diff}(x, y, z) = \frac{e^{ik(z + (x^2 + y^2)/2z)}}{i\lambda z} \int dx' dy' e^{-i(k_x x' + k_y y')} \tag{4.57}$$

Where $k_x = kx/z$ and $k_y = ky/z$. Thus the probability of detecting a particle on a screen a distance z from the slits is:

$$|\Psi_{diff}|^2 = \frac{4a^2}{(\lambda z)^2} \mathrm{sinc}^2 k_x a \cos^2 k_x d/2 \tag{4.58}$$

where $\mathrm{sinc}\, x \equiv \sin x/x$. If the distance between the diffracting slits is much larger than their width ($d \gg a$), the probability oscillates with a period: $\delta x = \lambda z/d$. While each particle is observed at a given position on the far-away screen, the overall probability distribution of the particles on the screen varies in space just as the diffraction pattern of EM radiation does.

There is an other way to derive that result, see Fig.4.12(b). The wave-function of the particle on the screen can be written as: $|\Psi\rangle = |I\rangle + |II\rangle$ where state $|I\rangle \sim e^{ikl_1} |O\rangle$ corresponds to the particle passing through the first slit and state $|II\rangle \sim e^{ikl_2} |O\rangle$ to the particle passing through the second slit. In other words, the wavefunction of the particle consists of a classically absurd superposition of a single particle following two different paths. If the slits are assumed to be thin, i.e. $a \ll d$, then

$$|\Psi_{diff}|^2 = \langle\Psi|\Psi\rangle \sim |1 + e^{ik(l_2-l_1)}|^2 \langle O|O\rangle = 4\cos^2(\phi_{int}/2)$$
$$\simeq 4\cos^2(kd\sin\theta/2) = 4\cos^2(k_x d/2)$$

with $\phi_{int} \equiv k(l_2 - l_1) \simeq kd\sin\theta \simeq kxd/z = k_x d$. Notice that the interference pattern is observed only if the phase-coherence between states $|I\rangle$ and $|II\rangle$ is maintained. For example if one wanted to see through which slit the particle has passed, one would have to use radiation with wavelength $\lambda_o \lesssim d$. Such radiation would impart on the detected particle a momentum uncertainty $\delta p_o \sim h/\lambda_o$ (i.e. wavenumber uncertainty $\delta k_o \sim 2\pi/\lambda_o$. In the far-field $d^2/\lambda z \ll 1$ the observation imparts on the observed state a phase uncertainty:

$$\delta\phi_o = \delta k_o z \sim 2\pi z/\lambda_o > 2\pi z/d \gg 2\pi d/\lambda > \phi_{int}$$

which is enough to destroy the interference pattern. This is the essence of Heisenberg's principle: if the particle interacts with radiation of small enough wavelength as required to determine its position the perturbation to its momentum will be so large as to destroy the coherence of the states. Trying to determine through which slit the particle has passed, namely to localize its position to better than the distance between the slits d, imparts on the particle a momentum uncertainty large enough to destroy the interference pattern, replacing it with an image of the two slits.

4.5.2 PARTICLE INTERFERENCE OBSERVED WITH BUCKYBALLS

In 1999, the group of Anton Zellinger observed QM interference effects in one of the largest molecule investigated so far[59, 60]: the fullerene or buckyball C_{60}, a molecule that can be seen[61] with a Scanning Tunneling Microscope (STM, see below), Fig.4.13(a). A beam of molecules at $\sim 900°C$ with a velocity distributed around 117 m/sec was shot at an array of slits of size $a \sim 40$nm and inter-distance $d \sim 100$nm. At this velocity the wavelength of the particles is $\lambda = 0.046$Å! They were observed on a detector positioned a distance $z \simeq 1.2$m from the slits (i.e. in the far-field $\lambda z \gg d^2$). Since the width of the slits a is only slightly smaller than d the intensity is strongly damped (see Eq.4.58) by the factor $\text{sinc}^2 k_x a$ when $x > \lambda z/2a$. Hence only a few oscillations in the probability of detection can be observed in Fig.4.13(c). In particular the first peaks (fringes) separated by a distance $x = \lambda z/d \sim 55\mu$m can be seen. The contrast is not as high as expected since the velocity of the molecules (hence their momentum and the distance between fringes) is thermally spread. It is interesting however that the coherence of the molecule passing through the slits can be maintained with molecules at such high temperatures. The reason for that coherence is that the particle is emitting black-body radiation at a wavelength of a few μm which is much larger than the distance between the slits $d \sim 0.1 \mu m$. Hence the uncertainty in the molecule's momentum imparted by its black-body radiation is not enough to destroy the coherence of the wave-function passing through contiguous slits.

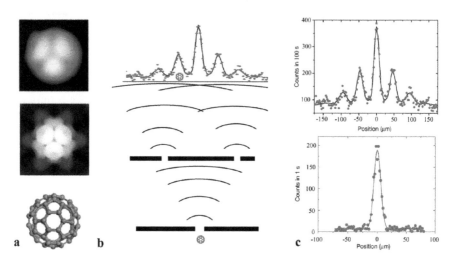

Figure 4.13 Diffraction experiment performed with C_{60} molecules (a) Top: scanning tunneling microscope (STM) images of a C_{60} molecule on a Si surface; middle: simulation of STM image; bottom: reconstructed theoretical image of the molecule. Reprinted from ref.[61] with the permission of AIP Publishing (b) Schematic drawing of a double slit experiment with C_{60} and the expected pattern, which is a cosine wave modulated by the diffraction pattern of a single slit ($sinc^2 k_x a$). (c) Top plot: results of the diffraction of the molecule from a grid (an array of slits 100nm apart; the continuous line is a best fit to the theoretical prediction). Notice that the distance between side peaks $\delta x = \lambda z / d \sim 55 \mu m$ is the one expected for the diffraction from a grid (see text). Bottom plot: data in absence of diffraction (no slits) where a single spot is observed (control experiment). Reprinted with permission from ref.[60] copyright (2003) by the American Physical Society.

4.5.3 PARTICLE-WAVE DUALITY: REFRACTION

Just as light impinging the interface between two media with different indices of refraction (i.e. different light velocities) is reflected and refracted, so is a particle when passing atop a potential barrier with resulting slow-down in momentum. The canonical case involves a free particle of momentum p and energy $E = p^2/2m$ entering at $x = 0$ a region with potential $V_0 < E$ and slowing down to momentum q: $q = \sqrt{2m(E - V_0)}$. On the left-handed side ($x < 0$), as for an EM wave, the particle's wave-function is the sum of an incident and a reflected wave:

$$\Psi_l(x) = Ae^{ipx/\hbar} + Be^{-ipx/\hbar}$$

while on the right-handed side the transmitted wavefunction is:

$$\Psi_r(x) = Ce^{iqx/\hbar}$$

Continuity of the wavefunction and its first derivative at $x = 0$ imply:

$$A + B = C$$

$$pA - pB = qC$$

Hence, the reflection coefficient R which is equal to the flux of reflected particles $p|B|^2/m$ divided by the flux of incident particles $p|A|^2/m$ is:

$$R = \left|\frac{B}{A}\right|^2 = \frac{(p-q)^2}{(p+q)^2} \tag{4.59}$$

And the transmission coefficient T which is the ratio of the flux of transmitted particles $q|C|^2/m$ divided by the flux of incident particles is:

$$T = \frac{q|C|^2}{p|A|^2} = \frac{4pq}{(p+q)^2} \tag{4.60}$$

As expected: $T + R = 1$. Notice that the reflection of a particle by a potential discontinuity is a purely quantum mechanical phenomena. Eq.4.59 is valid only if the potential varies on a scale smaller than the de Broglie wavelength of the particle $\lambda = h/p$, as we shall now show.

For a potential jump varying over a length-scale a:

$$V(x) = \frac{V_0}{1 + e^{-x/a}}$$

the reflection coefficient is given by[25]:

$$R = \frac{\sinh^2(\pi(p-q)a/\hbar)}{\sinh^2(\pi(p+q)a/\hbar)}$$

If $a/\lambda = ap/h \ll 1$ we recover Eq.4.59. In the other limit, that of a slowly varying potential (i.e. $ap/h > 1$, using $\sinh x \approx e^x/2$): $R \approx e^{-4\pi qa/\hbar}$ which tends to zero as a increases or $\hbar \to 0$, as expected in the classical limit.

This result can be understood qualitatively from Heisenberg uncertainty principle. In the region of size $\sim a$, where the potential is varying, a force $\vec{F}$ is exerted on the particle: $\vec{F} = -\vec{\nabla}V \sim -(V_0/a)\hat{x}$. In that region the uncertainty in the particle's energy is $\Delta E \sim V_0/2$, to which a time uncertainty $\Delta t \sim \hbar/\Delta E \sim 2\hbar/V_0$ is associated. From Newton's second law, the change in momentum Δp resulting from the application of a force F during a time Δt is: $\Delta p \sim F\Delta t = 2\hbar/a$. For the particle to be reflected by a potential jump the change in momentum Δp must be larger than twice the incident momentum p: $\Delta p > 2p$ which implies: $a < \hbar/p$.

---

[25]The solution of Schrödinger's equation with that potential requires some advanced mathematics. For the interested reader it is sketched in the Landau and Lifchitz course on Quantum Mechanics[41].

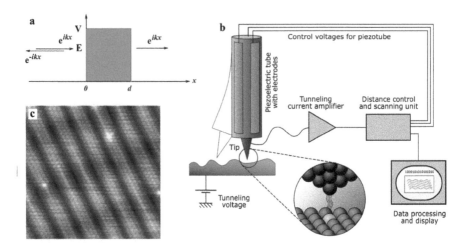

Figure 4.14 (a) An electron wave of energy E is incident on a barrier of energy V. Some of the wave is reflected and some tunnels through the barrier to be transmitted. (b) The principle of the Scanning Tunneling Microscope (STM). A conducting tip mounted on a piezo-electric drive is scanned across a surface. The current tunneling between the surface and the tip is kept constant via an appropriate feedback loop that maintains a constant distance between tip and surface. The feedback signal is used to generate an image of the surface with atomic resolution (reprinted from ref.[62]). (c) Image of the surface of a gold crystal taken in vacuum at 77°K. To minimize their surface energy the gold atoms deviate from the bulk crystal structure and arrange in columns several atoms wide with regularly-spaced furrows between them.

4.5.4 THE SCANNING TUNNELING MICROSCOPE

A consequence of Heisenberg's principle and a queer feature of Quantum Mechanics is the possibility for particles to pass through energetically forbidden regions, see Appendix A.8.2. Imagine a particle moving with momentum p along the x-axis and encountering a wall (an energy barrier of height $V > p^2/2m = E$ and thickness $x = d$, see Fig.4.14(a)). In classical mechanics the particle is reflected from such a wall. However in QM, the particle has a finite probability of passing through the wall and emerging on its other side! As shown in Appendix A.8.1, the probability of passing through the wall decays exponentially with the wall thickness:

$$T \approx \frac{16E}{V} e^{-d/\lambda_t}$$

where $\lambda_t = \hbar/2\sqrt{2mV}$. For an electron to pass through an energy barrier $V = 1\mathrm{eV}$: $\lambda_t \simeq$ 1Å which is the typical size of an atom. The transmission probability thus decreases exponentially with the size d of the barrier which can be but a few Å wide.

This sensitivity of QM tunneling to atomic size dimensions has led to the development in 1981 of the Scanning Tunneling Microscope (STM) by Gerd Binning

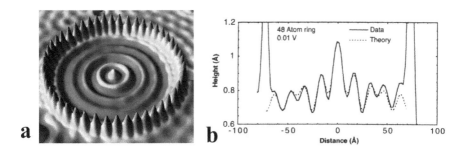

Figure 4.15 (a) Free electrons on the surface of copper (Cu) with effective mass $m = 0.38m_e$ are confined in a corral of radius $a = 71.3\text{Å}$ formed by 48 atoms of iron (Fe) and observed with a STM. (b) The probability of finding an electron at a given place within the corral is proportional to the intensity of the current tunneling into the scanning conducting tip of the STM. The observations are nicely fit by a linear superposition of just three eigen-functions, solutions of Schrödinger's equation for this problem. Reprinted from ref.[63] with permission from AAAS.

and Heinrich Rohrer. They used it to image surfaces with atomic resolution and were awarded for it the Nobel prize in Physics in 1986. In a STM a conducting tip is scanned above a conducting surface, see Fig.4.14(b). The electrons tunneling between the tip and the surface (across an insulating gap, usually vacuum) generate a current which amplitude is exponentially sensitive to the size of the gap and the voltage across it. A piezo-electric tube displaces the tip above the surface while maintaining (with an appropriate feedback loop) a constant current (i.e. a fixed distance) between the tip and the surface. Measuring the amplitude of the feedback signal while scanning the tip generates an image of the surface with atomic resolution, Fig.4.14(c). This invention revolutionized the study of matter at the nanoscale. It also led to the development of many other scanning probe microscopes, such as the Atomic Force Microscope (AFM) which uses the deflection of a small cantilever, like the needle in a gramophone, to map a surface with atomic resolution.

The STM also made possible the study of QM problems which for decades were assumed to be thought (gedanken) experiments, such as the problem of finding the eigen-energies of a particle confined in a 2D box. With the help of an STM such a situation could be constructed and studied[63]: consider a free particle (an electron on the surface of a conductor (Cu) with effective mass $m_c = 0.38m_e$, see section 4.3.4) confined by a two dimensional ring of radius $a = 71.3\text{Å}$ formed by 48 atoms of iron (Fe) deposited and manipulated with an STM to form a corral, Fig.4.15. The Schrödinger equation for such a particle is:

$$-\frac{\hbar^2}{2m_c}\nabla^2\Psi = E\Psi$$

for $r < a$ and $\Psi = 0$ for $r > a$. This is a typical Helmholtz equation (see appendix A.5) which solutions are: $|n, l\rangle \equiv \Psi_{n,l}(r, \phi) = J_l(\kappa_{n,l}r)e^{il\phi}$ where the eigen-energies $E_{n,l} = \hbar^2\kappa_{n,l}^2/2m_c$ are determined by the boundary condition $J_l(\kappa_{n,l}a) = 0$. Fig.4.15 presents the results from the group of Don Eigler at IBM who observed with an STM the wavefunction of an electron trapped in such a corral. The measured tunneling current is proportional to the probability of finding the electron under the scanning tip, which is remarkably fit by a linear combination of just three eigenstates: $|5, 0\rangle$, $|4, 2\rangle$ and $|2, 7\rangle$ with similar energies (since: $\kappa_{5,0}a = 14.931$, $\kappa_{4,2}a = 14.796$ and $\kappa_{2,7}a = 14.821$). In this corral the surface electrons with the highest energy are occupying the top of the conduction band from which they hop into the tip of the STM.

4.5.5 ◆ THE CORRESPONDENCE PRINCIPLE

One of the main self-consistency checks of the interpretation of QM is that its equations yield the classical results when the system is large and its wavefunction is localized so that one can replace the operators with their mean values. The correspondence principle therefore states that the time evolution of the mean of an observable should satisfy the equation of classical mechanics.

If $\hat{O}$ is an operator that does not depend explicitly on time, the evolution of its mean value is given by:

$$\frac{d}{dt} < \hat{O} > = \frac{d}{dt}\langle\Psi|\hat{O}|\Psi\rangle = (\frac{d}{dt}\langle\Psi|)\hat{O}|\Psi\rangle + \langle\Psi|\hat{O}\frac{d}{dt}|\Psi\rangle$$

$$= -\frac{1}{i\hbar}\langle\Psi|\hat{H}\hat{O}|\Psi\rangle + \frac{1}{i\hbar}\langle\Psi|\hat{O}\hat{H}|\Psi\rangle$$

where we used Eq.4.24 with $\hat{H}$ as the Hamiltonian operator. Hence the time evolution of the mean value of an operator obeys the so-called Ehrenfest theorem:

$$i\hbar\frac{d}{dt}\langle\hat{O}\rangle = \langle\Psi|[\hat{O}, \hat{H}]|\Psi\rangle = \langle[\hat{O}, \hat{H}]\rangle \qquad (4.61)$$

One can thus show that the angular moment $\vec{L} = \vec{r} \times \vec{p}$ is conserved during the motion of a free particle with Hamiltonian $\hat{H} = \hat{p}^2/2m$. Consider the x–component of the angular momentum: $\hat{L}_x = \hat{y}\hat{p}_z - \hat{z}\hat{p}_y$. If it is conserved, i.e. unchanged, during the motion then $d\langle\hat{L}_x\rangle/dt = 0$. Since only the spatial operators $\hat{y} = i\hbar\partial_{p_y}$ or $\hat{z} = i\hbar\partial_{p_z}$ do not commute with their associated momentum operators $\hat{p}_y$ or $\hat{p}_z$:

$$i\hbar\frac{d}{dt}\langle\hat{L}_x\rangle = \langle[\hat{L}_x, \hat{H}]\rangle = \frac{1}{2m}\{\langle[\hat{y}, \hat{p}_y^2]\hat{p}_z\rangle - \langle[\hat{z}, \hat{p}_z^2]\hat{p}_y\rangle\}$$

$$= \frac{i\hbar}{m}\langle\hat{p}_y\hat{p}_z - \hat{p}_z\hat{p}_y\rangle = 0$$

Consider further the problem of a particle moving in a potential $V(x)$. Its Hamiltonian is $\hat{H} = \hat{p}^2/2m + V(x)$. The mean velocity of the particle is

$$\langle v\rangle = \frac{d}{dt}\langle\hat{x}\rangle = \frac{\langle[\hat{x}, \hat{H}]\rangle}{i\hbar} = \frac{\langle[\hat{x}, \hat{p}^2]\rangle}{2i\hbar m} = \langle p\rangle/m$$

which is indeed the classical result $\vec{p} = m\vec{v}$. Similarly the evolution of the velocity operator yields:

$$m\frac{d}{dt}\langle\hat{v}\rangle = \frac{\langle[\hat{p},\hat{H}]\rangle}{i\hbar} = \frac{\langle[\hat{p},V(x)]\rangle}{i\hbar} = -\langle\frac{\partial V}{\partial x}\rangle = \langle F\rangle$$

Which is Newton's second law: $mdv/dt = -\partial_x V = F$. For a particle of charge q in a magnetic field $\vec{B} = \vec{\nabla}\times\vec{A}$ (where $\vec{A}$ is the magnetic vector potential, see section 3.2.4) the Hamiltonian is:

$$\hat{H} = \frac{(\hat{p}-(q/c)\hat{A})^2}{2m} \tag{4.62}$$

If the magnetic field $\vec{B}$ is constant along the z-axis, the vector potential is: $\vec{A} = (-yB, xB, 0)/2$. The mean velocity v_x is:

$$v_x \equiv \frac{d}{dt}\langle\hat{x}\rangle = \frac{\langle[\hat{x},\hat{H}]\rangle}{i\hbar} = \frac{\langle[\hat{x},(\hat{p}^2-(q/c)(\hat{A}\hat{p}+\hat{p}\hat{A}))]\rangle}{2i\hbar m}$$

$$= \frac{\langle[\hat{x},\hat{p}_x^2]\rangle - 2(q/c)\hat{A}_x\langle[\hat{x},\hat{p}_x]\rangle}{2i\hbar m}$$

$$= \frac{\langle p_x - (q/c)A_x\rangle}{m}$$

Which is the classical result: $m\vec{v} \equiv \vec{\pi} = \vec{p}-(q/c)\vec{A}$ (with Hamiltonian: $H = m\vec{v}^2/2$). Notice that the so-called canonical momentum $\vec{p}$ is different from the kinetic momentum $\vec{\pi}$. We will leave it as an exercise for the reader to derive the Lorentz force law in a uniform magnetic field, $\vec{B} = B_0\hat{z}$, namely $d\vec{\pi}/dt = (q/c)\vec{v}\times\vec{B}$, from the time evolution of the velocity operator[26]: $\hat{v} = (\hat{p}-(q/c)\hat{A})/m$.

---

[26] Hint: with the vector potential $\vec{A} = (-yB_0, xB_0, 0)/2$ write

$$m\langle\frac{dv_x}{dt}\rangle = \langle[p_x + \frac{qB_0 y}{2c}, H]\rangle/i\hbar$$

and show that it yields $m\dot{v}_x = (qB_0/c)v_y$

4.6 ◆ DIRAC'S EQUATION: ANTIPARTICLES AND SPIN

We have seen the power of the QM formalism in predicting effects that are not only at odds with classical mechanics but also counter-intuitive: molecules that can exist as superpositions of enantiomers, particles that behave like waves and are diffracted and electrons that can tunnel through walls. The most striking demonstration of the power and validity of QM has been the predictions of the existence anti-particles and spin (a property of electrons which has no classical equivalence) that came out of the successful attempt by P.A.M. Dirac to unify QM with special relativity. Since spin is crucial in understanding the properties of atoms and molecules, we shall see how it arises from Dirac's equation.

The Hamiltonian in Schrödinger's equation describes a non-relativistic particle for which the kinetic energy $E = p^2/2m \ll mc^2$. For a relativistic particle, Einstein showed (see appendix A.6) that the energy-momentum relation obey Eq.A.73: $E^2 = p^2c^2 + m^2c^4$. However writing the relativistic analogue of Eq.4.54:

$$i\hbar\frac{\partial\Psi}{\partial t} = \sqrt{-\hbar^2c^2\nabla^2 + m^2c^4}\Psi$$

led to insurmountable problems related to the interpretation of the square-root of the momentum operator. Similarly writing $H^2 = p^2c^2 + m^2c^4$ to derive the so-called Klein-Gordon equation:

$$\frac{\partial^2\Psi}{\partial(ct)^2} = [\nabla^2 - (mc/\hbar)^2]\Psi \tag{4.63}$$

yields an equation which is second order in time (Schrödinger's equation is first order) and which does not allow for the interpretation of $|\Psi|^2$ as a probability density. Dirac searched for a relativistic formulation that like Eq.4.24 would be first order in time, yet would treat time and space in a relativistic invariant way. He noticed that he could obtain the relativistic energy-momentum relation as the eigenvalue of a linear Schrödinger equation for a wavefunction Ψ that is a vector rather than a scalar.

Dirac's equation in 2D. For simplicity, let us consider first the case of a particle in two-dimensions (for example an electron moving on a graphene sheet[27]). Let the wave-function associated with the particle $|\Psi(x,y,t)\rangle$ be a two-component vector, obeying Schrödinger's equation $i\hbar\partial_t|\Psi\rangle = H|\Psi\rangle = E|\Psi\rangle$, where H is the 2×2 matrix:

$$H = \begin{pmatrix} mc^2 & p_xc - ip_yc \\ p_xc + ip_yc & -mc^2 \end{pmatrix} \equiv mc^2\sigma_z + cp_x\sigma_x + cp_y\sigma_y \tag{4.64}$$

and σ_k are the 2×2 Pauli matrices:

$$\sigma_x = \begin{pmatrix} 0 & 1 \\ 1 & 0 \end{pmatrix} \qquad \sigma_y = \begin{pmatrix} 0 & -i \\ i & 0 \end{pmatrix} \qquad \sigma_z = \begin{pmatrix} 1 & 0 \\ 0 & -1 \end{pmatrix} \tag{4.65}$$

---

[27] In reality the energy of an electron on a graphene sheet behaves (in certain conditions) as $E \propto |p|$ which corresponds to an extreme relativistic (or massless) Dirac particle[64].

which satisfy the following commutation relations:

$$[\sigma_x, \sigma_y] = 2i\sigma_z \qquad [\sigma_y, \sigma_z] = 2i\sigma_x \qquad [\sigma_z, \sigma_x] = 2i\sigma_y \qquad (4.66)$$

It is easy to verify that the eigenvalues E of this Hamiltonian satisfy the relativistic energy-momentum relation: $E^2 = m^2c^4 + p^2c^2$. For a particle at rest $p = 0$ there are two eigenvalues of the energy: $E = \pm mc^2$ with eigenstates: $|\phi_+\rangle = (1,0)$ and $|\phi_-\rangle = (0,1)$. The first eigenstate corresponds to an electron, while the second corresponds to a hole.

In the limit $0 < p \ll mc$ one can show that Eq.4.64 yields Schrödinger's equation for a non-relativistic particle. In that limit, we seek a solution that is a small perturbation on the eigenstate of a particle: $|\Psi\rangle = e^{-i(mc^2/\hbar)t}(\phi, \epsilon)$ (with $\epsilon \ll \phi$). Eq.4.64 yields the coupled equations:

$$mc^2\phi + i\hbar\frac{\partial\phi}{\partial t} = mc^2\phi + (p_xc - ip_yc)\epsilon$$

$$mc^2\epsilon + i\hbar\frac{\partial\epsilon}{\partial t} = (p_xc + ip_yc)\phi - mc^2\epsilon$$

from which we deduce that: $\epsilon \approx (p_x + ip_y)\phi/2mc \ll \phi$ (as assumed) and ϕ obeys Schrödinger's equation for a free particle in 2D:

$$i\hbar\frac{\partial\phi}{\partial t} = \frac{p^2}{2m}\phi$$

Hence in two dimensions a relativistic particle can be described by a two dimensional vector obeying a linear equation which is first order in time. For a particle at rest, the eigenstates corresponds to a particle and a hole. In the non-relativistic limit the 2×2 Hamiltonian reduces to Schrödinger's equation for the particle component of the wavefunction. In the following we shall see how Dirac generalized this approach to a particle moving in three dimensions with momentum: p_x, p_y, p_z.

Dirac's equation in 3D. Dirac noticed that in order for the eigenvalues E to satisfy the relativistic energy-momentum relation in 3D, he needed a 4 dimensional Hamiltonian[39] ("Principles of Quantum Mechanics", p.253ff):

$$H_D = mc^2\begin{pmatrix} I & 0 \\ 0 & -I \end{pmatrix} + cp_x\begin{pmatrix} 0 & \sigma_x \\ \sigma_x & 0 \end{pmatrix} + cp_y\begin{pmatrix} 0 & \sigma_y \\ \sigma_y & 0 \end{pmatrix} + cp_z\begin{pmatrix} 0 & \sigma_z \\ \sigma_z & 0 \end{pmatrix}$$

$$\equiv mc^2\beta + c\sum_k \alpha_k p_k \qquad (4.67)$$

Where the matrices β and α_k are the 4×4 Hermitian matrices (i.e. $\alpha_k^\dagger = \alpha_k$; $k = x, y$ or z):

$$\alpha_k = \begin{pmatrix} 0 & \sigma_k \\ \sigma_k & 0 \end{pmatrix} \qquad \beta = \begin{pmatrix} I & 0 \\ 0 & -I \end{pmatrix} \qquad (4.68)$$

and I is the 2×2 unit matrix. Hence:

$$
H_D = \begin{pmatrix}
mc^2 & 0 & p_z c & p_x c - i p_y c \\
0 & mc^2 & p_x c + i p_y c & -p_z c \\
p_z c & p_x c - i p_y c & -mc^2 & 0 \\
p_x c + i p_y c & -p_z c & 0 & -mc^2
\end{pmatrix}
$$

Notice the similarity of H_D with its 2D version, Eq.4.64.

With this Hamiltonian, H_D, Schrödinger's equation (now known as Dirac's equation) for a free particle becomes:

$$
i\hbar \partial_t |\Psi(\vec{x},t)\rangle = H_D |\Psi(\vec{x},t)\rangle = (mc^2 \beta + c \sum_k \alpha_k p_k) |\Psi(\vec{x},t)\rangle \tag{4.69}
$$

One can easily check that its eigenvalues satisfy: $(m^2 c^4 + p^2 c^2 - E^2)^2 = 0$. Notice that for each eigenvalue $E = \pm \sqrt{m^2 c^4 + p^2 c^2}$, there are two eigenstates. The fact that the Hamiltonian H_D, a 4×4 matrix, has two degenerate eigenstates signifies that the particle has some discrete internal degrees of freedom that commute with the particle's free energy operator. The two doubly degenerate eigenstates correspond (as we shall see) to a particle and an anti-particle each possessing two possible spin-states with the same energy.

In Dirac's equation as in Schrödinger's equation the wavefunction has a natural interpretation as a probability distribution. To see that, multiply Eq.4.69 on the left by $\langle \Psi(x,t)|$ and its complex conjugate

$$
-i\hbar (\langle \Psi(x,t)| \partial_t) = [\langle \Psi(x,t)| (c \sum_k \alpha_k^\dagger p_k^* + mc^2 \beta)]
$$

on the right by $|\Psi(x,t)\rangle$. Notice that $\alpha_k^\dagger = \alpha_k$ and $p_k = -i\hbar \partial_k$ and subtract the two to get:

$$
i\hbar \partial_t \langle \Psi|\Psi \rangle = -i\hbar c \sum_k \partial_k \langle \Psi|\alpha_k|\Psi \rangle
$$

which expresses a conservation law for the probability density $P(x,t) = \langle \Psi(x,t)|\Psi(x,t)\rangle$ with current density $J_k = c \langle \Psi|\alpha_k|\Psi\rangle$:

$$
\frac{\partial P}{\partial t} + \vec{\nabla} \cdot \vec{J} = 0
$$

Let us now investigate Eq.4.69. Let us first consider a free particle with zero momentum. Dirac's equation for its wavefunction $|\Psi^0\rangle$ is then:

$$
i\hbar \partial_t |\Psi^0\rangle = mc^2 \beta |\Psi^0\rangle
$$

which possesses two doubly degenerate solutions:

$$
|\Psi^0\rangle_{1,2} = e^{-i(mc^2/\hbar)t} \begin{pmatrix} \phi^0 \\ 0 \end{pmatrix} \qquad |\Psi^0\rangle_{3,4} = e^{i(mc^2/\hbar)t} \begin{pmatrix} 0 \\ \chi^0 \end{pmatrix}
$$

where ϕ^0 and χ^0 correspond to two normalized 2D vectors (such as $\phi_1^0 = (1,0)$ and $\phi_2^0 = (0,1)$; $\chi_1^0 = (1,0)$ and $\chi_2^0 = (0,1)$). The first solution corresponds to a doubly degenerate wavefunction for an electron with positive energy $E_e = mc^2$ and the second to a doubly degenerate wavefunction for a hole with energy $E_h = -mc^2$ on which we will have more to say shortly.

4.6.1 SPIN AND MAGNETIC DIPOLE

In the presence of an electro-magnetic field Dirac's equation is generalized just as Schrödinger's equation was generalized: by replacing the momentum p_k by $\pi_k = p_k - (q/c)A_k$ (see Eq.4.62) and adding the potential energy $q\Phi$

$$i\hbar\partial_t |\Psi(x,t)\rangle = [c\sum_k \alpha_k(p_k - (q/c)A_k) + q\Phi + mc^2\beta]|\Psi(x,t)\rangle \qquad (4.70)$$

Let us look for a solution in terms of the two-component vector: $|\Psi\rangle = (\tilde\phi, \tilde\chi)$. Dirac's equation can then be recast as:

$$i\hbar\partial_t \begin{pmatrix} \tilde\phi \\ \tilde\chi \end{pmatrix} = c\sum_k \sigma_k \pi_k \begin{pmatrix} \tilde\chi \\ \tilde\phi \end{pmatrix} + q\Phi\begin{pmatrix} \tilde\phi \\ \tilde\chi \end{pmatrix} + mc^2\begin{pmatrix} \tilde\phi \\ -\tilde\chi \end{pmatrix} \qquad (4.71)$$

To derive the low energy (non-relativistic, $pc \ll mc^2$) limit of Dirac's equation we look for a solution which is a perturbation of the previous solution for positive energy $E_e = mc^2$ at zero momentum, $|\Psi^0\rangle_{1,2}$:

$$|\Psi(x,t)\rangle = \begin{pmatrix} \tilde\phi \\ \tilde\chi \end{pmatrix} = e^{-i(mc^2/\hbar)t}\begin{pmatrix} \phi(x,t) \\ \chi(x,t) \end{pmatrix} \qquad (4.72)$$

where in the low energy limit, $\phi \approx \phi^0$ and $\chi \ll \phi$ are slowly time-varying fields (i.e. slower than $\omega_0 = mc^2/\hbar$), solutions of the coupled equations:

$$i\hbar\partial_t \begin{pmatrix} \phi \\ \chi \end{pmatrix} = c\sum_k \sigma_k \pi_k \begin{pmatrix} \chi \\ \phi \end{pmatrix} + q\Phi\begin{pmatrix} \phi \\ \chi \end{pmatrix} - 2mc^2\begin{pmatrix} 0 \\ \chi \end{pmatrix} \qquad (4.73)$$

In the low energy limit, i.e. when mc^2 is larger than any other energy, the terms $i\hbar\partial_t\chi$ and $q\Phi\chi$ are negligible with respect to the $2mc^2\chi$ term. The field χ is thus slaved to ϕ and given by:

$$\chi = \frac{\sum_k \sigma_k \pi_k \phi}{2mc}$$

which yields the following equation for ϕ:

$$i\hbar\partial_t\phi = \left[\frac{(\sum_k \sigma_k \pi_k)^2}{2m} + q\Phi\right]\phi \equiv \left[\frac{(\vec\sigma \cdot \vec\pi)^2}{2m} + q\Phi\right]\phi \qquad (4.74)$$

with $\vec\pi = -i\hbar\vec\nabla - (q/c)\vec{A}$. Using the following property of Pauli matrices:

$$(\vec\sigma \cdot \vec{A})(\vec\sigma \cdot \vec{B}) = \vec{A}\cdot\vec{B} + i\vec\sigma \cdot \vec{A}\times\vec{B}$$

one obtains:

$$(\vec{\sigma} \cdot \vec{\pi})^2 = \pi^2 + i\vec{\sigma} \cdot \vec{\pi} \times \vec{\pi}$$

$$= \pi^2 + i\vec{\sigma} \cdot (-i\hbar\vec{\nabla} - (q/c)\vec{A}) \times (-i\hbar\vec{\nabla} - (q/c)\vec{A}) \qquad (4.75)$$

$$= \pi^2 - (\hbar q/c)\vec{\sigma} \cdot \vec{\nabla} \times \vec{A} = \pi^2 - (\hbar q/c)\vec{\sigma} \cdot \vec{B}$$

From[28] which one derives the low energy version of Dirac's equation:

$$i\hbar\partial_t \phi = \left[\frac{(\hat{p} - (q/c)\hat{A})^2}{2m} - (\hbar q/2mc)\vec{\sigma} \cdot \vec{B} + q\Phi \right] \phi \qquad (4.76)$$

which resembles Schrödinger's equation except for the additional term on the right hand side which is the energy of a dipole $\vec{m} = (\hbar q/2mc)\vec{\sigma}$ in a magnetic field $\vec{B}$. Hence Dirac's equation implies that the electron possesses an intrinsic magnetic moment, a fact that was known from the spectrum of atoms in a magnetic field (see section 4.11.3). The value of the electron magnetic dipole, known as the Bohr magneton, is $\mu_B = \hbar e/2mc = 9.274 \ 10^{-21}$ erg/G. Exemplifying yet again the "unreasonable effectiveness of Mathematics"[1], it is amazing that this intrinsic property of the electron is a mathematical consequence of the unification of QM and special relativity!

Moreover, as we shall see below, Dirac was able to relate this magnetic moment to an intrinsic angular momentum of the particle: its spin. The electron being a point particle, with no internal structure (which is not the case for the proton or neutron), can classically have no moment of inertia and thus no angular momentum. Its spin is a quantum property for which we have no intuition[29].

Consider the time evolution of the orbital angular momentum $\vec{L} = \vec{r} \times \vec{p}$ for a free particle with Hamiltonian H_D (no magnetic or electric potential: $\vec{A} = \Phi = 0$). As we have seen in section 4.5.5, the time evolution of an observable mean (i.e. the classical observable) obeys:

$$i\hbar\partial_t \langle L_x \rangle = \langle [L_x, H_D] \rangle = \langle [yp_z - zp_y, (c \sum_k \alpha_k p_k + mc^2 \beta)] \rangle$$

$$= c\langle \alpha_y[y, p_y]p_z - \alpha_z[z, p_z]p_y \rangle = i\hbar c \langle \alpha_y p_z - \alpha_z p_y \rangle \neq 0$$

Thus in Dirac's equation $\langle L_x \rangle$ is not a constant of motion, in contrast with Schrödinger's equation for a free particle where it is, see section 4.5.5. Now consider the evolution of $\hat{\sigma}_x$ (the 4×4 matrix which diagonal elements are the Pauli σ_x matrices):

$$i\hbar\partial_t \langle \hat{\sigma}_x \rangle \equiv i\hbar\partial_t \left\langle \begin{pmatrix} \sigma_x & 0 \\ 0 & \sigma_x \end{pmatrix} \right\rangle = \langle [\hat{\sigma}_x, H_D] \rangle = \langle [\hat{\sigma}_x, (c \sum_k \alpha_k p_k + mc^2 \beta)] \rangle$$

$$= c\langle [\hat{\sigma}_x, \alpha_y]p_y + [\hat{\sigma}_x, \alpha_z]p_z \rangle = 2ic\langle \alpha_z p_y - \alpha_y p_z \rangle$$

---

[28] Use $(\vec{\nabla} + (q/i\hbar c)\vec{A}) \times (\vec{\nabla} + (q/i\hbar c)\vec{A})\phi = (q/i\hbar c)[\vec{\nabla} \times (\vec{A}\phi) + \vec{A} \times \vec{\nabla}\phi] = (q/i\hbar c)(\vec{\nabla} \times \vec{A})\phi$.

[29] An analogy with a spinning top is sometimes made but this is at odds with the point particle assumption.

where we used the following property of the Pauli matrices: $[\sigma_x, \sigma_y] = 2i\sigma_z$, $[\sigma_x, \sigma_z] = -2i\sigma_y$. Therefore[30], $L_x + \hbar\hat{\sigma}_x/2$ commutes with the relativistic Hamiltonian H_D, i.e. it is a constant of the motion. Dirac interpreted this result to mean that the particle has a spin angular momentum $\vec{S} = \hbar\vec{\sigma}/2$ which must be added to the orbital angular momentum $\vec{L}$ to get the total angular momentum, $\vec{J} = \vec{L} + \vec{S}$ which is the true constant of the motion.

According to Eq.4.76, the magnetic dipole moment of the electron is related to its spin by:

$$\vec{\mathbf{m}} = (\hbar e/2mc)\vec{\sigma} = \mu_B\vec{\sigma} = g(e/2mc)\vec{S} \qquad (4.77)$$

Where $g = 2$ is known as the electron gyromagnetic ratio. This is in contrast with the magnetic moment arising from the angular momentum $\vec{\mathbf{m}} = (e/2mc)\vec{L}$ (see Eq.3.66) for which $g = 1$. The prediction of the gyromagnetic ratio of the electron was another success of Dirac's equation. In fact the value of g has become the most stringent test of Quantum Electro-Dynamics (QED), the theory that generalized Dirac's approach and unified Electromagnetism and Quantum Mechanics. It is the most precisely known constant of Nature: its value has been tested to 12 decimal places $g = 2.0023193043617(!!)$, making QM and QED the most precisely tested theories ever proposed.

4.6.2 ANTIPARTICLES

Let us now return to the hole solution with energy $E_h = -mc^2$. We have already seen such solutions when studying the band theory of solids. In fact the original interpretation of Dirac was very similar. He assumed that all negative energy levels were occupied by a "sea" of electrons. Out of this sea, electrons could be excited to an energy level $E_e = mc^2$ leaving behind a hole (or positron) of energy $|E_h| = mc^2$. This process, electron-positron creation (and its reverse, annihilation), is routinely observed in particle accelerators once an energy $E > 2mc^2 \sim 1$MeV is achieved. The dynamic behavior of the positron is similar to that of an electron with opposite charge. To derive the positron wave-equation at low energies, we shall follow our previous study of the electron in a similar regime. We look for a low energy solution:

$$|\Psi\rangle = \begin{pmatrix} \tilde{\phi} \\ \tilde{\chi} \end{pmatrix} = e^{i(mc^2/\hbar)t} \begin{pmatrix} \phi \\ \chi \end{pmatrix} \qquad (4.78)$$

with $\phi \ll \chi$, see Eq.4.72, which leads to the following equation for the positron-wavefunction at low energies (setting $m \to -m$ in Eq.4.74):

$$i\hbar\partial_t\chi = \left(-\frac{[\sum_k \sigma_k \pi_k]^2}{2m} + q\Phi \right)\chi$$

---

[30]Using the relation: $[\hat{A} + \hat{B}, \hat{C}] = [\hat{A}, \hat{C}] + [\hat{B}, \hat{C}]$ and $[a\hat{A}, \hat{B}] = a[\hat{A}, \hat{B}]$.

Since the Pauli matrices are Hermitian, the equation for the complex conjugate of χ is:

$$i\hbar\partial_t\chi^* = \left(\frac{[\sum_k \sigma_k(i\hbar\partial_k - (q/c)A_k)]^2}{2m} - q\Phi\right)\chi^*$$
$$= \left(\frac{[\sum_k \sigma_k(-i\hbar\partial_k + (q/c)A_k)]^2}{2m} - q\Phi\right)\chi^* \qquad (4.79)$$

This is the same equation as the low energy limit of Dirac's equation for a particle of mass m and charge $-q$, Eq.4.74. Hence Dirac's hole solution describes the behavior of a particle with the same mass as the electron but of opposite charge (just like the holes in the band theory of solids, section 4.3.4). This prediction of Dirac was vindicated by the discovery of the positron three years later by Carl Anderson, for which they both shared the 1933 Nobel prize in Physics.

To summarize, the unification in 1929 of special relativity and QM by Dirac led to many unexpected predictions of which we have seen three:

1. The existence of anti-particles (e.g. the positron) with a mass identical to the particle but of opposite charge. These are created when an energy $2mc^2$ is available to create an electron-positron pair (raise the electron from the top of the "sea" at $E_h = -mc^2$ to the electron rest energy $E_e = mc^2$).

2. The association to both the electron and its anti-particle of an intrinsic angular momentum variable, the spin: $S = \hbar\sigma/2$.

3. The existence of a magnetic dipole moment linked to the electron's spin similar to the classical relation between magnetic moment and angular momentum but larger by a factor $g = 2$.

4.7 ANGULAR MOMENTUM WAVEFUNCTION

Many soluble Quantum Mechanical problems (e.g., the free particle, the Hydrogen atom, the freely rotating dipole) are spherically symmetric (they are invariant to arbitrary rotation about any axis). As a consequence of that symmetry they share the same angular wavefunctions. In spherical coordinates, the kinetic energy part of the Hamiltonian is:

$$H_k = -\frac{\hbar^2}{2m}\nabla^2 = -\frac{\hbar^2}{2m}\left(\frac{1}{r^2}\frac{\partial^2}{\partial r^2}r^2 + \frac{1}{r^2\sin\theta}\frac{\partial}{\partial\theta}(\sin\theta\frac{\partial}{\partial\theta}) + \frac{1}{r^2\sin^2\theta}\frac{\partial^2}{\partial\phi^2}\right)$$

$$= -\frac{\hbar^2}{2m}\left(\frac{\partial^2}{\partial r^2} + \frac{2}{r}\frac{\partial}{\partial r}\right) - \frac{\hbar^2}{2mr^2}\left(\frac{1}{\sin\theta}\frac{\partial}{\partial\theta}(\sin\theta\frac{\partial}{\partial\theta}) + \frac{1}{\sin^2\theta}\frac{\partial^2}{\partial\phi^2}\right)$$

$$\equiv -\frac{\hbar^2}{2m}\left(\frac{\partial^2}{\partial r^2} + \frac{2}{r}\frac{\partial}{\partial r}\right) - \frac{\hbar^2}{2mr^2}\mathcal{L}^2 \tag{4.80}$$

The operator : $\hat{L}^2 = -\hbar^2\mathcal{L}^2$ is known as the angular momentum operator, since its contribution to the Hamiltonian, i.e. $\hat{L}^2/2mr^2$ is the classical angular momentum kinetic energy: $E_L = L^2/2I$ (where $I = \int d^3r\,\rho(r)r^2$ is the moment of inertia), see section 2.1. The solution of Schrödinger's equation for spherically symmetric systems (e.g. in presence of a potential that depends only on the radial coordinate r) can be written as $\Psi(r,\theta,\phi) = R(r)Y_{lm}(\theta,\phi)$, where the spherical harmonics $Y_{lm}(\theta,\phi)$ are the eigenstates $|l,m\rangle$ of the angular momentum operator $\hat{L}^2$ with eigenvalue $\hbar^2 l(l+1)$ (with $l = 0,1,...$ and $-l \le m \le l$, see Appendix A.5):

$$\hat{L}^2 Y_{lm}(\theta,\phi) = -\hbar^2\left[\frac{1}{\sin\theta}\frac{\partial}{\partial\theta}(\sin\theta\frac{\partial}{\partial\theta}) + \frac{1}{\sin^2\theta}\frac{\partial^2}{\partial\phi^2}\right]Y_{lm}(\theta,\phi)$$

$$= \hbar^2 l(l+1)Y_{lm}(\theta,\phi) \tag{4.81}$$

$$\text{with :} \quad |l,m\rangle \sim Y_{lm}(\theta,\phi) = \sqrt{\frac{(2l+1)(l-m)!}{4\pi(l+m)!}}P_l^m(\cos\theta)e^{im\phi}$$

where $P_l^m(\cos\theta)$ are associated Legendre polynomials, see Appendix A.5. Thus, as shown in this Appendix, the solution of Schrödinger's equation for a free particle: $H_k\Psi = E\Psi$ is

$$\Psi_{lm}(kr,\theta,\phi) = R_l(kr)Y_{l,m}(\theta,\phi)$$

where the radial part of the wavefunction (with wavenumber $k = \sqrt{2mE}/\hbar$), $R_l(kr)$ are spherical Bessel functions given by the following formula (with $x = kr$):

$$R_l(x) = (-x)^l(\frac{1}{x}\frac{d}{dx})^l\frac{e^{ix}}{x}$$

Hence, as for the example of a particle in a box, the representation of the wavefunction of a free particle is arbitrary: it can either be expressed in the basis of plane waves $\exp\{i\vec{k}\cdot\vec{r}\}$, which are eigenstates of the linear momentum operator or in the basis of spherical waves $\Psi_{lm}(kr,\theta,\phi))$, which are eigenstates of the angular momentum operator.

When working in the spherical wave basis, the linear angular momentum operator $\hat{L} = \hat{r} \times \hat{p}$ can be written as:

$$\hat{L}_x = -i\hbar\left(y\frac{\partial}{\partial z} - z\frac{\partial}{\partial y}\right) = i\hbar\left(\sin\phi\frac{\partial}{\partial\theta} + \cot\theta\cos\phi\frac{\partial}{\partial\phi}\right) \tag{4.82}$$

$$\hat{L}_y = -i\hbar\left(z\frac{\partial}{\partial x} - x\frac{\partial}{\partial z}\right) = i\hbar\left(-\cos\phi\frac{\partial}{\partial\theta} + \cot\theta\sin\phi\frac{\partial}{\partial\phi}\right) \tag{4.83}$$

$$\hat{L}_z = -i\hbar\left(x\frac{\partial}{\partial y} - y\frac{\partial}{\partial x}\right) = -i\hbar\frac{\partial}{\partial\phi} \tag{4.84}$$

The spherical harmonics $Y_{lm}(\theta,\phi)$ are thus also eigenstates of $\hat{L}_z$ with eigenvalue $\hbar m$ (with $-l \le m \le l$):

$$\hat{L}_z|l,m\rangle = \hbar m|l,m\rangle$$

For a given value of the total angular momentum l, the matrix representation of L_z is diagonal of order $(2l+1) \times (2l+1)$. For example if $l = 1$:

$$L_z = \hbar\begin{pmatrix} 1 & 0 & 0 \\ 0 & 0 & 0 \\ 0 & 0 & -1 \end{pmatrix}$$

Notice that $\hat{L}_z$ does not commute with $\hat{L}_x$ or $\hat{L}_y$. In fact these operators satisfy the following commutation relations:

$$[\hat{L}_x, \hat{L}_y] = i\hbar\hat{L}_z \tag{4.85}$$

$$[\hat{L}_y, \hat{L}_z] = i\hbar\hat{L}_x$$

$$[\hat{L}_z, \hat{L}_x] = i\hbar\hat{L}_y$$

or $[\hat{L}_i, \hat{L}_j] = i\hbar\epsilon_{ijk}\hat{L}_k$, where i, j, k stands for x, y or z and $\epsilon_{ijk} = 1$ for an even permutation of the indices and -1 for an odd permutation[31]. Hence the eigenstates of $\hat{L}_z$ are not eigenstates of $\hat{L}_x, \hat{L}_y$ and any two components of the angular momentum cannot be measured simultaneously. Notice however that $\hat{L}_x$ and $\hat{L}_y$ do commute with the total angular momentum $\hat{L}^2 = \hat{L}_x^2 + \hat{L}_y^2 + \hat{L}_z^2$, see Appendix A.5:

$$[\hat{L}^2, \hat{L}_x] = [\hat{L}_y^2, \hat{L}_x] + [\hat{L}_z^2, \hat{L}_x] = \hat{L}_y[\hat{L}_y, \hat{L}_x] + [\hat{L}_y, \hat{L}_x]\hat{L}_y + \hat{L}_z[\hat{L}_z, \hat{L}_x] + [\hat{L}_z, \hat{L}_x]\hat{L}_z$$

$$= -i\hbar\hat{L}_y\hat{L}_z - i\hbar\hat{L}_z\hat{L}_y + i\hbar\hat{L}_z\hat{L}_y + i\hbar\hat{L}_y\hat{L}_z = 0$$

and therefore the total angular momentum and any component of the momentum along a defined axis can be measured simultaneously. These operators therefore do not mix eigenstates with different values of l. For any value of the total angular momentum l the possible eigenvalues of $\hat{L}_x, \hat{L}_y$ and $\hat{L}_z$ are the same: $\hbar m$ (with $-l \le m \le l$). But since $\hat{L}_x, \hat{L}_y$ do not commute with $\hat{L}_z$ they mix its eigenstates. Consequently the matrix representations of $\hat{L}_x, \hat{L}_y$ in the eigenstates of L_z (i.e. $|l,m\rangle$) are not diagonal, see Appendix A.8.3.

---

[31] The spin $S_i = \hbar\sigma_i/2$ obeys similar commutation rules, see section 4.6.1.

4.8 ◆ THE HYDROGEN ATOM AND ELECTRONIC ORBITALS

The first and major success of Quantum Mechanics was solving the energy levels of the hydrogen atom and explaining the properties of the elements as classified in Mendeleev's periodic table. Schrödinger's equation for a hydrogen-like atom is:

$$-\frac{\hbar^2}{2m_e}\nabla^2\Psi(\vec{x},t) - \frac{Ze^2}{r}\Psi(\vec{x},t) = E_n\Psi(\vec{x},t) \tag{4.86}$$

where the $-Ze^2/r$ term stands for the Coulomb interaction between the orbiting electron (of mass m_e) and a nucleus containing Z protons (in Hydrogen $Z = 1$). Since the potential is spherically symmetric, the spherical harmonics $Y_{lm}(\theta,\phi)$ are the eigenstates of the angular momentum operator $\hat{L}^2$ with eigenvalue $\hbar^2 l(l+1)$, see section 4.7. Thus, as for the free particle discussed above (see also Appendix A.5), we look for an eigenstate:

$$\Psi_{nlm}(r,\theta,\phi) = R_{nl}(r)Y_{lm}(\theta,\phi) \tag{4.87}$$

The equation for the radial function $R_{nl}(r)$ then becomes:

$$\left(-\frac{\hbar^2}{2m_e}\left(\frac{\partial^2}{\partial r^2} + \frac{2}{r}\frac{\partial}{\partial r}\right) + \frac{\hbar^2 l(l+1)}{2m_e r^2} - \frac{Ze^2}{r}\right)R_{nl} = E_n R_{nl}$$

The first term on the left is the radial part of the kinetic energy, the second term is the centrifugal potential (the angular part of the kinetic energy) and the third term is the Coulomb repulsion. Writing $R_{nl}(r) = u_{nl}(r)/r$ yields the following equation for $u_{nl}(r)$:

$$\frac{d^2 u_{nl}}{dr^2} + \left(\frac{2Zm_e e^2}{\hbar^2 r} - \frac{l(l+1)}{r^2}\right)u_{nl} = -\frac{2m_e E_n}{\hbar^2}u_{nl}$$

Defining $a = 2Zm_e e^2/\hbar^2 \equiv 2Z/r_0$ (where $r_0 = \hbar^2/m_e e^2$ is Bohr's radius, section 4.1.3) and $\lambda_n^2 = -2m_e E_n/\hbar^2 > 0$ (since the energy E_n of bound states is negative) one can recast the previous equation as:

$$u_{nl}'' + \left(\frac{a}{r} - \frac{l(l+1)}{r^2}\right)u_{nl} = \lambda_n^2 u_{nl}$$

In the limit $r \to \infty$, the terms in the bracket on the left hand side are negligible and the solution in that limit is $u_{nl}(r) \sim \exp(-\lambda_n r)$. Let us therefore look for a solution: $u_{nl}(r) = \mu_{nl}(r)\exp(-\lambda_n r)$. The equation that $\mu_{nl}(r)$ obeys is:

$$\mu_{nl}'' - 2\lambda_n \mu_{nl}' + \left(\frac{a}{r} - \frac{l(l+1)}{r^2}\right)\mu_{nl} = 0$$

Notice that close to $r = 0$, the equation becomes: $\mu_{nl}'' \approx l(l+1)\mu_{nl}/r^2$, which solution is: $\mu_{nl}(r) \sim r^{l+1}$. Let us therefore look for a polynomial solution: $\mu_{nl}(r) = \sum_{j=l+1}^{n} c_j r^j$. The coefficient of the term of order $j-1$ reads:

$$j(j+1)c_{j+1} - 2\lambda_n j c_j + a c_j - l(l+1)c_{j+1} = 0$$

which yields:

$$c_{j+1} = \frac{2\lambda_n j - a}{j(j+1) - l(l+1)} c_j$$

Notice that since $j \geq l+1$ the denominator is always positive. For the series to be finite, it must end at a value $j = n(> l)$ for which: $a = 2n\lambda_n$[32], i.e. $\lambda_n = 1/nr_0$ and a value of the eigen-energy, E_n (for Hydrogen, $Z = 1$):

$$E_n = -\frac{\hbar^2 \lambda_n^2}{2m_e} = -\frac{\hbar^2}{2m_e r_0^2} \frac{1}{n^2} \quad \text{with}: \quad n = 1, 2, 3 \cdots$$

which is the result obtained by Bohr, see Eq.4.8. Notice that while the radial function depends on the two quantum numbers (n, l) the eigen-energies depend only on n. This is a peculiarity of central $1/r$ potentials (such as the Coulomb potential between proton and electron). In non-hydrogen atoms that degeneracy is lifted due to the extra Coulomb repulsion between the electrons, see section 4.11.1.

It is customary to label the various orbitals (the Hydrogen atom's eigenstates) by their shell number $n = 1, 2, 3, \cdots$ and their main angular momentum number l ($0 \leq l < n$), such that $l = 0$ is called the s-orbital; $l = 1$ the p−orbital; $l = 2$ the d−orbital; $l = 3$ the f−orbital; $l = 4$ the g-orbital ... etc. (in alphabetical order skipping j). With this nomenclature we can proceed to write down the radial functions $R_{nl}(r) = \mu_{nl}(r)\exp(-r/nr_0)/r$ for the first orbitals:

1s orbital : $R_{10}(r) = 2r_0^{-3/2} e^{-r/r_0}$

2s orbital : $R_{20}(r) = (2r_0)^{-3/2}(2 - \frac{r}{r_0})e^{-r/2r_0}$

2p orbital : $R_{21}(r) = \frac{1}{\sqrt{3}}(2r_0)^{-3/2}\frac{r}{r_0}e^{-r/2r_0}$

In absence of other terms in the Hamiltonian, the orbitals of the electron in the Hydrogen atom are degenerate, i.e. for different values of the spin and angular quantum numbers (l, m) the energy of the electron is the same. Since there are 2 possible spin states ($S_z = \pm\hbar/2$) for every eigenstate Ψ_{nlm} and $2l + 1$ states with different values of m (and same value of l: $-l \leq m \leq l$) and since $0 \leq l \leq n - 1$, the degeneracy of a state with shell number n is: $2\sum_{l=0}^{n-1}(2l + 1) = 2n^2$. As we shall see below this degeneracy is partially lifted by the presence of other electrons, see section 4.11.1, by various relativistic effects (for example by the interaction between the electron spin and the magnetic field the electron experiences when orbiting the nucleus, the so-called spin-orbit coupling mentioned in section 4.11.3) or by the action of external fields, see sections 4.11.2 and 4.11.3.

---

[32] If the series is not finite it behaves for large j as $c_{j+1} = 2\lambda_n c_j/(j+1)$, i.e. it diverges as $\exp(2\lambda_n r)$, see Appendix A.3.1, resulting in an unphysical divergence of $R_{nl}(r)$.

4.9 ◆ MANY ELECTRON SYSTEMS

Until now we have mainly dealt with the problem of finding the energy level of a QM system consisting of a single electron (hoping between atoms, confined in a box or orbiting the nucleus). When studying a QM system consisting of two (or more) electrons one must consider their interaction (which can be treated using perturbation theory, section 4.11) and their identity. In classical mechanics while two particles (e.g. billiard balls) may look similar they are never identical: they can be identified from their different trajectories. In QM not only do all electrons "look" similar (for example all Hydrogen atoms have the same spectrum) but they cannot be told apart: they are identical. As we shall now see, this identity of QM particles has profound consequences on the properties of matter.

Consider the Hamiltonian of a two particle system $\hat{H}(1,2)$. Because of the identity of the particles the Hamiltonian is unchanged upon a permutation of the particles: $\hat{H}(1,2) = \hat{H}(2,1)$. Hence the permutation operator $\hat{P}$ commutes with the Hamiltonian:

$$\hat{P}\hat{H}(1,2)|1,2\rangle = \hat{H}(2,1)|2,1\rangle = \hat{H}(1,2)\hat{P}|1,2\rangle$$

and is a constant of motion. Two successive permutations brings us back to the initial state: $\hat{P}^2 = 1$ which implies that the eigenvalues of the permutation operator are ± 1: $|2,1\rangle = P|1,2\rangle = \pm|1,2\rangle$. These eigenvalues are constants of motion, namely the wavefunction of a many particles system is either even or odd under permutation of its (identical) particles and remains so for ever. Like spin this is a QM property of the particle. Particles known as fermions (electrons, protons, neutrons, etc.) are odd under permutations: $|2,1\rangle = -|1,2\rangle$, while particles which are even under permutation $|2,1\rangle = |1,2\rangle$ are known as bosons (e.g. photons). In 1940 Wolfgang Pauli, requiring QM to be invariant under a change of reference frames (relativistic invariance), related this permutation property to the spin of the particle: particles with half integer spin are fermions while particles with integer spin are bosons.

If the particles can occupy states $|\psi\rangle$ and $|\phi\rangle$, then for two bosons the joint wavefunction $|1,2\rangle$ (symmetric under exchange of the particles) is:

$$|1,2\rangle = \frac{|\psi(1)\rangle|\phi(2)\rangle + |\psi(2)\rangle|\phi(1)\rangle}{\sqrt{2}} = |2,1\rangle$$

For two fermions the joint wavefunction is:

$$|1,2\rangle = \frac{|\psi(1)\rangle|\phi(2)\rangle - |\psi(2)\rangle|\phi(1)\rangle}{\sqrt{2}} = -|2,1\rangle$$

which is an eigenstate of the permutation operator P with eigenvalue -1: $|2,1\rangle = P|1,2\rangle = -|1,2\rangle$. This wavefunction implies that in contrast to bosons, two fermions cannot share the same state: $|\psi\rangle \neq |\phi\rangle$. If their spatial wavefunction is identical then their spin-state must differ and if their spin-state is identical then their spatial wavefunction must differ. This extremely important property of fermions is known as Pauli's exclusion principle. It implies that the lowest energy level of an electronic system can be occupied by at most two electrons (with opposite spins). Contrast that

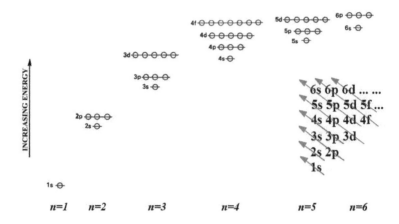

Figure 4.16 The orbital energies of many-electron atoms. The energy levels are ordered along the abscissa according to the main shell number n, which is also indicative of the size of the atom (the spread of the radial wavefunction increases roughly as nr_0). Notice that levels with different angular quantum number l (different colors) have different energies (see section 4.11.1), though they are still degenerate with respect to the azimuthal quantum number m. The number of circles indicate the level of degeneracy: $2l + 1$. Notice that in absence of an external field there is no preferred axis that would break that azimuthal symmetry. To minimize the energy electrons fill these levels from the bottom up while satisfying Pauli's exclusion principle: no more than two electrons (with opposite spin) per energy eigenstate. The first level to be filled is $1s$, then $2s$, $2p$, etc... Notice that the more outer orbitals $4s$ having lower energy is filled before $3d$; $5s$ before $4d$ and $6s$ before $4f$ and $5d$ following the empirical rule known as Madelung rule (see inset) which claims that states fill up following the sum of their quantum numbers $n + l$.

with the situation for bosons where the lowest energy state can be occupied by an arbitrary number of particles (for example photons of same energy as in Planck's black body radiation).

4.9.1 ◆ THE PERIODIC TABLE OF THE ELEMENTS

Pauli's exclusion principle and the solution of Schrödinger's equation for the hydrogen atom allows for a classifications of atoms according to the degree to which their orbitals are successively filled by the orbiting electrons: the so-called Aufbau principle. As pointed earlier due to electron-electron interactions the near angular degeneracy (neglecting spin-orbit coupling) existing in the hydrogen atom is lifted in multi-electron systems, see section 4.11.1. For the same value of the shell number n, orbitals with smaller values of l (more symmetric angular wavefunctions) have lower energy, see Fig.4.16. These energies cannot be calculated analytically, but numerical schemes have been developed to estimate them with great accuracy, see section 4.11 below. The orbitals are then filled from the bottom up with at most two electrons of opposite spins starting from the lowest energy level. An empirical rule due to Madelung summarizes the sequence in which orbitals are filled up by noticing that the successive orbital occupation correlates with the sum of the orbital quantum numbers $n + l$. Hence orbital $4s$ fills up before $3d$ and orbital $5p$ before $4f$, etc., see Fig.4.16. The agreement between the periodic table organized by Mendeleev according to the chemical properties of the elements and the filling up of the energy levels (with elements in the same column sharing the same number of outer shell electrons, also known as valence electrons) is one of the greatest success of Quantum Mechanics, see Fig.4.17. The universality of the organization of the elements in that periodic table was suggested as a possible Rosetta stone for the deciphering of advanced alien civilizations (see "Omnilingual" by H.Beam Piper[65]).

Hydrogen (H) has one electron in the lowest orbital: $1s$. Its spin $S = 1/2$ imply that there are $2s + 1 = 2$ possible states in a s−orbital. In terms of Dirac's eigenstates $|j, l, s, m_J>$, see Appendix A.8.3.1 with total angular momentum $j = l + 1/2 = 1/2$ that state can also be designated in so-called spectroscopic notation as: $^{2S+1}L_J = {}^2S_{1/2}$.

Helium (He) has two electrons occupying the lowest orbital (written as $(1s)^2$) with anti-parallel spin, i.e with total spin $S = 0$. The spectroscopic notation of that state is $^{2S+1}L_J = {}^1S_0$.

Lithium (Li) has three electrons, two occupying with opposite spin the lowest orbital $(1s)^2$ and one electron occupying the next outer orbital: $2s$, hence its electron occupancy is : $(1s)^2(2s)$. Since the chemical properties of atoms are determined by their outer-shell electrons (so called valence electrons which can engage in bonding interactions, see section 4.10), Lithium with one valence electron in the same s−state as the electron in Hydrogen share similar chemical properties with Hydrogen, as indeed all the alkalines (the elements in the first column of Mendeleev's table, see Fig.4.17: Sodium (Na), Potassium (K), Rubidium (Rb), Cesium (Cs), Francium (Fr)) which all have a single electron in an outer s−state. In spectroscopic notation the state of the electron in the unfilled shell is: $^{2S+1}L_J = {}^2S_{1/2}$, which is the same as Hydrogen.

Beryllium (Be) has four electrons, two occupying with opposite spin the lowest orbital $(1s)^2$ and two electrons occupying the next orbitals $(2s)$ with opposite spin. Its electron occupancy is thus: $(1s)^2(2s)^2$ which is similar to Helium. Beryllium is however more reactive than Helium since its $2p$ orbital has an energy only slightly higher than $2s$ which result in the electron easily hoping to the $2p$ orbital, with resulting occupancy: $(1s)^2(2s)(2p)$. Its outer shell electron occupancy is identical to that of Magnesium (Mg), Calcium (Ca), Strontium (Sr), Barium (Ba) and Radium (Ra) which have similar chemical properties.

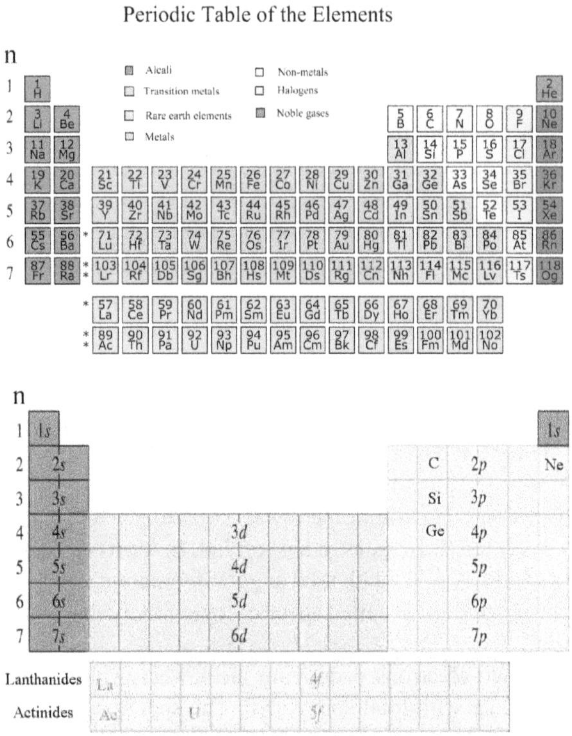

Periodic Table of the Elements

Figure 4.17 The periodic table of the elements. The periodic table (top) devised by Mendeleev and ordered according to increasing atomic number (number of electrons) and similar chemical properties. In comparison (bottom) the table is redrawn as deduced from the filling up of the orbitals from the bottom up, taking into account Pauli's exclusion principle and ordering the columns according to the number of valence (outer shell) electrons: the two $(l = 0)$ s-orbitals in violet; the six $(l = 1)$ p-orbitals in yellow; the 10 $(l = 2)$ d-orbitals in blue; the 14 $(l = 3)$ f-orbitals in green. The agreement between the two is one of the greatest success of QM.

Boron (B) has 5 electrons, two in orbital $1s$, two in orbital $2s$ and one in orbital $2p$: $(1s)^2(2s)^2(2p)$. Its outer shell occupancy and chemical properties are similar to that of Aluminium (Al), Gallium (Ga), Indium (In), and Thalium (Tl). These elements are often used as p-type dopants (electron acceptors) in semiconductors, see section 5.9.2.2. The value of the total angular momentum is $j = l - s = 1/2$. Its outer electronic state is thus $^2P_{1/2}$.

Carbon (C) has 6 electrons, two in orbital $1s$, two in orbital $2s$ and two in orbital $2p$: $(1s)^2(2s)^2(2p)^2$. Since orbital $2p$ is triply degenerate, the question arises as to where the two electrons in $2p$ settle. Because of Pauli's principle, electrons with identical wavefunction (i.e. orbitals) have opposite spin, while electrons in different orbitals may have similar spin. Since electrons in different orbitals overlap less and thus repel each other less, that situation is energetically favorable giving rise to Hund's rule: "the state of highest spin has the lowest energy". Hence in Carbon the electrons in the $2p$ orbitals have similar (parallel) spin and occupy orbitals with different azimuthal number m. In fact, it is energetically favorable for one of the electrons in the $2s$ orbital to occupy one of the empty $2p$-orbitals, giving rise to a electron configuration known as sp_3 where the four orbitals are occupied with 4 electrons (each characterized by a different wavefunction): $(1s)^2(2s)(2p)^3$. The outer shell properties of Carbon and its chemical properties are similar to Silicon (Si) and Germanium (Ge), i.e. semiconductors. In the same column, Tin (Sn) and Lead (Pb) have similar outer shells, but smaller band gaps (see section 4.3.4) and are in fact conductors at room temperatures.

Nitrogen (N) has 7 electrons, two in orbital $1s$, two in orbital $2s$ and three in orbital $2p$: $(1s)^2(2s)^2(2p)^3$. These outer electrons have parallel spin because of Hund's rule and occupy the three $2p$ orbitals with different azimuthal number m. The outer shell occupancy and chemical properties of Nitrogen are similar to that of Phosphorus (P), Arsenic (As), Antimony (Sb) and Bismuth (Bi). These elements are often used as n-type dopants (electron donors) in Si and Ge semiconductors, see section 5.9.2.2.

Oxygen (O) has 8 electrons, two in $1s$, two in $2s$ and four in $2p$: $(1s)^2(2s)^2(2p)^4$. Two of those are paired (they share the same orbital with opposite spin) and two have parallel spins and share orbitals with different m. Its outer shell electron occupancy is identical to that of Sulphur (S), Selenium (Se), Tellurium (Te) and Polonium (Po).

Fluorine (F) has nine electrons, the last five of which occupy the three $2p$ orbitals, four are paired (they share the same orbital with opposite spin) and one occupies the third orbital: $(1s)^2(2s)^2(2p)^5$. The outer shell occupancy of Fluorine is similar to that of other Halogens: Chlorine (Cl), Bromine (Br), Iodine (I) and Astatine (At).

Neon (Ne) has ten electrons which occupy all the $1s$, $2s$ and $2p$ orbitals: $(1s)^2(2s^2)(2p)^6$. All electrons are paired and therefore not disponible for bonding (see section 4.10 next). Neon like Helium and the other Noble gases: Argon(Ar), Krypton (Kr), Xenon (Xe), Radon (Rn) have a similar outer electron shell and share a very low chemical reactivity.

An interesting situation occurs when orbitals $4s$ and $5s$ fill up (for elements Calcium (Ca) and Strontium (Sr)). The next ten elements have their electrons fill the more inner $3d$ or $4d$ shells. These elements are known as transition metals. Similarly when orbitals $6s$ fill up (for Barium (Ba)), the next shell to fill up is the inner $4f$ shells which defines a set of 14 rare-earth elements (Lanthanides) with similar chemical properties. Once this shell is filled the next one, $5d$ is also an inner shell and the next ten elements share similar properties with transition metals.

4.10   THE CHEMICAL BOND

The role that the Hydrogen atom has played in the understanding of the periodic table of the elements has been played by the Hydrogen molecule ion H_2^+ in understanding the chemical bond. Setting the protons at positions $\pm \vec{d} = \pm a\hat{z}$, the problem is to solve Schrödinger's equation for a single electron:

$$\hat{H}\Psi(\vec{x},t) = -\frac{\hbar^2}{2m_e}\nabla^2\Psi(\vec{x},t) - \left(\frac{e^2}{|\vec{x}+\vec{d}|} + \frac{e^2}{|\vec{x}-\vec{d}|} - \frac{e^2}{2a}\right)\Psi(\vec{x},t) = E_n\Psi(\vec{x},t) \qquad (4.88)$$

The constant term $-e^2/2a$ accounts for the repulsion between the nuclei (protons). This equation can be solved by separation of variables in prolate spheroidal coordinates ($\mu \geq 0; 0 \leq \nu \leq \pi; 0 \leq \phi \leq 2\pi$):

$$x = a\sinh\mu\sin\nu\cos\phi$$
$$y = a\sinh\mu\sin\nu\sin\phi$$
$$z = a\cosh\mu\cos\nu$$

Looking for a solution $\Psi(\mu,\nu,\phi) = M(\mu)N(\nu)\Phi(\phi))$ (as was done for the hydrogen atom in spherical coordinates) and noticing that : $|\vec{x}\pm\vec{d}| = \sqrt{x^2+y^2+(z\pm a)^2} = a(\cosh\mu\pm\cos\nu)$, we leave it as an exercise to the reader[33] to show that Eq.4.88 then yields the following equations:

$$\frac{\partial^2\Phi}{\partial\phi^2} + m^2\Phi = 0$$

$$\frac{\partial^2 M}{\partial\mu^2} + \coth\mu\frac{\partial M}{\partial\mu} + (2\alpha\cosh\mu - \frac{m^2}{\sinh^2\mu} + E_n'\sinh^2\mu - \lambda)M = 0$$

$$\frac{\partial^2 N}{\partial\nu^2} + \cot\nu\frac{\partial N}{\partial\nu} + (-\frac{m^2}{\sin^2\nu} + E_n'\sin^2\nu + \lambda)N = 0 \qquad (4.89)$$

Where $\alpha \equiv 2a/r_0$ (where $r_0 = \hbar^2/m_e e^2 = 0.53$Å is the Bohr radius of Hydrogen), $E_n' \equiv (2m_e a^2/\hbar^2)(E_n - e^2/2a)$ and $m = 0, \pm 1, \pm 2,...$ and λ are quantum numbers (playing the role that m, l played in the hydrogen atom). The energy levels E_n (with $m = 0$) have been solved in 1927 (only one year after Schrödinger solved the Hydrogen atom) by O.Burrau who then looked for the distance $2a$ between the atoms that would minimize the molecule ground state energy E_0. He found that energy to be minimal with value: $E_0^{min} = -16.4$ eV when the distance between the atoms is twice their Bohr

---

[33] Hint:

$$\nabla^2\Psi = \frac{1}{a^2(\sinh^2\mu + \sin^2\nu)}[\frac{\partial^2\Psi}{\partial\mu^2} + \frac{\partial^2\Psi}{\partial\nu^2} + \coth\mu\frac{\partial\Psi}{\partial\mu} + \cot\nu\frac{\partial\Psi}{\partial\nu}] + \frac{1}{a^2\sinh^2\mu\sin^2\nu}\frac{\partial^2\Psi}{\partial\phi^2}$$

$$\frac{1}{|\vec{x}+\vec{d}|} + \frac{1}{|\vec{x}-\vec{d}|} = \frac{2\cosh\mu}{a(\sinh^2\mu + \sin^2\nu)}$$

radius: $2a = 2r_0$ ($\alpha = 2a/r_0 = 2$), see Fig.4.18. This exact solution provided a test-bed for approximation methods that we will present next.

4.10.1 VARIATIONAL APPROACH TO MOLECULAR ENERGY LEVELS

Since most problems in physics are not exactly soluble, as for example the determination of the energy spectra of complex molecules, the existence of an exact solution for H_2^+ was helpful in the development and testing of approximate methods to compute eigen-energies and eigenstates (orbitals) for more complex QM systems. One of these approximations is known as the variational method (for details see chap.7 in L.P.Pauling and E.B.Wilson, "Introduction to quantum mechanics"[42]). The idea is pretty simple: one assumes that in some limit a given problem can be solved exactly, i.e. a set of eigenstates can be computed for the Hamiltonian in that limit. One then uses a linear superposition of these states as a trial function and one estimates the energy of that state for the complete Hamiltonian. This energy is then minimized with respect to the parameters of the superposition (e.g. the coefficients of the linear superposition). Since one does not expect the trial function to yield the exact solution of the full problem, the estimated minimal energy E_{min} is an upper bound to the true ground state energy of the system: $E_{min} > E_0$. In general the more complete the set of functions used as a trial, the better the approximation. However, the main problem with the variational approach is that there is no way of estimating how good an approximation it is. For that one needs to resort to perturbation expansions, which we will study later.

In the case of the hydrogen molecule ion H_2^+ its Hamiltonian can be written as[66, 43]:

$$\hat{H} = \hat{H}_H(r_1) - e^2/r_2 + e^2/2a = \hat{H}_H(r_2) - e^2/r_1 + e^2/2a$$

where $\hat{H}_H$ is the Hamiltonian of the Hydrogen atom and $\vec{r}_1 = \vec{r} - \vec{a}$; $\vec{r}_2 = \vec{r} + \vec{a}$. When the two protons are far apart ($a \gg r_0$) the solutions for the electron wavefunction are simply those computed for a hydrogen atom centered on either proton, i.e. $|\Psi_{nlm}(r_1)\rangle$ or $|\Psi_{nlm}(r_2)\rangle$. When the atoms are brought in proximity, we shall use for simplicity a trial wavefunction $|\Psi_T\rangle$ which is a superposition of hydrogen $1s$ eigenstates (with energy $E_1 = -e^2/2r_0 = -13.6\text{eV}$) : $|\Psi(r)\rangle = R_{10}(r)Y_{00}(\theta,\phi) = \exp(-r/r_0)/\sqrt{\pi r_0^3}$:

$$|\Psi_T\rangle = c_1|\Psi(r_1)\rangle + c_2|\Psi(r_2)\rangle$$

Where the coefficients c_1, c_2 are parameters to be determined such as to minimize the energy E of that trial wavefunction[66, 42, 43]. Inserting that ansatz into the full Schrödinger equation yields:

$$\langle \Psi_T|\hat{H}|\Psi_T\rangle = E\langle \Psi_T|\Psi_T\rangle$$

which —defining $H_{11} = \langle\Psi(r_1)|\hat{H}|\Psi(r_1)\rangle$; $H_{12} = \langle\Psi(r_1)|\hat{H}|\Psi(r_2)\rangle = H_{21}$; $H_{22} = \langle\Psi(r_2)|\hat{H}|\Psi(r_2)\rangle$ and the overlap function: $\Delta \equiv \langle\Psi(r_2)|\Psi(r_1)\rangle$ —is equivalent to:

$$c_1^2(H_{11} - E) + 2c_1c_2(H_{12} - \Delta E) + c_2^2(H_{22} - E) = 0 \tag{4.90}$$

We now have to determine the coefficients c_1, c_2 that minimize E, i.e. for which $\partial E/\partial c_1 = \partial E/\partial c_2 = 0$. Taking the derivative of the previous equation with respect to c_1 and c_2 yields at the minimum of the energy a set of coupled homogeneous equations:

$$(H_{11} - E)c_1 + (H_{12} - \Delta E)c_2 = 0$$
$$(H_{12} - \Delta E)c_1 + (H_{22} - E)c_2 = 0$$

Which has a solution when the determinant:

$$\begin{vmatrix} H_{11} - E & H_{12} - \Delta E \\ H_{12} - \Delta E & H_{22} - E \end{vmatrix} = 0$$

Since $H_{11} = H_{22}$ we obtain:

$$E_s = \frac{H_{11} + H_{12}}{1 + \Delta}$$
$$E_a = \frac{H_{11} - H_{12}}{1 - \Delta}$$

The first solution corresponds to a symmetric trial wavefunction: $c_1 = c_2$:

$$|\Psi_s\rangle = \frac{1}{\sqrt{2 + 2\Delta}}(|\Psi(r_1)\rangle + |\Psi(r_2)\rangle)$$

The second solution corresponds to an anti-symmetric wavefunction: $c_2 = -c_1$:

$$|\Psi_a\rangle = \frac{1}{\sqrt{2 - 2\Delta}}(|\Psi(r_1)\rangle - |\Psi(r_2)\rangle)$$

The various integrals can then be evaluated numerically or in this case even exactly[34] to yield (with $\alpha = 2a/r_0$):

$$H_{11} = -\frac{e^2}{2r_0} + \frac{e^2}{2a} + J < 0$$
$$H_{12} = \left(-\frac{e^2}{2r_0} + \frac{e^2}{2a}\right)\Delta + K < 0$$

where J known as the Coulomb integral, K the resonance integral, and Δ the overlap integral are:

$$J = \int d^3 r \Psi(r_1)(-e^2/r_2)\Psi(r_1) = -\frac{e^2}{2a}\left[1 - e^{-2\alpha}(\alpha + 1)\right] < 0$$
$$K = \int d^3 r \Psi(r_1)(-e^2/r_2)\Psi(r_2) = -\frac{e^2}{r_0}(1 + \alpha)e^{-\alpha} < 0$$
$$\Delta = \int d^3 r \Psi(r_1)\Psi(r_2) = \left[1 + \alpha + \frac{\alpha^2}{3}\right]e^{-\alpha} > 0$$

---

[34] The integrals are evaluated in prolate spheroidal coordinates with $|\vec{r} \pm \vec{d}| = a(\cosh\mu \pm \cos\nu)$ and the volume element: $d^3 r = a^3 \sinh\mu \sin\nu(\sinh^2\mu + \sin^2\nu)\, d\mu\, d\nu\, d\phi$

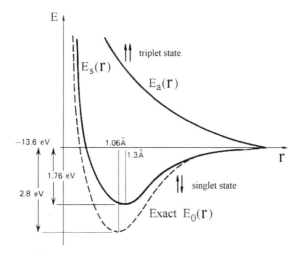

Figure 4.18   The ground state energy level of the Hydrogen molecule ion H_2^+ as a function of the distance $R = 2a$ between the two protons. The exact solution $E_0(R)$ is shown in dashed and compared to the results obtained from the variational method for a symmetric and an anti-symmetric electronic wavefunction. Only the symmetric solution describes a bound state (energy E_s smaller than the hydrogen ground state energy E_1). For two-electrons the bound state is a spin-singlet state (anti-symmetric spin wavefunction). Adapted with permission from ref.[66] copyright (1928) by the American Chemical Society.

The energies of the symmetric and antisymmetric wavefunctions can then be re-cast as:

$$E_s = -\frac{e^2}{2r_0} + \frac{e^2}{2a} + \frac{J+K}{1+\Delta} \tag{4.91}$$

$$E_a = -\frac{e^2}{2r_0} + \frac{e^2}{2a} + \frac{J-K}{1-\Delta} \tag{4.92}$$

The first term on the right is the ground state ($n = 1$) energy of an electron in a Hydrogen atom. The second term is the Coulomb repulsive energy between the Hydrogen nuclei (protons). The third term is the contribution from electron sharing between the two protons. Only the symmetric solution has a minimal energy $E_s = -15.36\text{eV}$ lower than the Hydrogen ground state: $E_1 = -e^2/2r_0 = -13.6\text{eV}$ at an internuclear distance $2a \sim 1.3\text{Å}$. The anti-symmetric solution has an energy E_a always larger than E_1 and thus cannot represent a bound state of the H_2^+ ion. Comparing these values to the exact solution ($E_0 = -16.4\text{eV}$ at $2a = 1.06\text{Å}$), see Fig.4.18, suggests that in this case the simple variational approach yields a reasonable approximation to the ground state of the system.

Molecular Orbitals. Just as we constructed the atomic orbitals, by filling up the (perturbed) Hydrogen orbitals from the bottom up taking into consideration Pauli's exclusion principle we may arrange the two electrons in the H_2 molecule by filling up the molecular orbitals just determined taking into account Pauli's principle. One might worry that the electron-electron repulsion $(e^2/|\vec{r}_1 - \vec{r}_2|)$ may not be negligible in comparison to the Coulomb or exchange energies (J, K) introduced above. However because the electron charges are spread over a region of size $\sim r_0$, their repulsive interaction turns out to alter only slightly the ground state energy which is still stable for the symmetric wavefunction and unstable for the antisymmetric one. The two electrons in H_2 thus share the same ground state molecular orbital with opposite spin (they are paired). This is a general result: **a chemical bond between two elements is formed by the sharing of an electron pair with opposite spin (a spin singlet state) whose spatial wavefunction is symmetric with respect to particle exchange** (to satisfy Pauli's exclusion principle).

4.10.1.1 Hückel's Molecular Orbital Theory

The variational approach used to estimate the ground state of H_2^+ has been used by Eric Hückel in 1930 to estimate the eigen-energies and orbitals of hydro-carbon chains and rings of n carbon atoms. As we shall now show this Hückel Molecular Orbital (HMO) theory justifies *a-posteriori* our initial approach to multi-state systems (the ammonia and aromatic molecules discussed in section 4.3).

Hückel's idea was to use a trial molecular wavefunction $|\Psi_{MO}\rangle$ which is a superposition of atomic wavefunctions $|\phi\rangle$ (p-orbitals) centered on the individual Carbon atoms ($i = 1, 2, \cdots, n$):

$$|\Psi_{MO}\rangle = \sum_{i=1}^{n} c_i |\phi_i\rangle$$

Just as we have done for H_2^+ the idea is to determine the coefficients $\{c_i\}$ that minimize the energy of the full Hamiltonian:

$$\langle \Psi_{MO}|\hat{H}|\Psi_{MO}\rangle = E\langle \Psi_{MO}|\Psi_{MO}\rangle$$

Neglecting the direct overlap between neighboring atomic orbitals, i.e. assuming: $\Delta = \langle \phi_i|\phi_j\rangle \approx 0$ one gets:

$$\sum_{ij} c_i c_j H_{ij} - E c_i^2 \delta_{ij} = 0$$

Where $H_{ij} = \langle \phi_i|\hat{H}|\phi_j\rangle$. H_{ii} and H_{ij} are respectively the Coulomb and resonance integrals we saw above. If only the overlap between nearest neighbors $j = i \pm 1$ contributes significantly to the resonance integral, minimization of the energy with respect to the vector of coefficients $\{c_i\}$ yields (for a n-atoms linear carbon chain):

$$\begin{pmatrix} E_0 & -A & 0 & \cdots & 0 \\ -A & E_0 & -A & \cdots & 0 \\ \vdots & & & & \vdots \\ 0 & \cdots & 0 & -A & E_0 \end{pmatrix} \begin{pmatrix} c_1 \\ \vdots \\ c_n \end{pmatrix} = E \begin{pmatrix} c_1 \\ \vdots \\ c_n \end{pmatrix} \tag{4.93}$$

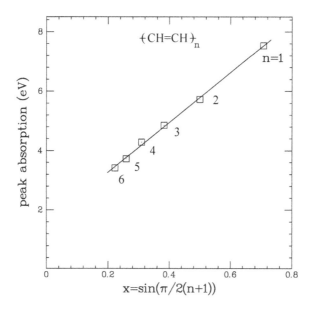

Figure 4.19 The band gap ΔE of a few alkenes with $n = 1,...6$ measured by the absorption of photons of energy $h\nu = \Delta E$. Notice that the measured energies vary linearly with $\sin \pi/2(n+1)$ as predicted from Hückel's molecular orbital theory. The fact that the gap doesn't go to zero as $n \rightarrow \infty$, is due to finite size effects, i.e. slight variation of the overlap integral A with n [67].

where we defined: $E_0 \equiv H_{ii}$ and $-A \equiv H_{i,i+1} = H_{i-1,i}$. Notice the similarity with the model studied earlier for aromatic molecules (circular carbon chains), Eq.4.34. The eigen-energies can then be computed as follows:

$$\Delta_n = \begin{vmatrix} 2\cos\theta & 1 & 0 & \cdots & 0 \\ 1 & 2\cos\theta & 1 & \cdots & 0 \\ \vdots & & & & \\ 0 & & \cdots & 0 & 1 & 2\cos\theta \end{vmatrix} = 0$$

with $2\cos\theta \equiv (E - E_0)/A$. Computing the determinant Δ_n yields a recursion relation: $\Delta_n = 2\cos\theta\Delta_{n-1} - \Delta_{n-2}$ with $\Delta_0 = 1$ and $\Delta_1 = 2\cos\theta$. Its solution is $\Delta_n = \sin(n+1)\theta/\sin\theta$ and the eigenvalues are consequently determined by: $\Delta_n - 0$, i.e. $\theta = m\pi/(n+1)$ $(m = 1, 2,...,n)$:

$$E = E_0 + 2A\cos\frac{m\pi}{n+1} = E_0 - 2A\cos\frac{l\pi}{n+1} \tag{4.94}$$

with $l = n + 1 - m$, compare with Eq.4.36. The filling up of these molecular orbitals is similar to the filling up of atomic orbitals, each eigenstate with eigenvalue E_l accommodating an electron pair with opposite spin, contributed by neighboring carbon atoms. As each energy level is occupied by two electrons, the energy gap ΔE between the Highest Occupied Molecular Orbital (HOMO, the top of the valence band) with $l_{homo} = n/2$ (or $(n+1)/2$ for odd n) and the Lowest Unoccupied Molecular Orbital (LUMO, bottom of the conduction band) with $l_{lumo} = l_{homo} + 1$ is:

$$\Delta E = 4A \sin \pi/2(n+1)$$

The energy gap of alkene chains of different lengths (i.e. different n) has been measured[68] and compares well with this prediction, see Fig.4.19. The small gap in long alkene chains such as poly-acetylene and progress in their doping (with electron donors or acceptors) have led to the development of conducting polymers for which Alan J. Heeger, Alan MacDiarmid and Hideki Shirakawa were awarded the Nobel prize in 2000. These conducting polymers are the basic blocks of Organic Light Emitting Diodes (OLED), a technology used in flexible screen displays.

4.10.2 MOLECULAR ROTATIONAL SPECTRUM

Consider a diatomic molecule consisting of two atoms of mass m_1, m_2 (as for example in the molecule CO) at a distance $\vec{r}_{12} = \vec{r}_1 - \vec{r}_2$ (where $\vec{r}_1, \vec{r}_2$ are the atoms coordinates). The Hamiltonian describing this molecule is

$$H = -\left(\frac{\hbar^2}{2m_1} \frac{\partial^2}{\partial \vec{r}_1^2} + \frac{\hbar^2}{2m_2} \frac{\partial^2}{\partial \vec{r}_2^2} \right)$$

Defining the center of mass (CM) coordinate : $\vec{R}_{CM} = (m_1 \vec{r}_1 + m_2 \vec{r}_2)/(m_1 + m_2)$ allows to rewrite the previous Hamiltonian as: $H = H_{CM} + H_o$, where H_{CM} is the Hamiltonian of the center of mass (describing the motion of a free particle of mass $m_1 + m_2$) and H_o is the Hamiltonian describing the relative motion of the bound atoms in the CM coordinate system[35]. If the distance between the atoms is fixed $|\vec{r}_{12}| = d$, the rotational Hamiltonian is (see Eq.4.80 with fixed r):

$$H_{rot} = H_o \Big|_{r_{12}=d} = -\frac{\hbar^2}{2md^2} \left[\frac{1}{\sin \theta} \frac{\partial}{\partial \theta} (\sin \theta \frac{\partial}{\partial \theta}) + \frac{1}{\sin^2 \theta} \frac{\partial^2}{\partial \phi^2} \right] \qquad (4.95)$$

$$\equiv \frac{\hbar^2 \vec{f}^2}{2I} \qquad (4.96)$$

where $I = m_1 (\vec{r}_1 - \vec{R}_{CM})^2 + m_2 (\vec{r}_2 - \vec{R}_{CM})^2 = m_1 m_2 d^2/(m_1 + m_2) = md^2$ is the moment of inertia of the di-atomic molecule ($m \equiv m_1 m_2/(m_1 + m_2)$ is known as the reduced

---

[35] Using the fact that $\partial \Psi/\partial \vec{r}_i = [m_i/(m_1 + m_2)]\partial \Psi/\partial \vec{R}_{CM} - (-1)^i \partial \Psi/\partial \vec{r}_{12}$ (with $i = 1, 2$) from which one derives the following equality:

$$-\frac{\hbar^2}{2m_1} \frac{\partial^2}{\partial \vec{r}_1^2} - \frac{\hbar^2}{2m_2} \frac{\partial^2}{\partial \vec{r}_2^2} = -\frac{\hbar^2}{2(m_1 + m_2)} \frac{\partial^2}{\partial \vec{R}_{CM}^2} - \frac{\hbar^2(m_1 + m_2)}{2m_1 m_2} \frac{\partial^2}{\partial \vec{r}_{12}^2} = H_{CM} + H_o$$

mass). Notice that the angular momentum operator $\hat{J}^2$ is the same as the angular momentum operator $\hat{L}^2$ for a single particle, section 4.7. The eigen-energies of the rotational Hamiltonian are therefore:

$$E_j = \frac{\hbar^2 j(j+1)}{2I} \quad \text{with:} \quad j = 0, 1, 2, \cdots \tag{4.97}$$

The typical energies associated with molecular rotational levels are on the order of 0.001eV (meV). Transitions between rotational levels corresponds to absorption or emission of microwave radiation. The eigenstates are the spherical harmonics (see appendix A.5.3): $|j,m\rangle = Y_{jm}(\theta, \phi)$ and as in the Hydrogen atom, each eigenstate is $2j+1$-fold degenerate (since $m = -j, -j+1, ..., j$). For more complex molecules with different moments of inertia, the analysis becomes more complicated and will not be presented here.

4.10.3 MOLECULAR VIBRATIONAL SPECTRUM

We have seen that the nuclei of diatomic molecules such as H_2 are stabilized by the exchange of a pair of electrons at the bottom of an effective potential $V(r)$, see Fig.4.18. Near its equilibrium position $\vec{r}_{12} = \vec{d}$ the binding potential $V(r_{12})$ can be expanded in a Taylor series:

$$V(r_{12}) = V(d) - \frac{1}{2} \frac{\partial^2 V}{\partial r_{12}^2}\bigg|_{r_{12}=d} (\vec{r}_{12} - \vec{d})^2 + \ldots$$

Defining $k \equiv \partial^2 V / \partial r_{12}^2 \big|_{r_{12}=d}$ and looking at the deviations from equilibrium: $\vec{r} = \vec{r}_{12} - \vec{d}$ yields the Hamiltonian of a harmonic oscillator:

$$H_{osc} = H_o + V(r_{12}) = -\frac{\hbar^2}{2m}\nabla^2 + \frac{1}{2}kr^2 + V(d) \tag{4.98}$$

The determination of the eigen-energies[36] and eigenstates of a harmonic oscillator turns out to be one of those rare examples of an exactly soluble problem in QM (the hydrogen atom is another). As done earlier, we shall look for a separable solution $\Psi(x, y, z) = X(x)Y(y)Z(z)$ of Schrödinger's equation:

$$H_{osc}\Psi(x, y, z) = E\Psi(x, y, z)$$

Since $r^2 = x^2 + y^2 + z^2$ we obtain:

$$\frac{\partial^2}{\partial x^2}X(x) + (\lambda_x - \alpha^2 x^2)X(x) = 0 \tag{4.99}$$

---

[36] As only energy differences matter we shall set $V(d) = 0$

with $\alpha = m\omega/\hbar$, $\omega = \sqrt{k/m}$ and similar equations for $Y(y)$ and $Z(z)$. The eigen-energy is given by:

$$E = \frac{\hbar^2(\lambda_x + \lambda_y + \lambda_z)}{2m}$$

In the limit $x \to \infty$, Eq.4.10.3, simplifies to $X'' = \alpha^2 x^2 X$, for which the asymptotic bounded solution is $X_\infty \sim \exp(-\alpha x^2/2)$. As we have done for the solution of the radial equation of the Hydrogen atom, let us then look for a solution:

$$X(x) = f(x)\exp(-\alpha x^2/2)$$

where $f(x)$ is a finite polynomial. Inserting that Ansatz into Eq.4.10.3 yields:

$$f'' - 2\alpha x f' + (\lambda_x - \alpha)f = 0$$

Looking for a polynomial solution: $f(x) = \sum_n a_n x^n$ yields the recursion relation:

$$(n+2)(n+1)a_{n+2} + [\lambda_x - (2n+1)\alpha]a_n = 0$$

The polynomial $f(x)$ is finite[37] if $\lambda_x = (2n+1)\alpha$ with $n = 0, 1, 2, ...$ Thus for a one dimensional oscillator the energy levels are:

$$E_n^{1D} = \frac{\hbar^2 \lambda_x}{2m} = \hbar^2(n+1/2)\frac{\alpha}{m} = (n+1/2)\hbar\omega \qquad (4.100)$$

which (except for the $1/2$) is the result obtained using the Bohr-Sommerfeld rule, Eq.4.10. The energy levels in the harmonic oscillator are equally spaced by $\hbar\omega$.

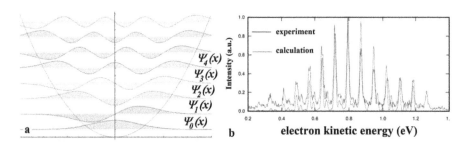

Figure 4.20 (a) The first eigenstates of the harmonic oscillator. Notice the inversion symmetry of the even function ($\Psi_{2n}(-x) = \Psi_{2n}(x)$) and the antisymmetry of the odd ones ($\Psi_{2n+1}(-x) = -\Psi_{2n+1}(x)$). (b) The computed and measured kinetic energies of the photoelectrons (section 4.1.2) emitted from the vibrational levels of $C_6H_4^-$ (data courtesy of A.I.Krylov[69]). Notice the equal spacing of the spectral lines and their value ($\sim 0.1eV$) typical of vibrational energies in molecules.

---

[37] If the polynomial $f(x)$ is not finite, its series expansion behaves for large $n = 2m$ as $a_{2(m+1)} = \alpha a_{2m}/(m+1)$, i.e. it diverges as $\exp\{\alpha x^2\}$ (from the result on the expansion of e^x in Appendix A.3.1) which results in an unphysical divergence of the wavefunction $X(x)$ for large x.

For molecules the spacing is typically $\hbar\omega \sim 0.1\text{eV}$, in the infrared (see Fig.4.20). The polynomials $f(x)$ are known as Hermite polynomials of the variable $\xi = \sqrt{\alpha}x$:

$$H_n(\xi) = (-1)^n e^{\xi^2} \frac{d^n}{d\xi^n}\left(e^{-\xi^2}\right)$$

so that the eigenstates read:

$$X_n(x) = A_n e^{-\alpha x^2/2} H_n(\sqrt{\alpha}x) \qquad \text{with}: \qquad A_n = \frac{(\alpha/\pi)^{1/4}}{\sqrt{2^n n!}} \qquad (4.101)$$

The first Hermite polynomials describing the lowest eigenstates are:

$$H_0(\xi) = 1 \ ; \ H_1(\xi) = 2\xi \ ; \ H_2(\xi) = 4\xi^2 - 2 \ ; \ H_3(\xi) = 8\xi^3 - 12\xi \qquad (4.102)$$

The first eigenstates of the 1D-harmonic oscillator are shown in Fig.4.20. The even functions in n are symmetric with respect to inversion $x \to -x$, whereas the odd functions are antisymmetric. Therefore, for identical di-atoms (such as H_2) the spin of the nuclei in the vibrational ground state $(n = 0)$ must be opposite (by Pauli's exclusion principle).

The 3D-harmonic oscillator can be decomposed into three components with resulting energy levels: $E_{n_1,n_2,n_3} = (n_1 + n_2 + n_3 + 3/2)\hbar\omega$. Note that if the molecule both vibrates and rotates its energy levels are only approximately a sum of rotational and vibrational levels (in fact the system does not separate nicely into pure vibrational and pure rotational modes).

4.11 TIME INDEPENDENT PERTURBATION THEORY

Most of the problems in physics and in QM in particular are not exactly soluble. Therefore the development of methods that yield approximate solutions is of particular importance. We have seen earlier one such method (the variational approach) and have pointed out its main drawback: the absence of a systematic way of estimating the error made. In the following we will study a systematic method to get an approximate solution and to estimate the error made within this approximation. The idea is to express the Hamiltonian H as the sum of a known exactly soluble Hamiltonian H_0 plus a perturbation λH_1, where λ is a small parameter ($\lambda \ll 1$):

$$H = H_0 + \lambda H_1$$

One seeks the eigenstates $|\Psi_n >$ of the full Schrödinger equation:

$$H|\Psi_n\rangle = E_n|\Psi_n\rangle$$

knowing the eigenstates $|\phi_n\rangle$ of the soluble problem:

$$H_0|\phi_n\rangle = \epsilon_n^0|\phi_n\rangle$$

If these eigenstates span the Hilbert space of H (the space of its solutions) we may look for an eigenstate of H as:

$$|\Psi_n\rangle = \sum_m a_{nm}|\phi_m\rangle$$

Inserting that ansatz into the full Schrödinger equation yields:

$$\sum_m a_{nm}H|\phi_m\rangle = \sum_m a_{nm}(H_0 + \lambda H_1)|\phi_m\rangle = \sum_m a_{nm}(\epsilon_m^0 + \lambda H_1)|\phi_m\rangle = \sum_m a_{nm}E_n|\phi_m\rangle$$

multiplying this equation on the left by $\langle\phi_l|$ yields:

$$\sum_m a_{nm}(E_n - \epsilon_m^0)\langle\phi_l|\phi_m\rangle = \lambda\sum_m a_{nm}\langle\phi_l|H_1|\phi_m\rangle$$

Since $\langle\phi_l|\phi_m\rangle = \delta_{lm}$ $\qquad (E_n - \epsilon_l^0)a_{nl} = \lambda\sum_m a_{nm}\langle\phi_l|H_1|\phi_m\rangle$ (4.103)

When $\lambda = 0$: $E_n = \epsilon_n^0$ and $|\Psi_n\rangle = |\phi_n\rangle$, i.e. $a_{nl} = \delta_{nl}$. When $\lambda \ll 1$ we will therefore look for a solution as an expansion in λ:

$$a_{nl} = \delta_{nl} + \lambda a_{nl}^{(1)} + \lambda^2 a_{nl}^{(2)} + \dots$$
$$E_n = \epsilon_n^0 + \lambda E_n^{(1)} + \lambda^2 E_n^{(2)} + \dots$$

Inserting that ansatz into Eq.4.103 yields:

$$(\epsilon_n^0 - \epsilon_l^0 + \lambda E_n^{(1)} + \lambda^2 E_n^{(2)} + \dots)(\delta_{nl} + \lambda a_{nl}^{(1)} + \lambda^2 a_{nl}^{(2)} + \dots)$$
$$= \lambda\sum_m\langle\phi_l|H_1|\phi_m\rangle(\delta_{nm} + \lambda a_{nm}^{(1)} + \lambda^2 a_{nm}^{(2)} + \dots)$$

Equating terms order by order in λ yields:

$$\lambda: \quad E_n^{(1)}\delta_{nl} + (\epsilon_n^0 - \epsilon_l^0)a_{nl}^{(1)} = \langle\phi_l|H_1|\phi_n\rangle$$

$$\lambda^2: \quad E_n^{(2)}\delta_{nl} + E_n^{(1)}a_{nl}^{(1)} + (\epsilon_n^0 - \epsilon_l^0)a_{nl}^{(2)} = \sum_m \langle\phi_l|H_1|\phi_m\rangle a_{nm}^{(1)} \qquad (4.104)$$

$$\vdots$$

At order λ when $n = l$ we obtain the first order correction to the eigen-energy ϵ_n^0:

$$E_n^{(1)} = \langle\phi_n|H_1|\phi_n\rangle \qquad (4.105)$$

When $n \neq l$ we obtain the first order correction to the eigenstate $|\phi_n\rangle$:

$$a_{nl}^{(1)} = \frac{\langle\phi_l|H_1|\phi_n\rangle}{\epsilon_n^0 - \epsilon_l^0} \qquad (4.106)$$

the value of the first order correction amplitude $a_{nn}^{(1)}$ is obtained from the normalization condition $\langle\Psi_n|\Psi_n\rangle = 1$ (to first order in λ). It turns out that $a_{nn}^{(1)}$ is purely imaginary and can be adsorbed into a phase shift of $|\Psi_n\rangle$ and thus chosen to be zero. In any case, we shall be here interested in the corrections to the eigen-energies, not the eigenstates. To second order in λ, Eq.(4.104) yields when $n = l$ the second-order correction to the eigen-energy:

$$E_n^{(2)} + E_n^{(1)}a_{nn}^{(1)} = \sum_m \langle\phi_n|H_1|\phi_m\rangle a_{nm}^{(1)}$$

$$= \sum_{m\neq n} \langle\phi_n|H_1|\phi_m\rangle a_{nm}^{(1)} + \langle\phi_n|H_1|\phi_n\rangle a_{nn}^{(1)}$$

Using the first order results, Eq.4.105 and Eq.4.106 yields:

$$E_n^{(2)} = \sum_{m\neq n} \frac{\langle\phi_n|H_1|\phi_m\rangle\langle\phi_m|H_1|\phi_n\rangle}{\epsilon_n^0 - \epsilon_m^0}$$

$$= \sum_{m\neq n} \frac{|\langle\phi_n|H_1|\phi_m\rangle|^2}{\epsilon_n^0 - \epsilon_m^0} \qquad (4.107)$$

As before the second order equality when $n \neq l$ yields the correction to the eigen-states, which won't interest us here. To summarize, the eigen-energy valid to second order in the perturbation strength λ can be read directly from the matrix representation of the perturbation Hamiltonian: $\lambda H_{1,nm} \equiv \lambda\langle\phi_n|H_1|\phi_m\rangle$.

$$E_n = \epsilon_n^0 + \lambda E_n^{(1)} + \lambda^2 E_n^{(2)} + O(\lambda^3)$$

$$= \epsilon_n^0 + \lambda\langle\phi_n|H_1|\phi_n\rangle + \lambda^2 \sum_{m\neq n} \frac{|\langle\phi_n|H_1|\phi_m\rangle|^2}{\epsilon_n^0 - \epsilon_m^0} + O(\lambda^3) \qquad (4.108)$$

One may compute the higher order terms in this expansion, but usually only the first non-zero contribution is computed. The advantage of this perturbation expansion over the variational approach seen earlier is that the error in a perturbation expansion up to order n is known to be of $O(\lambda^{n+1})$. In the following we shall see how to implement this approach in various cases.

4.11.1   ENERGY LEVELS IN NON-HYDROGEN ATOMS

We have seen that in Hydrogen, the energy levels have an accidental degeneracy in the quantum number l associated with the angular momentum: while the radial wavefunctions R_{nl} depend on the quantum numbers n and l, the associated energy levels depends only on n. To see how this degeneracy is lifted in non-Hydrogen atoms consider the alkalines (the first column in the Periodic Table, Fig.4.17) which have their inner shells filled and one electron on a new shell. For an alkaline with Z protons in its nucleus and $Z-1$ electrons in its inner shells, the electrostatic potential experienced by the outer electron is:

$$V(r) = -\frac{e^2}{r} - \frac{(Z-1)e^2}{r}(1-p(r)) \equiv -\frac{e^2}{r} - \delta V(r)$$

The term $\delta V(r)$ represents the Coulomb interaction of the outer electron with the $Z-1$ partially screened charges of the nucleus; $p(r)$ represent the effective screening of the nuclear charge by the $Z-1$ inner electrons within a sphere of radius r. Close to the nucleus, there is no screening of the nuclear charges $p(0) = 0$, while far from it there is complete screening by the inner electrons: $\lim_{r\to\infty} p(r) = 1$. To take into account that screening effect we shall assume for simplicity that $\delta V = \delta V_0 r_s \exp(-r/r_s)/r$ (with $\delta V_0 = (Z-1)e^2/r_s$), where r_s is a typical distance over which the inner electrons wave-functions are spread. To first order in the perturbation δV, the shift in the energy levels of the outer electron is given by Eq.4.105:

$$\Delta E_{nl}^{(1)} = -\langle \Psi_{nlm}|\delta V|\Psi_{nlm}\rangle$$

Where $|\Psi_{nlm}\rangle$ is the Hydrogen eigenstate of the outer electron, Eq.4.87. Since δV is maximal at $r = 0$ (the nuclear charge is less screened the closer one gets to the nucleus), the energy shift is larger the greater the probability of finding the electron near the nucleus at $r = 0$, i.e. for the lower l-orbitals. To see how this comes about let us compute the first order perturbation in the outer electron's energy. When solving Schrödinger's equation for the Hydrogen atom, section 4.8, we found that near $r = 0$: $R_{nl} \sim (r/r_n)^l/r_n^{3/2}$ with $r_n = nr_0$. Due to the orthogonality of the angular eigenstates

$Y_{lm}(\theta, \phi)$ the first order correction to the energy becomes[38]:

$$\Delta E_{nl}^{(1)} = -\int_0^\infty dr\, r^2 R_{nl}^2(r)\delta V(r) \approx -\frac{\delta V_0 r_s}{r_n} \int_0^\infty dx\, x^{2l+1} e^{-r_n x/r_s}$$

$$= -\delta V_0 (2l+1)! (r_s/r_n)^{2l+3}$$

Since the inner electrons are spread over a distance $r_s < r_n$ (less than the distance of the outer electron orbit) the negative energy split is maximal for $l = 0$ and decreases as l increases. This explains the lift in the accidental degeneracy of the Hydrogen energy levels in multi-electron atoms which we used when filling-up their electronic orbitals (see Aufbau principle above and Fig.4.17).

4.11.2 THE STARK EFFECT

The Stark effect is the name given to the differential shift in the eigen-energies of a QM system in a constant electric field. This effect has applications in communication and in the measurement of local electric fields, for example across a cell membrane or in deep space.

4.11.2.1 Quantum Wells

Consider first the case of a particle of mass μ and charge e in a box (or quantum well, section 4.2.1) of size a. We have seen that the eigenstates of the particle are $|\psi_n\rangle = \sqrt{2}\sin n\pi x$ (with $0 \le x = z/a \le 1$) with eigen-energies: $E_n^0 = \hbar^2 \pi^2 n^2/2a^2\mu$. In presence of an electric field $\mathcal{E}$ across the box, we need to add a perturbation due to the electric potential $V(x) = e\mathcal{E}ax$ to the Hamiltonian H_0 of the free particle: $H = H_0 + V(x)$. To first order the energy shift, Eq.4.105 is:

$$\Delta E_n^{(1)} = \langle \psi_n | e\mathcal{E}ax | \psi_n \rangle = \frac{e\mathcal{E}a}{2}$$

which results in an inconsequential constant shift of all energy levels. To second order the energy shift is[39], see Eq.4.107:

$$\Delta E_n^{(2)} = \sum_m \frac{|\langle \psi_n | e\mathcal{E}ax | \psi_m \rangle|^2}{E_n^0 - E_m^0}$$

$$= \frac{\mu a^4 e^2 \mathcal{E}^2}{\hbar^2} \sum_m \frac{128 n^2 m^2}{\pi^6 (n^2 - m^2)^5} \quad \text{with}: \quad n \pm m \text{ odd} \qquad (4.109)$$

---

[38]Define $x \equiv r/r_n$ and use:

$$\int_0^\infty x^n e^{-\alpha x} = (-1)^n \frac{d^n}{d\alpha^n} \int_0^\infty e^{-\alpha x} = n! \alpha^{-n-1}$$

[39]Using the identity:

$$2\int_0^1 dx\, x\sin n\pi x \sin m\pi x = \int_0^1 dx\, x[\cos(n-m)\pi x - \cos(n+m)\pi x] = -\frac{8nm}{\pi^2(n^2-m^2)^2}$$

if $n \pm m$ is odd and zero otherwise.

For a typical quantum well of size $a = 10$nm in a field $\mathcal{E} = 3 \ 10^6$ V/m $= 10^2$ statV/cm (i.e. a voltage across the well of 30 mV), the Stark shift in the ground state energy E_1^0 is $\Delta E_1^{(2)} \sim -0.41 \ 10^{-14}$ erg (neglecting contributions of $m > 2$ in Eq.4.109), a magnitude similar to the unperturbed ground state energy $E_1^0 \sim 0.6 \ 10^{-14}$ erg, section 4.2.1. The energy spectrum of a quantum well can thus be significantly affected by the voltage across the well which allows for a strong modulation of the output intensity of near infrared laser diodes, a very useful feature for optical communication.

Notice that the sum in Eq.4.109 is well defined only if the denominator is nonzero ($E_n^0 \neq E_m^0$), namely if the eigenstates are non-degenerate. That condition is not fulfilled in Hydrogen where electronic eigenstates with different quantum number l have the same energy. The investigation of the Stark effect in Hydrogen necessitates a perturbation analysis specific for degenerate eigenstates that is presented in Appendix A.8.4.

4.11.2.2 Diatomic Molecules

Next, let us consider a diatomic molecule with dipole moment $\vec{\mu}$ in an electric field $\vec{\mathcal{E}}$ defining the z-axis. To the unperturbed rotational Hamiltonian H_{rot} of the diatomic molecule with eigen-energy $E_j^0 = \hbar^2 j(j+1)/2I$, section 4.10.2 one needs to add the perturbation due to the alignment of the molecular dipole with the electric field: $H = H_{rot} + \mu \mathcal{E} \cos \theta$.

From the eigenstates of the rotational Hamiltonian $|j,m\rangle = Y_{jm}(\theta,\phi)$, using the mathematical identities:

$$\cos \theta \, P_j^m(\cos \theta) = \frac{j+1-m}{2j+1} P_{j+1}^m(\cos \theta) + \frac{j+m}{2j+1} P_{j-1}^m(\cos \theta)$$

$$\cos \theta \, Y_{jm}(\theta,\phi) = \sqrt{\frac{(j+1+m)(j+1-m)}{(2j+3)(2j+1)}} Y_{j+1,m}(\theta,\phi) + \sqrt{\frac{(j+m)(j-m)}{(2j+1)(2j-1)}} Y_{j-1,m}(\theta,\phi)$$

$$\equiv C_{j+1,m} Y_{j+1,m}(\theta,\phi) + C_{j-1,m} Y_{j-1,m}(\theta,\phi) \tag{4.110}$$

one can show that the first order energy shift is null:

$$\Delta E_j^{(1)} = \langle j,m|\mu \mathcal{E} \cos \theta|j,m\rangle = \int d\Omega Y_{jm}^*(\theta,\phi)\mu \mathcal{E} \cos \theta Y_{jm}(\theta,\phi)$$

$$= \mu \mathcal{E} \int d\Omega Y_{jm}^*(\theta,\phi)[C_{j+1,m} Y_{j+1,m}(\theta,\phi) + C_{j-1,m} Y_{j-1,m}(\theta,\phi)] = 0$$

where we used the orthogonality of the spherical harmonics (see appendix A.5.3).

The second order shift in the energy, Eq.4.107 is then:

$$\Delta E_j^{(2)} = \sum_{j',m'} \frac{|\langle j,m|\mu \mathcal{E}\cos\theta|j',m'\rangle|^2}{E_j^0 - E_{j'}^0}$$

$$= \sum_{j'=j\pm 1} \frac{\mu^2 \mathcal{E}^2 C_{j',m}^2}{E_j^0 - E_{j'}^0}$$

$$= -\frac{\mu^2 \mathcal{E}^2 I}{\hbar^2}\left[\frac{(j+1)^2 - m^2}{(j+1)(2j+3)(2j+1)} - \frac{j^2 - m^2}{j(2j+1)(2j-1)}\right] \tag{4.111}$$

Notice that the Stark shift in energy varies with both j and $|m|$, see Fig.4.21. For the ground state $j = m = 0$ the shift is $\Delta E_0^{(2)} \equiv \hbar\Delta\omega = -\mu^2\mathcal{E}^2 I/3\hbar^2$: it increases (in absolute value) as the square of the electric field as indeed observed.

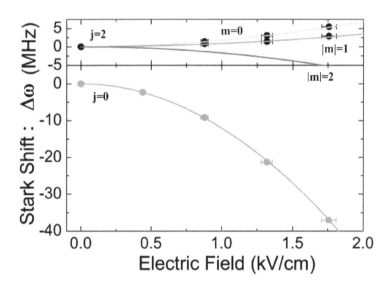

Figure 4.21 The split and shift of the energy $\Delta E_j^{(2)} = \hbar\Delta\omega$ in presence of an electric field of the $j = 0$ and $j = 2$ rotational lines of the diatomic molecule KRb with electric dipole moment $\mu = 0.566$D. Notice the parabolic dependence (different for different j and $|m|$, see Eq. 4.111) as a function of electric field. Reprinted from ref.[70] with permission from AAAS.

The Stark shift of atomic absorption or emission lines can be used to measure the local electric field experienced by the molecule. If the molecule is a fluorescent dye the Stark shift will affect the fluorescent emission of the molecule which can be used to deduce the local electric field or if the dye is embedded in a biological membrane the voltage across the membrane. Such dyes are used to monitor the activity of neurons in live animals.

4.11.3 THE ZEEMAN EFFECT

The Zeeman effect is the name given to the differential shift in the eigen-energies of a QM system in a constant magnetic field. This effect has applications in magnetic resonance imaging (MRI) and in the measurement of local magnetic fields, as for example at the Sun's surface. In the following we shall study first the response of electrons in an atom to the presence of a (strong) magnetic field. We shall then study the behavior in a magnetic field of atomic nuclei which possess a magnetic dipole moment associated with their spin.

To investigate the response of an electron of mass m_e (and charge $q = -e$) in a Hydrogen atom to a constant magnetic field we shall use the low energy limit of Dirac's equation, see Eq.4.76 and Eq.4.77:

$$\left\{ \frac{(\vec{p} + (e/c)\vec{A})^2}{2m_e} - \frac{e^2}{r} + \frac{eg}{2m_e c} \vec{S} \cdot \vec{B} \right\} \Psi = E\Psi \tag{4.112}$$

For magnetic fields below about 1000 Gauss, one needs to consider the so-called spin-orbit coupling between the electron spin and the internal magnetic field it experiences in its frame of reference, see section 3.2.2,:

$$\vec{B}_{int} = -\vec{v} \times \vec{E}/c \sim -\frac{\vec{p} \times \vec{r}}{m_e c} \frac{e}{r^3} = \frac{e}{m_e c r^3} \vec{L} \sim 1000 \ \text{Gauss}$$

In the following we shall assume that the external magnetic field is large enough, so as to allow us to neglect this spin-orbit interaction. In the Coulomb gauge: $\vec{\nabla} \cdot \vec{A} = 0$ (i.e. $\vec{A} = (\vec{B} \times \vec{r})/2$) Eq.4.112 becomes:

$$\left\{ -\frac{\hbar^2 \nabla^2}{2m_e} - \frac{e^2}{r} + \frac{e}{2m_e c}(\vec{B} \times \vec{r}) \cdot \vec{p} + \frac{e^2}{2m_e c^2}\vec{A}^2 + \frac{eg}{2m_e c}\vec{S} \cdot \vec{B} \right\} \Psi = E\Psi$$

For magnetic fields $B \sim 10^4$ Gauss the term in A^2 of order $(e^2 r_0^2/2m_e c^2)B^2 \sim 10^{-9} \text{eV}$ is negligible in comparison with the terms linear in B (of order (10^{-4}eV). The Hamiltonian can thus be written as:

$$H = H_0 + \frac{e}{2m_e c}(\vec{B} \times \vec{r}) \cdot \vec{p} + \frac{egB}{2m_e c} S_z \equiv H_0 + \delta H$$

where $H_0 = -(\hbar^2/2m_e)\nabla^2 - e^2/r$ is the Hamiltonian in absence of magnetic field, i.e. the Hydrogen atom Hamiltonian. The eigenstates of this Hamiltonian are the Hydrogen atom eigenstates: $|n, l, m, s, s_z\rangle$. Since: $(\vec{B} \times \vec{r}) \cdot \vec{p} = \vec{B} \cdot (\vec{r} \times \vec{p}) = \vec{B} \cdot \vec{L}$, the magnetic perturbation Hamiltonian δH becomes (with the magnetic field B defining the z-axis):

$$\delta H = \frac{eB}{2m_e c}(L_z + gS_z) \tag{4.113}$$

where $g = 2$ is the gyromagnetic ratio of the electron (see Eq.4.77). Notice that δH commutes with H_0. Therefore the perturbed Hamiltonian (in this approximation) has the same eigenstates as the unperturbed one. Given that $S_z = \pm 1/2$ the perturbation to the energy is therefore:

$$\Delta E_s = \frac{e\hbar B}{2m_e c}(m \pm 1) \equiv \mu_B B(m \pm 1) \equiv \hbar \omega_L(m \pm 1) \tag{4.114}$$

Where $\mu_B = e\hbar/2m_e c$ is the Bohr magneton, see section 4.6.1, and the frequency $\omega_L = eB/2m_e c$, known as the Larmor frequency, is the frequency of precession of the magnetic dipole of the electron about the direction of the magnetic field. Notice that the magnetic field lifts the degeneracy in the azimuthal quantum number m, just as the electric field in the Stark effect was lifting the degeneracy in the angular momentum l. As we shall see below (section 4.12.1) in presence of a magnetic field, transitions are only possible between states with identical spin, angular momentum l differing by one unit and quantum number m differing by 0 or ± 1. Hence in presence of a strong magnetic field three emission (or absorption) lines corresponding to transitions between state $2p$ and state $1s$ with $\Delta l = 1; \Delta m = 0, \pm 1$ can be observed while in absence of magnetic field only one line is seen ($\Delta m = 0$), see Fig.4.22. The difference in frequency between these lines $\Delta \nu = \omega_L$ allows for an estimate of the strength of the local magnetic field, as for example in a Sun's spot which can be as high as thousands of Gauss.

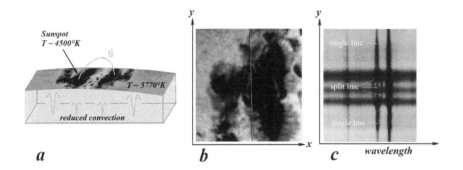

Figure 4.22 (a –b) Strong magnetic field lines pierce the surface of the Sun at its spots. These regions on the Sun's surface are cooler by about 1000°K than their surrounding due to the damping of the underlying convective motion by the magnetic field. (c) The magnetic field of about 4000 Gauss causes the splitting of the spectral emission lines measured along the vertical white line passing through the Sun's spot shown in (b). Images courtesy of NSO/AURA/NSF[71].

4.11.3.1 Magnetic Resonance Imaging (MRI)

One of the main applications of the Zeeman effect is in nuclear Magnetic Resonance Imaging which operating principle is based on the use of a magnetic field B to align the nuclear spin of atoms, mainly Hydrogen with magnetic dipole moment $\mu_p = 2.82 \ 10^{-23}$ erg/G (= $2.82 \ 10^{-26}$ J/T) in other words to split their **nuclear** (not electronic) energy levels according to Eq.4.114: $\Delta E = \mu_p B \sigma_z = \hbar \omega_L^p \sigma_z$ (where $\omega_L^p = \mu_p B/\hbar$ is the Larmor frequency of a proton with $\sigma_z = \pm 1/2$ which for the typical fields of MRI, i.e. 1 Tesla=10^4 Gauss is in the radio-frequency range: $v_L^p = \omega_L^p/2\pi = 42$MHz). The low energy state (spin aligned with the field) will be more occupied than the high energy one. As seen earlier in the study of the ammonia molecule, radiation at the resonant (Larmor) frequency ($\omega = \omega_L^p = \mu_p B/\hbar$) is absorbed by the protons, exciting them to the high energy state (spin anti-parallel to the magnetic field) from which they return to the ground state by spontaneous emission.

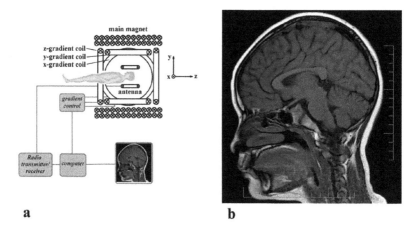

Figure 4.23 (a) A Magnetic Resonance Imaging machine consists of an apparatus in which a strong permanent magnet and various coils allow for the generation of strong magnetic field gradients along a pre-determined direction: $\vec{B}(x)$. This field gradient results in a splitting of the proton energy levels with a gap that varies with space: $\Delta E(x) = \mu_p B(x)$ (where μ_p is the proton's magnetic dipole moment). By measuring the amplitude and phase of the radio-waves emitted at given frequency ($\omega = \Delta E(x)/\hbar$) from protons excited by a pulse of EM radiation, one can deduce the concentration of Hydrogen atoms at the corresponding location and their local chemical environment. The result is a very detailed image (b) of various cuts through the body, here the head (reprinted from ref.[72]).

In MRI one applies a strong magnetic field **gradient** on a patient: $B = B_0 + \beta x$. As a result excited protons (e.g. the nuclei of hydrogen in H$_2$O)) will display a Larmor frequency that varies with distance along the gradient $\omega_L^p = \omega_0^p + \alpha x$ (with $\alpha = \mu_p \beta/\hbar$. These protons are excited with a pulse of EM radiation around ω_0^p and one mesures the radio signal $f(t)$ received from the protons as they decay back to their ground state within a time $\tau_x \sim 1\text{sec}$. The Fourier transform of that signal is (see section 4.12.3 below):

$$\tilde{f}(\omega) = \int dx \frac{A(\omega_L^P)}{i(\omega - \omega_L^P) - \gamma_x}$$

with $\gamma_x = 1/\tau_x$. The amplitude of the Fourier transform $\propto A(\omega_L^P)$ informs on the density of hydrogen atoms emitting at frequency ($\omega_L^P \approx \omega$), i.e. at position x. The phase ($\propto \gamma_x$) reveals the local chemical environment of the Hydrogen nuclei, e.g. whether they are bound to Oxygen as in water or Carbon as in fatty acids). One can thus distinguish between different tissues (e.g. richer in fat or water). The result is a high resolution tissue specific image of various cuts (along the magnetic field gradient) of a patient's body, see Fig.4.23.

4.12 TIME DEPENDENT PERTURBATION THEORY

Until now, in our investigations of Schrödinger's equation, we have dealt mostly with time independent problems (the few exceptions were our study of the ammonia maser, section 4.3.2 and our description of the MRI principle, see previous section), namely we looked at systems for which the Hamiltonian did not depend explicitly on time and for which the goal was to determine the eigenstates and eigen-energies. Very often though to investigate these states, one drives the system with a time dependent external perturbation, such as an electro-magnetic field, and studies the induced transitions between the various eigenstates. In such a situation the Hamiltonian of the system can be written as: $H(t) = H_0 + \lambda H_1(t)$, where H_0 is the time independent unperturbed Hamiltonian for which the eigenstates and eigen-energies are known:

$$H_0 |\phi_n\rangle = \epsilon_n^0 |\phi_n\rangle$$

and $\lambda H_1(t)$ is a small time dependent perturbation. To solve the time dependent Schrödinger equation:

$$i\hbar \partial_t |\Psi(t)\rangle = H |\Psi(t)\rangle = H_0 |\Psi(t)\rangle + \lambda H_1(t) |\Psi(t)\rangle \qquad (4.115)$$

with initial condition: $|\Psi(0)\rangle = |\phi_i\rangle$, we shall look for a solution as an expansion in terms of the eigenstates of H_0:

$$|\Psi(t)\rangle = \sum_n a_n(t) |\phi_n\rangle$$

Inserting that ansatz into Eq.4.115 yields:

$$i\hbar \sum_n \frac{da_n}{dt} |\phi_n\rangle = \sum_n [\epsilon_n^0 + \lambda H_1(t)] a_n(t) |\phi_n\rangle$$

Multiplying this equation on the left by $\langle \phi_m |$ and recalling that the eigenstates are orthonormal: $\langle \phi_m | \phi_n \rangle = \delta_{mn}$ yields:

$$i\hbar \frac{da_m(t)}{dt} = \epsilon_m^0 a_m(t) + \lambda \sum_n \langle \phi_m | H_1(t) | \phi_n \rangle a_n(t)$$

Define: $a_m(t) = \mu_m(t)e^{-i\epsilon_m^0 t/\hbar}$ to obtain the following equation for $\mu_m(t)$:

$$i\hbar\frac{d\mu_m(t)}{dt} = \lambda\sum_n \langle\phi_m|H_1(t)|\phi_n\rangle e^{i(\epsilon_m^0 - \epsilon_n^0)t/\hbar}\mu_n(t)$$

Integrating yields:

$$\mu_m(t) - \delta_{mi} = \frac{\lambda}{i\hbar}\sum_n \int_0^t dt\, \langle\phi_m|H_1(t)|\phi_n\rangle e^{i\omega_{mn}t}\mu_n(t)$$

with $\omega_{mn} = (\epsilon_m^0 - \epsilon_n^0)/\hbar$ and the initial condition $\mu_n(0) = \delta_{ni}$. For short enough times, the system remains essentially in its initial state $|\phi_i\rangle$ and for states $m \neq i$ we may write:

$$\mu_m(t) \simeq \frac{\lambda}{i\hbar}\int_0^t dt\, \langle\phi_m|H_1(t)|\phi_i\rangle e^{i\omega_{mi}t}$$

Now consider a periodic perturbation, such as an electro-magnetic standing wave: $\lambda H_1(t) = (Ve^{-i\omega t} + V^* e^{i\omega t})/2 = |V|\cos(\omega t + \theta)$:

$$\mu_m(t) \simeq \frac{1}{i\hbar}\int_0^t dt\, \frac{\langle\phi_m|V|\phi_i\rangle}{2}e^{i[\omega_{mi}-\omega]t} + \frac{\langle\phi_m|V^*|\phi_i\rangle}{2}e^{i[\omega_{mi}+\omega]t}$$

$$= -[V_{mi}F(\omega) + V_{mi}^*F(-\omega)]$$

with $V_{mi} = \langle\phi_m|V|\phi_i\rangle/2$ and :

$$F(\omega) = \frac{1}{i\hbar}\int_0^t dt\, e^{i[\omega_{mi}-\omega]t} = -\frac{e^{i(\omega_{mi}-\omega)t} - 1}{\hbar(\omega_{mi}-\omega)} = \frac{te^{i(\omega_{mi}-\omega)t/2}}{i\hbar}\text{sinc}[(\omega_{mi}-\omega)t/2]$$

The probability that the system is in state $|\phi_m\rangle$ at time t is:

$$P_m(t) = |a_m(t)|^2 = |\mu_m(t)|^2 = |V_{mi}F(\omega) + V_{mi}^*F(-\omega)|^2$$

This transition probability will be large when either $F(\omega)$ or $F(-\omega)$ are large, which occurs when $\epsilon_m^0 - \epsilon_i^0 = \hbar\omega$ or $\epsilon_m^0 - \epsilon_i^0 = -\hbar\omega$. The former case (with $\epsilon_m^0 > \epsilon_i^0$) corresponds to photon absorption, the latter to stimulated emission of radiation from a high energy initial state ($\epsilon_i^0 > \epsilon_m^0$).

In case of absorption the probability of finding the system in the excited state $|\phi_m\rangle$ at times t is thus:

$$P_m(t) = |V_{mi}|^2|F(\omega)|^2 = |V_{mi}|^2(t/\hbar)^2\text{sinc}^2[(\omega_{mi}-\omega)t/2]$$

$$\approx \frac{2\pi|V_{mi}|^2}{\hbar^2}t\delta(\omega_{mi}-\omega) \tag{4.116}$$

where we used the approximation: $\text{sinc}^2 ax \approx (\pi/a)\delta(x)$, see Eq.4.31. Hence the transition rate from state $|\phi_i\rangle$ into state $|\phi_m\rangle$: $T_{mi} = dP_m/dt$ is:

$$T_{mi} = \frac{2\pi|V_{mi}|^2}{\hbar^2}\delta(\omega_{mi}-\omega) = \frac{2\pi|V_{mi}|^2}{\hbar}\delta(\epsilon_m^0 - \epsilon_i^0 - \hbar\omega)$$

If there exists a distribution of excited state energies (which is the case due to the radiative lifetime of the excited state, to Doppler broadening, collisions, etc; see section 4.12.3) then the δ-function in the above equation has to be replaced by the probability density $\rho(\epsilon_m)$ of occupation of the excited states (known as the density of excited states)[40]:

$$T_{mi} = \frac{2\pi|V_{mi}|^2}{\hbar}\rho(\epsilon_m^0) \tag{4.117}$$

This equation is known as Fermi's Golden rule. It relates the transition rate from the ground state $|\phi_i\rangle$ to the excited state $|\phi_m\rangle$ to the coupling between initial and final states and the density of excited states. This is the result we obtained when studying the ammonia maser, see Eq.4.32.

In case of stimulated emission, the probability of finding the system in a lower energy state $|\phi_m\rangle$ after a time t is computed exactly as before:

$$T_{mi} = \frac{2\pi|V_{mi}^*|^2}{\hbar}\delta(\epsilon_m^0 - \epsilon_i^0 + \hbar\omega) = \frac{2\pi|V_{mi}|^2}{\hbar}\rho(\epsilon_i^0)$$

which is the same result as that for absorption (notice that here it is the density of the initial (excited) states that appears in Fermi's Golden rule, as the ground state energy is generally much sharper). This is also in agreement with Einstein treatment of radiation, see section 4.1.4.

4.12.1 TRANSITIONS BETWEEN ELECTRONIC ENERGY LEVELS

Let us consider the case of an atom illuminated with a plane electromagnetic wave:

$$\vec{E} = \vec{E_0}\cos(kx - \omega t) = \frac{\vec{E_0}}{2}(e^{i(kx-\omega t)} + e^{-i(kx-\omega t)})$$

propagating along the x-axis and let us choose the z-axis as parallel to $\vec{E_0}$ (which is always possible unless some external field exist which defines the z-axis). The average intensity of this electromagnetic wave is, see Eq.3.113:

$$I = \frac{c}{8\pi}Re[\hat{k}\cdot\vec{E}\times\vec{B^*}] = \frac{c}{8\pi}\vec{E_0^2}$$

When an atom is illuminated its electronic states are perturbed by the EM field: the perturbation Hamiltonian results from the interaction of the electric-field with the dipole formed by the nucleus at the center and the electron (with charge $-e$)) swirling around at position $\vec{r}$.

$$\lambda H_1(t) = -\vec{p}\cdot\vec{E} = e\vec{r}\cdot\vec{E} = \frac{eE_0z}{2}(e^{i(kx-\omega t)} + e^{-i(kx-\omega t)})$$

---

[40]Such that $1 = \int d\epsilon\delta(\epsilon - \epsilon_i^0 - \hbar\omega) = \int d\epsilon\rho(\epsilon)$

The perturbation coupling matrix element V_{fi} between the initial state of the atom $|i\rangle$ and its final one $|f\rangle$ is thus:

$$V_{fi} = \langle \phi_f | \lambda H_1 | \phi_i \rangle = \frac{eE_0}{2} \int d^3 x \phi_f z \phi_i e^{\pm ikx}$$

Since the electromagnetic wavelength λ is much larger than atomic scales we may assume that $kx \ll 1$ and approximate $e^{\pm ikx} \simeq 1$, hence:

$$V_{fi} \simeq \frac{eE_0}{2} \int d^3 x \phi_f z \phi_i = eE_0 z_{fi}/2$$

where the matrix element z_{fi} is similar to the one computed when studying the reponse of a diatomic molecule to the presence of an electric field, see Eq.4.111, $z_{fi} = \langle \phi_f | z | \phi_i \rangle = \langle \Psi_{nlm} | r \cos\theta | \Psi_{n'l'm'} \rangle$. Since $|\Psi_{nlm}\rangle \sim R_{nl}(r)Y_{lm}(\theta,\phi)$ this element is null except when $m = m'$ and $l' = l \pm 1$, see Eq.4.110:

$$z_{fi} = \langle \Psi_{n,l,m} | r \cos\theta | \Psi_{n',l\pm1,m} \rangle \tag{4.118}$$

And by Eq.4.117 the transition rate is then:

$$T_{fi} = \frac{2\pi |V_{fi}|^2}{\hbar}\rho(\epsilon_f) = \frac{\pi e^2}{2\hbar}E_0^2 |z_{fi}|^2 \rho(\epsilon_f)$$

$$= (\frac{4\pi^2 e^2}{\hbar c})I |z_{fi}|^2 \rho(\epsilon_f) = 4\pi^2 \alpha I |z_{fi}|^2 \rho(\epsilon_f) \tag{4.119}$$

Where $\rho(\epsilon_f)$ is the density of excited states, see Eq.4.117 and $\alpha = e^2/\hbar c \approx 1/137$ is known as the fine structure constant. The Einstein coefficient associated to that transition can be derived as in section 4.3.2.1, see Eq.4.33:

$$B_{fi} = 4\pi^2 \alpha v c |z_{fi}|^2 \tag{4.120}$$

In this analysis we have assumed that the electric field defined the z-axis. In presence of a static field that defines that axis (e.g. a constant magnetic field, as in a Sun's spot, section 4.11.3)), the EM field couples to the three components of the dipole moment:

$$V_{fi} = \langle \phi_f | eE_{0,x} r \sin\theta \cos\phi + eE_{0,y} r \sin\theta \sin\phi + eE_{0,z} r \cos\theta | \phi_i \rangle$$

Using the identities (see also Eq.4.110):

$$\cos\theta \, P_l^m(\cos\theta) = \frac{l-m+1}{2l+1}P_{l+1}^m(\cos\theta) + \frac{l+m}{2l+1}P_{l-1}^m(\cos\theta)$$

$$\sin\theta \, P_l^m(\cos\theta) = \frac{(l-m+1)(l-m+2)}{2l+1}P_{l+1}^{m-1}(\cos\theta) - \frac{(l+m+1)(l+m)}{2l+1}P_{l-1}^{m-1}(\cos\theta)$$

$$= \frac{-1}{2l+1}P_{l+1}^{m+1}(\cos\theta) + \frac{1}{2l+1}P_{l-1}^{m+1}(\cos\theta) \tag{4.121}$$

and the orthogonality of the spherical harmonics: $Y_{lm}(\cos\theta,\phi) \sim P_l^m(\cos\theta)e^{im\phi}$ (see Eq.4.82), one notices that the perturbation coupling elements V_{fi} are null unless $m' - m \equiv \Delta m = -1,0,1$ and $l' - l \equiv \Delta l = \pm 1$.

Therefore in absence of a magnetic field (i.e. the electric field defines the z-axis), transitions with $\Delta l = \pm 1$ and $\Delta m = 0$ are possible. Thus transitions between s and p states ($\Delta l = \pm 1$) are allowed but (to leading order) not between two s or two p states.

In presence of a magnetic field transitions between states differing in their angular momentum l by ± 1 and in m by 0, ± 1 are possible. This explains the splitting in three of the emission lines from a Sun's spot (for the three possible values of Δm) described in section 4.11.3 and shown in Fig.4.22.

Moreover, since the coupling of the spin to the magnetic field $((e/m_e c)\vec{S} \cdot \vec{B}(t))$ is much weaker than the electric dipole coupling (eEr)[41], spin is conserved during the interaction of an atom with the EM field: $\Delta S = 0$. These considerations set the selection rules for the transitions between the electronic energy levels in an atom.

4.12.2 MOLECULAR ROTATIONAL-VIBRATIONAL TRANSITIONS

Rotational transitions. Let us consider the interaction of an EM wave with a diatomic molecule possessing a permanent electric dipole $\vec{p}$ (which may be the case if the two atoms are different as in CO but not if they are identical as in O_2) . As done in the previous section we shall assume that the wave interacts with the electric dipole of the molecule:

$$\lambda H_1(t) = -\vec{p} \cdot \vec{E} = -\frac{pE_0 \cos\theta}{2}(e^{-i\omega t} + e^{i\omega t})$$

where θ is the angle between the dipole and the electric field. For a diatomic molecule with moment of inertia I we have seen, section 4.10.2, that the rotational eigenenergies are given by $E_j = \hbar^2 j(j+1)/2I$, while the rotational eigenstates $|j,m\rangle = Y_{jm}(\theta,\phi)$ are the same as the Hydrogen atom angular wavefunctions. Consequently the same selection rules hold for dipole transitions in a diatomic molecule as in the Hydrogen atom. Transitions between an initial state $|i\rangle \equiv |j,m\rangle$ and a final state $|f\rangle \equiv |j',m'\rangle$ are possible only if their angular quantum numbers j, j' differ by ± 1 and their quantum number m, m' are identical.

This can be checked using the identities Eq.4.110 to compute the matrix element V_{fi} between initial and final states:

$$V_{fi} = -\frac{pE_0}{2} \langle j',m'|\cos\theta|j,m\rangle = -\frac{pE_0}{2}[C_{j+1,m}\delta_{j',j+1} + C_{j-1,m}\delta_{j',j-1}]\delta_{m',m}$$

That value can then be substituted in Eq.4.119 to yield the dipole transition rate for excitations between rotational energy levels.

Vibrational transitions. If vibrational states with energy $E_n = (n+1/2)\hbar\omega$ are excited by an EM wave, the dipole coupling is:

$$\lambda H_1(t) = -\vec{p} \cdot \vec{E} = -p_{0,z}E - q_{eff}zE$$

---

[41]The magnetic dipole coupling is weaker than the electric dipole coupling by the ratio $(e\hbar/m_e c)/er_0 = e^2/\hbar c \equiv \alpha \approx 1/137$, where $r_0 = \hbar^2/m_e e^2$ is Bohr's radius.

(where $\vec{E}$ defines the z-axis and the effective coupling charge is $q_{eff} \equiv \partial p/\partial z|_0$). The first term on the right is constant and doesn't couple between initial and final vibrational eigenstates, $|n\rangle$ and $|m\rangle$ of the harmonic oscillator. Transition are only possible if the dipole moment depends on the vibrational state, i.e. if the second term is non-zero:

$$\langle m|q_{eff}z|n\rangle = \int dz X_m(z) q_{eff} z X_n(z) \equiv q_{eff} z_{mn}$$

Where $X_n(z)$ is an eigenstate of the 1D-harmonic oscillator, see Eq.4.101:

$$X_n(z) = (\alpha')^{1/4} A_n e^{-\alpha' x^2/2} H_n(\sqrt{\alpha'}z)$$

with $A_n = 1/\sqrt{\pi^{1/2} 2^n n!}$ and $\alpha' \equiv m\omega/\hbar$. Using the property of the Hermite polynomials: $2\xi H_n(\xi) = H_{n+1}(\xi) + 2nH_{n-1}(\xi)$ (with $\xi \equiv \sqrt{\alpha'}z$) and the orthogonality of the Harmonic oscillator eigenstates, it is easy to see that the matrix element z_{mn} coupling the initial and final eigenstates is :

$$\begin{aligned}
z_{mn} &= \frac{A_m A_n}{2\sqrt{m\omega/\hbar}} \int d\xi e^{-\xi^2} H_m(\xi)[H_{n+1}(\xi) + 2nH_{n-1}(\xi)] \\
&= \frac{A_m A_{n+1} \sqrt{2(n+1)}}{2\sqrt{m\omega/\hbar}} \int d\xi e^{-\xi^2} H_m(\xi)H_{n+1}(\xi) + \frac{A_m A_{n-1} \sqrt{2n}}{2\sqrt{m\omega/\hbar}} \int d\xi e^{-\xi^2} H_m(\xi)H_{n-1}(\xi) \\
&= \frac{1}{\sqrt{2m\omega/\hbar}}[\sqrt{n+1}\delta_{m,n+1} + \sqrt{n}\delta_{m,n-1}] \quad\quad\quad (4.122)
\end{aligned}$$

which is nonzero if $m = n \pm 1$. The transition rates between these states are then given by Eq.4.119 with an effective charge q_{eff} and a transition element $z_{fi} \equiv z_{mn}$ determined in Eq.4.122.

Diatomic molecules can both vibrate and rotate. Neglecting the possible coupling between vibrational and rotational states, the gap between successive vibrational energy levels is:

$$\Delta E_n = E_{n+1} - E_n = \hbar\omega \quad \text{of } O(0.1 \text{ eV}).$$

Similarly the gap between successive rotational energy levels increases linearly with j:

$$\Delta E_{j+1,j} \equiv E_{j+1} - E_j = \hbar^2(j+1)/I \quad \text{of } O(1 \text{ meV})$$

which is much smaller than the constant energy gap between successive vibrational energy level, ΔE_n.

Light-induced transitions (absorption) between vibrational-rotational energy levels in diatomic molecules occur between vibrational levels with $\Delta n = 1$ (energy difference $\Delta E_n = \hbar\omega$) and $\Delta j = \pm 1$. When $\Delta j = 1$: $\Delta E_{j+1,j} = \hbar^2(j+1)/I$. When $\Delta j = -1$: $\Delta E_{j-1,j} = -\hbar^2 j/I$ Hence the rotational-vibrational absorption spectrum of diatomic molecules exhibits two bands known as R- and P-bands, see Fig.4.24.

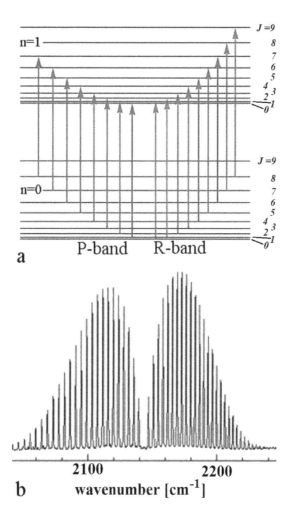

Figure 4.24   (a) Transitions between vibrational energy level $n = 0$ and $n = 1$ and rotational energy levels differing by $\Delta j = \pm 1$ in a diatomic molecule such as CO (reprinted from ref.[73]). These transitions define two characteristic spectral bands known as P-band when $\Delta j = -1$ and R-band when $\Delta j = 1$.

The R-band corresponds to transitions with $\Delta j = +1$ at frequencies:

$hv_j = \Delta E = \hbar\omega + \hbar^2(j+1)/I.$

The P-band corresponds to transitions with $\Delta j = -1$ at frequencies:

$hv_j = \Delta E = \hbar\omega - \hbar^2 j/I.$

4.12.3 ABSORPTION AND FLUORESCENCE EMISSION.

The emission and absorption spectra of a molecule display a unique finger-print, as shown in Figs.4.24 and 4.25, that is widely used to identify the components of a mixture and to estimate their concentration. Thus the composition of far away stars is deduced from the observation of the specific emission lines of certain elements. Similarly the search for Earth-like planets orbiting those stars relies of the observation of water-specific absorption lines in the light reflected from those planets. Closer to us, the measurement of the concentration of pollutants in air or water relies on the detection of their specific spectral finger-print and on the quantification of their absorption at these specific frequencies. Some cancer photo-therapies are based on light absorption by appropriate molecules (cancer drugs) which trigger a chemical reaction producing poisonous oxygen radicals. Understanding the absorption and emission of EM radiation by an atom/molecule is thus a fundamental question with many applications[74, 75]. We shall in the following revisit that issue, first sketched in section 4.1.4, using the results from time dependent perturbation theory, Eq.4.117 and Eq.4.119.

Consider the absorption of a coherent light beam (e.g. a laser beam) of frequency v by a molecule in its ground state $|i\rangle$ (of energy ϵ_i), resulting in its excitation to an electronic and vibrational energy level $|f\rangle$ (of energy $\epsilon_f > \epsilon_i$, see sections 4.10 and 4.10.3). The molecules in the excited state decay back to the ground state within a typical time τ_f via a variety of pathways: direct emission of radiation is one possibility (within a radiative lifetime $\tau_s > \tau_f$, see section 4.1.4), but de-excitation via non-radiative channels, such as collisions, energy transfer via dipole-dipole interactions, Doppler-broadening, etc. are often competing alternatives[42]. Typically, the probability of de-excitation obeys a Poisson distribution: $P(t > 0) = e^{-t/\tau_f}/\tau_f$, which implies that the excited state wavefunction vary in time (for $t > 0$) as:

$$\Psi(t) = \frac{e^{-t/2\tau_f - i\omega_{fi}t}}{\sqrt{\tau_f}}$$

where $\omega_{fi} = (\epsilon_f - \epsilon_i)/\hbar = 2\pi v_{fi}$. In the frequency (or Fourier) space, see Appendix A.4 the wavefunction is:

$$\tilde{\Psi}(\omega) = \left(\sqrt{2\pi\tau_f}[i(\omega_{fi} - \omega) + 1/2\tau_f] \right)^{-1}$$

from which we derive the density of states $\rho(v)$ at frequency v around v_{fi}:

$$\rho(v) = 2\pi|\tilde{\Psi}(\omega)|^2 = 4\tau_f \frac{1}{1 + [4\pi\tau_f(v_{fi} - v)]^2} = \frac{\Delta v_f}{\pi(\Delta v_f^2 + (v_{fi} - v)^2)} \qquad (4.123)$$

The spectral width of the excited state, $\Delta v_f = 1/4\pi\tau_f$ determines the probability of excitation[43] $\rho(v)$ of the ground state $|i\rangle$ by light at frequency v close to the resonant

---

[42]The overall decay rate $k_f = 1/\tau_f$ is a sum over the decay rates from all possible de-excitation pathways: $k_f = k_s + k_{nr} > k_s$, where k_{nr} is the decay rate via all non-radiative pathways. Hence: $\tau_s > \tau_f$.

[43]Notice that as expected: $\int |\Psi(t)|^2 dt = \int |\tilde{\Psi}(\omega)|^2 d\omega = \int \rho(v)dv = 1$.

frequency $\nu_{fi} = (\epsilon_f - \epsilon_i)/h$. As a light beam propagates in a media containing such molecules at concentration n_i particles/cm^3, it is absorbed at a rate: $B_{fi}\rho(\nu)n_i$ (see section 4.1.4 and ignoring stimulated emission processes which are negligible at low light intensities). As a consequence of absorption the density of photons $\eta(x,\nu)$ at position x and frequency ν (or equivalently the beam intensity $I(x,\nu) = ch\nu\eta(x,\nu)$, see Eq.4.3) decreases along the beam path x as:

$$\frac{1}{c}\frac{d\eta(\nu,x)}{dt} = \frac{d\eta(\nu,x)}{dx} = -B_{fi}\rho(\nu)n_i\eta(\nu,x)/c \equiv -\sigma_{fi}(\nu)n_i\eta(x,\nu) \qquad (4.124)$$

Where $\sigma_{fi}(\nu) \equiv B_{fi}\rho(\nu)/c$ is the absorption cross-section at frequency ν. The beam intensity then decays as: $I(\nu,x) = I(\nu,0)e^{-\sigma_{fi}(\nu)n_ix}$. From a measurement of the transmittance at frequency ν over a distance l: $\mathcal{T}_l(\nu) = I(\nu,l)/I(\nu,0)$, see Eq.3.139 one deduces the concentration n_i of molecules absorbing at frequency ν (e.g. pollutants in air or DNA in water):

$$n_i = -\frac{\log \mathcal{T}_l(\nu)}{l\sigma_{fi}(\nu)}$$

In a given sample, measurements of $\mathcal{T}_l(\nu)$ at various frequencies ν (or wavelengths λ) allow to quantify the concentrations of different molecules (absorbing at different frequencies). For example the concentration of DNA in a given solution is assessed by the transmittance at $\lambda = 260$nm (with $\sigma_{DNA} = 5.07 \ 10^{-17}$ cm^2 or an extinction coefficient[44] $\epsilon^{-1} = 50(\mu g/ml)$cm), while the purity of a DNA sample is assessed by measuring the ratio of transmittance at 260nm and 280nm (a ratio close to 1.8 in absence of protein contamination). Notice that absorption is maximal at resonance ($\nu = \nu_{fi}$):

$$\sigma_{max} = 4B_{fi}\tau_f/c = \lambda^2\tau_f/2\pi\tau_s \qquad (4.125)$$

Where we used Einstein's relation, Eq.4.17. At low temperatures and in a solid medium, where decay via non-radiative channels (e.g. collisions) is less probable ($\tau_f \sim \tau_s$), the absorption cross-section in the visible (about 10^{-10} cm^2) is much larger than the atomic size. In more common situations (e.g. room temperature and molecules in air or water), the excited state lifetime τ_f can be $10^5 - 10^6$ times smaller than the radiative lifetime $\tau_s \sim 6$ nsec and the absorption cross-section is closer to molecular dimensions: $\sigma_{max} \sim 10^{-16}$ cm^2.

Fluorescence. An interesting class of absorbing molecules are fluorophores, molecules that fluoresce (glow) when excited by light (at frequency ν_{ex}). They are often used by biologists to label non-fluorescent proteins, DNA or RNA, thus allowing for the observation of their localization and traffic within a cell or an organism. Upon excitation to high electronic and vibrational energy levels, fluorophores rapidly decay via non-radiative channels (within a time $\tau_f \sim 10^{-15}$sec) to the lowest vibrational energy level of the electronic excited state. Within a time τ_s they decay from

---

[44]One sometimes uses the Optical Density, defined as: $OD \equiv -\log_{10}\mathcal{T}_l(\nu) = -\log\mathcal{T}_l(\nu)/\log(10) = \epsilon c_i x$, with: c_i is the molar concentration of particles (see section 5.7.3): $c_i = 10^3 n_i/N_A$, where $N_A = 6 \ 10^{23}$ particles/l is Avogadro's number and the extinction coefficient $\epsilon \equiv 10^{-3}\sigma_{fi}N_A/\log(10)$.

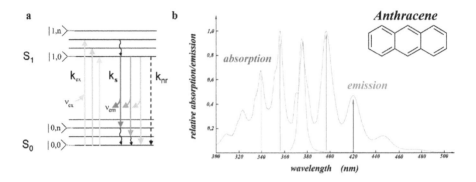

Figure 4.25 (a) Schematic diagram of absorption and emission between a molecular electronic and vibrational ground state S_0 ($|0,0\rangle$) and the electronic excited states $|1,n\rangle$. Illumination with light of frequency ν_{ex} of the molecule in its ground state results in its excitation to high electronic and vibrational state $|1,n\rangle$. From these vibrational excited states the molecule decays quickly (within $\sim 10^{-15}$ sec) via non-radiative channels (symbolized by the wiggly arrow) to the lowest excited state S_1 ($|1,0\rangle$). It then decays back to the ground state (or to one of its vibrational excited states $|0,n\rangle$) at a rate k_s via the spontaneous emission of light at frequency ν_{em} or at a rate k_{nr} via some non-radiative channel (e.g. transfer of energy to an other molecule via dipole-dipole interaction). (b) Absorption and emission spectra of anthracene[76]. Due to energy conservation and dissipation effects, emission is always at higher wavelength, lower frequency (i.e. lower energy) that absorption.

that state back to the ground state via emission of light of frequency $\nu_{em} \lesssim \nu_{ex}$, see Fig.4.25. This emitted light is the cause of fluorescence.

From our analysis of the electronic transitions, section 4.12.1, we can relate many of the parameters relevant to fluorescent absorption and emission. From Eq.4.120:

$$B_{fi} = 4\pi^2 \alpha \nu_{ex} c |z_{fi}|^2$$

and Einstein relation, Eq.4.17:

$$B_{fi}\tau_s = c\lambda^2/8\pi$$

one derives a relation between the radiative lifetime, τ_s and the transition matrix element for absorption, z_{fi}:

$$\tau_s = \frac{\lambda^3}{32\pi^3 \alpha c |z_{fi}|^2} \tag{4.126}$$

Inserting typical values for the wavelength of visible light ($\lambda \sim 0.5\mu m$) and for the transition matrix element $z_{fi} \sim 1\text{Å}$ (a typical atomic size), yields $\tau_s \sim 6$ nsec, in agreement with the typical radiative lifetimes of fluorophores. Notice that alteration in the dipole matrix element z_{fi}, due for example to the presence of an external field, will affect the radiative lifetime.

On the other hand, from Eq.4.125 one finds:

$$\sigma_{max} = 4B_{fi}\tau_f/c = 16\pi^2 \alpha v_{ex}|z_{fi}|^2 \tau_f \tag{4.127}$$

From Eq.4.119, one deduces[45] that the rate of light absorption at the resonant frequency v_{ex} is $k_{ex} = \sigma_{max}I/hv_{ex}$. At steady state that rate equals the rate of light emission. As the illumination intensity I is increased the excitation rate increases untill it reaches the maximal rate set by the decay rate of the emission state: $k_{ex} \leq k_s = 1/\tau_s$. At intensities $I > I_s \equiv hv_{ex}/\sigma_{max}\tau_s \sim 500$ kW/cm^2, stimulated emission from the excited state effectively saturates the absorption capacity of the molecules. This regime while easily achieved with a laser beam of a few mW focused by a microscope objective on a spot of diameter $\sim 0.5\mu$m, can be detrimental to biological samples and is almost never used. Indeed to limit photo-damage to biological tissues observed under a microscope, illumination intensities in the visible range are usually kept below 1 W/cm^2 (which is already 10 times more intense than the sun-light at noon) and illumination times are kept as short as possible.

4.12.3.1 Super-Resolution Microscopy

The preceding description of light-matter interaction is the basis of Stefan Hell's revolutionary proposal[77, 78] for the increase of the optical resolution of microscopes, see Appendix A.7.2, for which he was awarded the Nobel prize in 2014. S.Hell suggested to use a stimulated (STED) beam of frequency $v_{STED} \lesssim v_{em}$ to force excited molecules to emit light coherently with that beam, thus depleting fluorescence emission at frequency v_{em} in the area where the STED beam intensity is large, see Fig.4.26(a,b). By using a doughnut-shaped high intensity STED beam superposed on a Gaussian excitation beam at frequency $v_{ex} \gtrsim v_{em}$, the area of fluorescent emission could be made as small as desired thus increasing the resolution of microscopes beyond the diffraction limit. The optical images obtained with that technique, see fig.4.26(c,d), allow for real time observations of biological processes with a resolution that was until now achievable only with electron microscopes on fixed (i.e. dead) samples.

---

[45] At the resonant frequency $(v = v_{fi})$: $\rho(\epsilon_f) = \rho(v_{fi})/h = 1/\pi h \Delta v_f$. Since $4\pi\tau_f \Delta v_f = 1$:

$$k_{ex} \equiv T_{fi} = 4\pi^2 \alpha I |z_{fi}|^2 \rho(\epsilon_f) = \frac{\sigma_{max}I}{hv_{ex}4\pi\tau_f\Delta v_f} = \frac{\sigma_{max}I}{hv_{ex}}$$

This equation can be derived simply by considering the different ways of computing the power absorbed P_{ex}: either from the cross section of absorption $P_{ex} = \sigma_{max}I$ or from the rate of excitation $P_{ex} = k_{ex}hv_{ex}$.

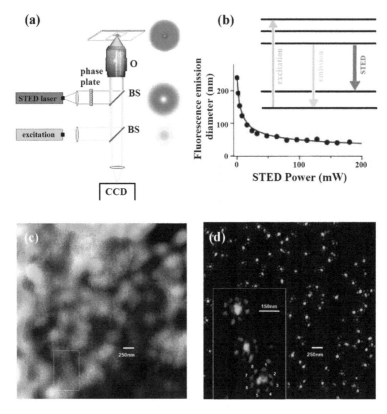

Figure 4.26 (a) STimulated Emission Depletion (STED) microscopy is a super-resolution microscopy approach that breaks the diffraction limit by relying on the strong non-linearity of stimulated emission. Two co-linear beams combined with appropriate dichroic mirrors (BS) are used to illuminate the sample via the objective (O): a gaussian shaped excitation beam (at frequency v_{ex} whose intensity is maximal at the center) and a longer wavelength doghnut shaped stimulation (STED) beam (at frequency $v_{STED} \lesssim v_{em} \lesssim v_{ex}$. Due to the non-linear dependence of stimulated emission on the stimulating intensity, see (b), the stronger the STED beam the more the excited states are driven to emit light in phase with the stimulating beam. Where the intensity of the doghnut beam is close to zero, i.e. at its center, spontaneous fluorescence emission at frequency v_{em} is possible. This happens in an area which size decreases as the intensity of the STED beam increases (see (b)), allowing for a theoretically unlimited optical resolution (in practice resolutions down to a few tens of nanometers have been achieved). (c) Regular diffraction limited confocal image of the nuclear pore complex which controls transport of molecules in and out of the cell nucleus. (d) Same image using STED microscopy. Notice the improved resolution which allows to identify the 8 protein sub-units of the complex (labeled with a red fluorescent molecule). Adapted from ref.[79] with permission from Elsevier.

A related super-resolution technique, known as Single Molecule Localization Microscopy (SMLM), uses molecules which can be switched into a fluorescent state by illumination with UV light (with wavelength around 390nm)[80, 81]. Using a low UV intensity, only a few molecules are switched on and their localization determined from the center of their isolated fluorescent image spots of radius $r \approx \lambda/2$, see Appendix A.7.2.1, with a precision[46] $\delta r = r/\sqrt{N}$ (where N is the number of detected photons emitted by a fluorescent molecule before it photobleaches and switches off). By successive rounds of UV-illumination and photobleaching, a pointillist image of a fluorescently labeled biological object (membrane, chromosome, axon, etc.) is obtained with resolution δr.

4.12.3.2 The Laser

As we have seen previously Einstein phenomenological description of light-matter interaction has served as the basis for the understanding of absorption and emission of EM radiation by atoms and molecules[74]. It has also made possible decades later the invention of the laser, a device now found ubiquitously in CD players, laser pointers and surgical devices and used extensively for communication through fiber-optics cable.

A laser consists of a light amplifying medium inside a highly reflective optical cavity, which usually consists of two mirrors arranged such that light bounces back and forth, each time passing through the gain medium, see Fig.4.27. Typically one of the two mirrors is partially transparent to let part of the beam exit the cavity. To achieve light amplification the excited state $|f\rangle$ in the medium of a laser cavity emitting at frequency v_{if} has to be more populated than the lower energy state $|i\rangle$. Since at thermodynamic equilibrium low energy states are always more populated than higher energy ones, energy must be injected in the medium to "pump" (i.e. excite) atoms from the ground state $|i\rangle$ into the light emitting state $|f\rangle$.

Let us therefore consider light of intensity $I(x,v) = chv\eta(x,v)$ $(0 < x < l)$ propagating in a cavity of length l. Due to stimulated emission the increase in the photon density $\eta(x,v)$ (or intensity $I(x,v)$) is, see Eq.4.124:

$$\frac{d\eta}{dx} = (n_f - n_i)B_{if}\rho(v)\eta(x,v)/c$$
$$= (n_f - n_i)\sigma_{fi}(v)\eta(x,v)$$

In a laser, light absorption by the ground state is exceeded by stimulated emission from the more populated excited state $(n_f > n_i)$. As it propagates in the amplifying medium, the light intensity grows exponentially as : $I(x,v) = I(0,v)\exp\left[(n_f - n_i)\sigma_{fi}x\right]$. However, the light in a laser cavity experiences also losses (due to unrelated absorption, scattering, etc.) of magnitude α per cm. Moreover since

---

[46]The standard deviation (error) on the mean position is smaller by $\sqrt{N}$ than the error $\approx \lambda/2$ on the individual photon positions, see section 5.1.3.

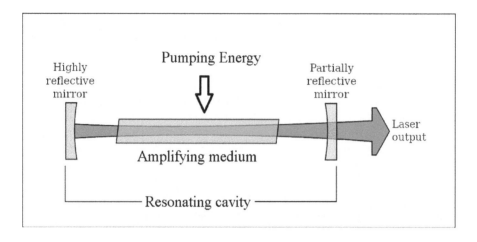

Figure 4.27 Principle of operation of a laser. An amplifying medium is pumped by an external energy source (e.g. a flash lamp) to generate a higher density of excited states than of ground states (population inversion). The medium is placed in a cavity with reflecting mirrors, one of which lets a small fraction of the light out (adapted from ref.[82]). The light reflected in the cavity is amplified by the stimulated emission of the excited states. When the threshold for lasing is achieved the losses in the cavity are balanced by the gain from the amplifying medium.

light has to come out of the laser there are also losses due to the fact that only a portion R of the intensity (at $x = l$) is reflected back into the cavity. For a laser to operate the losses must equal the gain, i.e.: $R\exp\{[\sigma_{fi}(n_f - n_i) - \alpha]l\} = 1$, which implies that the population inversion at threshold has to satisfy at best (when $\sigma_{fi} = \sigma_{max}$):

$$(n_f - n_i)_{thresh} = (\alpha - \frac{1}{l}\log R)/\sigma_{max} = \frac{2\pi\tau_s}{\lambda^2\tau_f}(\alpha - \frac{1}{l}\log R)$$

Hence the longer the wavelength (λ) the smaller the required population inversion for lasing. That is one of the reasons that masers (lasers in the microwave range) were the first to be invented while X-ray lasers, potentially of great utility (to resolve the structure of big molecules), have been difficult to develop. Notice that at steady-state the balance of losses and gain imply that the population inversion remains at threshold: the more the ground state is pumped, the more the excited state is induced to emit by the increased light density in the cavity, thus keeping the difference between the density of the two states fixed at its threshold.

5 Statistical Mechanics

Statistical mechanics is the branch of physics that deals with the behavior of large ensembles of particles (molecules or atoms). As such it addresses almost all the phenomena of our daily experience, from the characterization of the different phases of matter (solid, liquid, gas, glasses, polymers, etc.) to their dynamical properties (heat, flow, diffusion, conduction, etc.). By being the most accessible to our senses, these phenomena have been known and studied ever since mankind discovered the use of fire (heat). In fact the most ancient classification of our daily experience has been laid in terms of the four elements: earth, water, air and fire (to which some added aether (or quintessence) to describe the world inaccessible to our senses). Today these elements are referred to as phases of matter: solid, liquid, gas and plasma. These elements had attributes: hot, cold, wet and dry. Nowadays these are called thermodynamic properties: temperature, humidity, pressure, volume, etc. The mix of the various elements, their weight, extant and shape defined the objects around us. Hence pottery could be described as the breathing of fire into a mix of water and earth shaped into a vessel of certain form.

Since historical pre- and mis-conceptions have so influenced the evolution of statistical mechanics (also known as thermodynamics), following the historical narrative (as I have mostly done for the presentation of electromagnetism and quantum mechanics) is more confusing than illuminating. In the following therefore I will approach the development of the concepts of statistical mechanics from a modern perspective based on the theory of information. This approach has been pioneered by E.T.Jaynes in two seminal papers (*"Information theory and Statistical Mechanics"*[83, 84] and nicely expanded in a monograph by A.Katz: *"Principles of Statistical Mechanics: the information theory approach"*[85].

After a brief introduction to probability theory and its use in the solution of the random walk problem (a problem encountered in many contexts: diffusion, polymer theory, stock market fluctuations, etc.), I will introduce the concept of missing information or entropy and will argue that all problems in equilibrium Statistical Mechanics can be reduced to maximizing the entropy under certain constraints (energy, number of particles, etc.). I will exemplify the use of this approach in the estimation of the probability of appearance of the various faces of fair and loaded die. I will then use it systematically to study three statistical mechanics systems: magnetism, polymers and ideal gases which can be classical (atoms, molecules, ions) or quantum (electrons, photons, phonons, bosons). We will study these systems first by neglecting the interactions between their components (e.g. spins or particles) and then discuss various approximations used when these interactions cannot be neglected (ferromagnetism, large polymer deformations, fluids). I will end up with a discussion of the most difficult and unsolved problems in physics: the behavior of out of equilibrium systems, which is to say almost all the systems we face in our everyday life, such as biological, ecological, geological, meteorological, economical and social systems.

DOI: 10.1201/9781003218999-5

Along the way we will see how the foundations established by Statistical Mechanics have led to the development of vast fields of fundamental and applied research: astrophysics (star evolution), physical-chemistry (polymers, chemical reactions, etc.), micro-electronics (diodes, transistors, etc.), hydrodynamics, etc.

The major message I would like to convey here, is that a simple mathematical concept (maximizing the missing information) has enormous power. It unifies the approach to all the many particles systems that we can solve, those for which Nature happens to be "fair".

5.1 ◆ PROBABILITY THEORY

Probabilities play a central role in Statistical Mechanics, thus our first task will be to review some concepts of probability theory. By definition the probability of observing an event A is the ratio known as the event frequency between the number of occurrences of that event to the total number of events in the limit where the total number of events tends to infinity.

$$P(A) = \frac{\text{number of events A}}{\text{total number of events} \to \infty} \tag{5.1}$$

As that limit can never be reached in practice, the definition of probabilities is problematic at the outset. In situations were there is no reason to favor one event over another, it has been conjectured for a long time that the probability of all events should be the same. For example, when tossing an unloaded die the probability of one of its faces (1, 2,...6) coming up should be 1/6. This issue of the estimation of probabilities is central to Statistics and to Statistical Mechanics. We shall delve on it extensively later on. For the time being let us assume that the probability can be defined and measured as accurately as desired.

If two events are independent, i.e. if the occurrence of one doesn't affect the occurrence of the other, then the probability of observing the two events is the product of the probability of observing each event separately:

$$P(A, B) \equiv P(A \text{ AND } B) \equiv P(A \bigcap B) = P(A) \cdot P(B) \tag{5.2}$$

For example the probability of observing two successive heads (H) when throwing a coin is $P(H,H) = 1/2 \cdot 1/2 = 1/4$. If two events are mutually exclusive (i.e. they cannot be observed simultaneously) then the probability of observing either of them is, see Fig.5.1(a):

$$P(A \text{ OR } B) \equiv P(A \bigcup B) = P(A) + P(B) \tag{5.3}$$

For example, the probability of observing an even number when throwing a die is: $P(\text{even}) = P(2) + P(4) + P(6) = 1/2$. If the events are not mutually exclusive then, see Fig.5.1(b):

$$P(A \bigcup B) = P(A) + P(B) - P(A \bigcap B) \tag{5.4}$$

For example, the probability of drawing an Ace or a Diamond in a pack of 52 cards is: $P(\text{Ace OR Diamond}) = 4/52 + 13/52 - 1/52 = 16/52$. The conditional probability

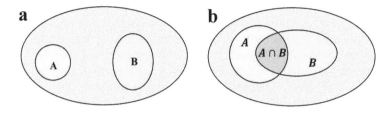

Figure 5.1 The probability of events A or B shown in these Venn diagrams is proportional to their area relative to that of the whole set (light brown). (a) mutually exclusive events (Kings and Aces in a card game for example); (b) mutually non-exclusive events (Hearts and Queens for example))

expresses then the probability of observing event B conditioned on the observation of event A:

$$P(B|A) = \frac{P(A \cap B)}{P(A)} \tag{5.5}$$

For example, the probability in a card game of drawing an Ace if the card drawn is a Spade is: $P(\text{Ace}|\text{Spade}) = P(\text{Ace AND Spade})/P(\text{Spade}) = (1/52)/(13/52) = 1/13$. In making these estimates we assumed that the pack of cards, the die or the coin are not "loaded", so that similar events have the same a-priori probability. One of the basic assumptions of Statistical Mechanics is that Nature is not "loaded" so that the probability of observing a certain state/event can be determined a-priori by counting the number of ways that state/event can be achieved. If the system is fair, all outcomes/ways are equiprobable. For example one can guess the probability of getting two heads when throwing a coin 3 times by counting the number of outcomes/ways that yield two heads: HHT,HTH,THH. Since the total number of possible outcomes is $\Omega = 2^3 = 8$, the probability of observing two heads is expected to be 3/8. Counting the number of ways of obtaining a given state/event is thus an essential part of probability theory (and statistical mechanics).

Consider N different objects (e.g. billiard balls identified by their numbers). How many ways are there for ordering them? There are N possibilities for the first place, N-1 for the second, etc. So overall there are $N(N-1)(N-2)\cdots 1 \equiv N!$ ways of ordering N different objects. If however among the N objects, m are identical (say m black billiard balls), then one cannot distinguish between the permutations of these m objects and therefore the number of ways of ordering the N balls becomes $\Omega = N!/m!$. Since all ordering are a-priori equivalent, the probability of observing a particular one is $P(\Omega) = 1/\Omega$. In general if there are M classes of identical objects, e.g. balls of identical colour, with m_i objects in each class ($i = 1, \cdots, M$ such that $\sum_{i=1}^{M} m_i = N$) then the probability of observing one particular ordering (configuration or state) is

$$P(\Omega) = 1/\Omega = m_1!m_2!\cdots m_M!/N! \equiv \Pi_{i=1}^{M} m_i!/N!$$

In these examples the ordering was important. Often though it is not the ordering of the objects that is important, but the kind of objects one has. For example in a card game, such as Poker, it is the cards one has that matter, not the order in which they were distributed. Selecting m objects out of N different ones, yields a total number of combinations: $\Omega = N!/(N-m)!m!$. For example, in Poker the number of possible hands is $\Omega_{poker} = 52!/47!5!$. The number of hands with 4 aces is 48 (the 4 aces + one of the remaining 48 cards). The probability of having 4 aces is thus:

$$\mathcal{P}(4 \text{ aces}) = \frac{\text{number of hands with 4 aces}}{\text{total number of hands}} = \frac{48}{\Omega_{poker}} = \frac{48!5!}{52!} = \frac{1}{54145}$$

5.1.1 ◆ RANDOM WALK

The random walk plays a central role in many processes. In the original experiments of Brown and later Perrin, it described the motion of a pollen particle subjected to the random shocks of the environment (the water molecules), but it is nowadays also used to describe the fluctuations in the values of stocks (subjected to the random whims of investors), the disappearance of species (subjected to stochastic competition), the stochastic earning and ultimate ruin of a gambler, the foraging habits of ants, etc. Even though Einstein's analysis of Brownian motion convinced physicists of the existence of atoms, it was the Roman philosopher Lucretius who 2000 years earlier linked their existence to the random motion of dust particles in the air[13]:

"Observe what happens when sunbeams are admitted into a building and shed light on its shadowy places. You will see a multitude of tiny particles mingling in a multitude of ways... their dancing is an actual indication of underlying movements of matter that are hidden from sight... It originates with the atoms which move of themselves. Then those small compound bodies that are least removed from the impetus of the atoms are set in motion by the impact of their invisible blows and in turn cannon against slightly larger bodies. So the movement mounts up from the atoms and gradually emerges to the level of our senses, so that those bodies are in motion that we see in sunbeams, moved by blows that remain invisible".

Because of the importance of that model we will, in the following pages, study it from a variety of perspectives. Let us first consider a random walk in one dimension, such as a drunk walking up or down a staircase (or a gambler whose earnings go up or down), see Fig.5.2. Let the probability of moving up be p that of moving down be $q = 1 - p$. Let N be the total number of steps (number of games in the case of a gambler), m the number of steps up (or gains) and $N - m$ the number of steps down (i.e. losses).

The total number of paths ending at position: $l = m - (N - m) = 2m - N$ (i.e. with a net of l gains) is: $N!/m!(N-m)!$ (= number of combinations with m steps up and $N - m$ steps down). Each of these paths can occur with probability $p^m q^{N-m}$, hence the probability of ending at position l is by virtue of Eq.5.3:

$$\mathcal{P}(m) = \frac{N!}{m!(N-m)!} p^m q^{N-m} \tag{5.6}$$

Notice that the right-hand side is the binomial expansion of $(p+q)^N$, Appendix A.3.1. Therefore the sum of the probabilities of being anywhere along a path of length N: $\sum_m \mathcal{P}(m) = (p+q)^N = 1$, as it should. The mean number of steps up: $\langle m \rangle$ is by

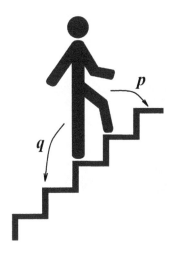

Figure 5.2 A drunk is climbing a staircase (adapted from ref.[86]). It has probability p of going up and probability q of going down.

definition:

$$\langle m \rangle \equiv \sum_m m \mathcal{P}(m) = p \frac{\partial}{\partial p} \sum_m \mathcal{P}(m) = p \frac{\partial}{\partial p}(p+q)^N = Np \qquad (5.7)$$

Similarly the average of m^2 is :

$$\langle m^2 \rangle \equiv \sum_m m^2 \mathcal{P}(m) = p^2 \frac{\partial^2}{\partial p^2} \sum_m \mathcal{P}(m) + p \frac{\partial}{\partial p} \sum_m \mathcal{P}(m) \qquad (5.8)$$

$$= (p^2 \frac{\partial^2}{\partial p^2} + p \frac{\partial}{\partial p}) (p+q)^N$$

$$= N(N-1)p^2 + pN = (Np)^2 + Npq$$

The variance σ_m^2 is by definition the average of the square deviation from the mean:

$$\sigma_m^2 \equiv \langle (m - <m>)^2 \rangle = \langle m^2 \rangle - \langle m \rangle^2 = Npq \qquad (5.9)$$

σ_m is also known as the standard deviation of the random variable m. It quantifies the mean fluctuations of m about its average value, $\langle m \rangle$. For a random walker the mean position after N steps is:

$$\langle l \rangle = 2 \langle m \rangle - N = N(2p-1) = N(p-q)$$

while the mean fluctuations about that position are

$$\sigma_l = 2\sigma_m = 2\sqrt{Npq}$$

If the probabilities of moving up or down are equal ($p = q = 1/2$): $\langle l \rangle = 0$ and $\sigma_l = \sqrt{N}$. In other words after N steps a population of drunks (or pollen particles in Perrin's experiments) starting from the same initial position would have spread about that position by a typical distance $\sigma_l = \sqrt{N}$. The probability of finding a particle (or the density of particles) at a distance $l = 2m - N$ from the origin can be deduced from Eq.5.6. An interesting limit of that probability distribution is the many steps (or long time) limit: $N \gg 1$ and $m/N = p + \mu$ (with $\mu \ll 1$ since the probability p of moving up is by the definition Eq.5.1: $p = \lim_{N \to \infty} m/N$). To compute the distribution $\mathcal{P}(m)$ in that limit we will use Stirling's formula:

$$N! = N^N e^{-N} \sqrt{2\pi N}(1 + O(1/N)) \tag{5.10}$$

Eq.5.6 can then be rewritten as:

$$\log \mathcal{P}(m) = \log N! - \log m! - \log(N-m)! + m\log p + (N-m)\log q$$
$$= N\log N - m\log m - (N-m)\log(N-m) + \frac{1}{2}\log \frac{N}{2\pi m(N-m)}$$
$$+ m\log p + (N-m)\log q + O(1/N)$$
$$= N[-\frac{m}{N}\log \frac{m/N}{p} - (1-\frac{m}{N})\log \frac{(1-m/N)}{q}] - \frac{1}{2}\log 2\pi N(m/N)(1-m/N)$$
$$= -N[(p+\mu)\log(1+\mu/p) + (q-\mu)\log(1-\mu/q)] - \frac{1}{2}\log 2\pi Npq$$
$$= -\frac{(N\mu)^2}{2Npq} - \frac{1}{2}\log 2\pi Npq + N \cdot O(\mu^3)$$

where we used the expansion $\log(1+x) = x - x^2/2 + x^3/3 + ...$, see Appendix A.3.1. Hence in the limit $N \gg 1$:

$$\mathcal{P}(m) = \frac{1}{\sqrt{2\pi Npq}}e^{-\frac{(m-Np)^2}{2Npq}} = \frac{1}{\sqrt{2\pi}\sigma_m}e^{-(m-\langle m \rangle)^2/2\sigma_m^2} \tag{5.11}$$

This distribution is known as a Gaussian probability distribution. It is maximal at $m = \langle m \rangle$ and its width is proportional to σ_m. By shifting m by its mean and dividing by the standard deviation, one obtains the so-called normal distribution $\mathcal{P}(\xi)$ (with $\xi = (m - \langle m \rangle)/\sigma_m$) shown in Fig.5.3. It is helpful for computing the probability of finding a value of m between two bounds m_1, m_2:

$$\mathcal{P}(m_1 < m < m_2) = \int_{m_1}^{m_2} \mathcal{P}(m)dm = \int_{\xi_1}^{\xi_2} \mathcal{P}(\xi)d\xi \tag{5.12}$$

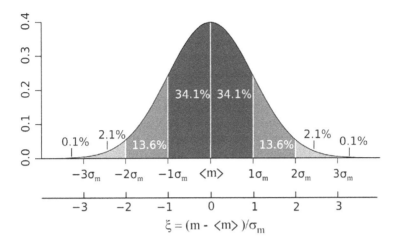

Figure 5.3 The Gaussian (Normal) distribution. The normal distribution $\mathcal{P}(\xi) = \frac{1}{\sqrt{2\pi}}e^{-\xi^2/2}$ is plotted. It corresponds to a Gaussian distribution $\mathcal{P}(m)$ (see text) which has been shifted by its mean and rescaled by its standard deviation: $\xi = (m - \langle m \rangle)/\sigma_m$. The integral below the curve in the various bands is shown. The probability of finding a value of m within a standard deviation from its mean (i.e. $-1 < \xi < 1$) is $\mathcal{P}(-1 < \xi < 1) = \int_{-1}^{1} \mathcal{P}(\xi)d\xi = 0.682$.

Freely jointed chain model of Polymers. The random walk model is a practical model for the description of polymers, such as Poly-Vinyl-Chloride (PVC), Poly-Ethylene-Glycol (PEG), ssDNA, RNA, etc.. In that approach the polymer is modeled as a chain of N freely jointed segments of size b (known as the Kuhn length) and of contour length: $L_0 = Nb$, see Fig.5.4. As these segments can adopt any orientation in space the chain is similar to a random walk of N steps of size b.

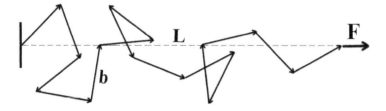

Figure 5.4 A polymer can be described as a chain of freely jointed segments of size b, whose orientation in space is uncorrelated but biased by the force F pulling on it. Notice the similarity with a random walk of step size b.

One is then interested in estimating the mean end-to-end distance L of the chain under tension or its average mean-square extension. Clearly the end-to-end distance is: $L = \sum_i \vec{b}_i$. From the investigation of the random walk we know that: $\langle L \rangle = Nb(p - q) = 2Nb\epsilon$, where the bias ϵ in this random walk is due to the tension pulling on the polymer (an issue we shall address below, see section 5.4). For a free (unstretched) polymer ($p = q = 1/2$) the average mean square extension will be $\sigma_L^2 \equiv \langle L^2 \rangle = Nb^2 = L_0 b$.

Consequently, for man-made polymers whose Kuhn length is about the size of the monomers i.e. $O(1\text{nm})$, a chain consisting of a few hundred monomers will occupy a region of size $\sigma_L \sim O(10 \text{ nm})$. Due to its double helical structure, a DNA molecule is however much stiffer. Its Kuhn length is very large: $b = 100\text{nm}$. A bacterial chromosome such as that of E.coli (of contour length $L_0 \approx 1.4$ mm) spontaneously occupies a region of about 12 μm in size, which is an order of magnitude larger than the size of an E.coli. To squeeze the chromosome inside the bacterial capsule, the circular DNA is twisted and compactified by specific DNA binding proteins. An even more acute problem of DNA compaction exists in Eukaryotes (e.g. multi-cellular organisms such as ourselves, where the length of one chromosome can be a few cm and $\sigma_L \sim 0(100\mu\text{m})$). In that case to squeeze it into the nucleus the molecule is first wrapped around specific proteins assemblies, known as histones, forming a bead on a string structure known as chromatine which is then further compacted via the formation of various highly coiled structures.

5.1.2   DIFFUSION

For reasons that will become clearer below, the Gaussian distribution is the most frequently encountered probability distribution in Statistics and Statistical Mechanics. As argued, it describes the spreading of particles from an initial seed due to the random "invisible blows" from the atoms invoked by Lucretius, a process known as diffusion. We can use our previous results on the discrete motion of a random walker to deduce the probability distribution $\mathcal{P}(x,t)$ associated to the diffusion of particles to position x and time t if initially localized at position $x = 0$: $\mathcal{P}(x,0) = \delta(x)$.

In that instance $N = t/\Delta t$ is the number of successive time points (taken at interval Δt), a is the step size (the mean distance traveled during the interval Δt), so that the overall distance x traveled by a particle is $x = la = (2m - N)a$. The probability of going up/down (or left/right) will be the same $p = q = 1/2$ unless the movement is biased by a flow which carries the particles in a given direction, in which case $p = 1/2 + \epsilon$ and $q = 1/2 - \epsilon$. The relation between ϵ and the velocity of the flow v can be deduced from the mean distance traveled: $vt \equiv \langle x \rangle = a\langle l \rangle = aN(p-q) = (2a\epsilon/\Delta t)t$, namely: $v = 2a\epsilon/\Delta t$. If the flow rate is small (i.e. if $\epsilon \ll 1$) then the fluctuations about that mean distance will be $\sigma_x = 2a\sqrt{Npq} = \sqrt{(a^2/\Delta t)t} \equiv \sqrt{2Dt}$, where $D \equiv a^2/2\Delta t$ is known as the diffusion constant. The probability distribution $\mathcal{P}(x,t)$ of the particle can then be deduced from Eq.5.11:

$$\mathcal{P}(x,t) = \frac{1}{\sqrt{4\pi Dt}} e^{-(x-vt)^2/4Dt} \tag{5.13}$$

This probability distribution describes the diffusion of particles in 1D. In presence of a flow the particles are advected at velocity v, but due to random collisions with the molecules of the fluid they also disperse to occupy a region of size $\sigma_x(t) = \sqrt{2Dt}$ ($\sqrt{4Dt}$ in 2D and $\sqrt{6Dt}$ in 3D). For small molecules (e.g. sugars) the coefficient of diffusion in water is $D = 5 \ 10^{-6}$ cm^2/sec, while for a small protein it is $D = 10^{-6}$ cm^2/sec (as we shall see below, D decreases as the inverse of the radius of the diffusing particle (see section 5.11.3).

Eq.5.13 is the solution to the diffusion equation:

$$\frac{\partial P}{\partial t} = D\frac{\partial^2 P}{\partial x^2} - v\frac{\partial P}{\partial x} \tag{5.14}$$

This can be verified directly by plugging $P(x,t)$ from Eq.5.13 into Eq.5.14. One can also see that by deriving Eq.5.14 from the 1D random walk process: let $P(x,t+\Delta t)$ be the probability of finding a particle at position x at time $t+\Delta t$. The particle can reach that position by moving up from position $x-a$ at time t with probability $p = 1/2+\epsilon$ or moving down from position $x+a$ with probability $q = 1/2-\epsilon$. We can thus write the so-called master equation describing the conservation of probability in a random walk:

$$P(x,t+\Delta t) = pP(x-a,t) + qP(x+a,t)$$
$$= (1/2+\epsilon)[P(x,t) - a\frac{\partial P}{\partial x} + \frac{a^2}{2}\frac{\partial^2 P}{\partial x^2} + ...]$$
$$+ (1/2-\epsilon)[P(x,t) + a\frac{\partial P}{\partial x} + \frac{a^2}{2}\frac{\partial^2 P}{\partial x^2} + ...] \tag{5.15}$$
$$= P(x,t) - (2a\epsilon)\frac{\partial P}{\partial x} + \frac{a^2}{2}\frac{\partial^2 P}{\partial x^2}$$

Expanding $P(x,t+\Delta t) = P(x,t) + \Delta t \cdot \partial P/\partial t + ...$ and identifying as before the flow velocity $v = 2a\epsilon/\Delta t$ and the diffusion constant $D = a^2/2\Delta t$ yields Eq.5.14. In 3D the diffusion equation becomes:

$$\frac{\partial P}{\partial t} = D\nabla^2 P - \vec{v} \cdot \vec{\nabla}P \tag{5.16}$$

which solution is the generalization of Eq.5.13:

$$P(\vec{x},t) = (\frac{1}{4\pi Dt})^{3/2} e^{-(\vec{x}-\vec{v}t)^2/4Dt} \tag{5.17}$$

Hence in 3D the mean displacement is $\langle \vec{x} \rangle = \vec{v}t$ and the variance: $\sigma_r^2 = \langle (\vec{x}-\vec{v}t)^2 \rangle = \sigma_x^2 + \sigma_y^2 + \sigma_z^2 = 6Dt$.

Just as Eq.3.43 expresses charge conservation, the diffusion equation is a conservation law for the probability P. Eq.5.16 can be recast as:

$$\frac{\partial P}{\partial t} + \vec{\nabla} \cdot \vec{J} = 0$$

where the current density is $\vec{J} = -D\vec{\nabla}P + \vec{v}P$, see also section 5.11.1.1.

Sometimes one is interested in the probability that a random walker or a diffusing particle has not reached a certain distance x_0 within a time t, namely one is interested in the integrated probability:

$$\mathcal{P}(x < x_0, t) = \int_{-\infty}^{x_0} \mathcal{P}(x, t)dx = \frac{1}{2}[1 + \mathrm{erf}(x_0 - vt)] \qquad (5.18)$$

where the error function $\mathrm{erf}(\xi)$ is defined as:

$$\mathrm{erf}(\xi) \equiv \frac{2}{\sqrt{\pi}} \int_0^{\xi} e^{-z^2}dz$$

The diffusion equation is encountered in many different contexts (thermodynamics, fluid dynamics, economics, ecology, etc.), see below and section 5.11. The initial condition $\mathcal{P}(\vec{x}, 0)$ maybe less trivial than assumed here (i.e. $\mathcal{P}(\vec{x}, 0) = \delta(\vec{x})$) and the boundary conditions more restrictive (for example that on some surface Γ, $\mathcal{P}(\vec{x}, t)\big|_{\Gamma} = \mathrm{const}$). In such instances our solution, Eq.5.17, may be less relevant and one may often have to resort to numerical solutions of Eq.5.16 to obtain an estimate of $\mathcal{P}(\vec{x}, t)$.

Stock-market fluctuations. In recent years, Statistical Mechanics methods have been used in other contexts, in particular in Economics and Finance. For example the fluctuations of the value of stocks have been modeled by a random walk. Specifically, the value of a stock has been assumed to fluctuate due to the uncorrelated random whims of investors. This assumption is not always valid: during a stock-market crash, for example, panic selling imply highly correlated impulses (a phenomenon resulting in non-Gaussian behavior and the presence of so-called fat-tails in the probability distribution of gains and losses). As the change in stock value is proportional to the price S of the stock (it is measured in percents, not in absolute value), the "randomly walking" variable (which is going up or down by an average percentage, i.e. a step of size a) is $u = \log S/S_0$, where S_0 is the price of the stock at time $t = 0$. After a time t the distribution $\mathcal{P}(u, t)$ (of the various stocks in an index such as the Dow Jones or the SP 500) will be given by Eq.5.13:

$$\mathcal{P}(u, t) = \frac{1}{\sqrt{2\pi\sigma_u^2 t}} e^{-(u - vt)^2/2\sigma_u^2 t} \qquad (5.19)$$

where σ_u is the standard deviation in the logarithmic fluctuations of the stock price (known as the volatility) and v is related to the rate of increase (or decrease) μ of the mean stock value $\langle S \rangle$:

$$\langle S \rangle = S_0 \int e^u \mathcal{P}(u, t)du = \frac{S_0}{\sqrt{2\pi\sigma_u^2 t}} e^{(v + \sigma_u^2/2)t} \int du e^{-[u - (v + \sigma_u^2)t]^2/2\sigma_u^2 t}$$

$$= S_0 e^{(v + \sigma_u^2/2)t} \equiv S_0 e^{\mu t}$$

hence $v = \mu - \sigma_u^2/2$. Notice that even if $v = 0$ (i.e. no bias in the random walk) $\mu > 0$, i.e. the average value of the stocks' portfolio increases (which is a reflection of the fact that the stock price is bounded from below: $S > 0$)). One can also show that $\langle S^2 \rangle / \langle S \rangle^2 = e^{\sigma_u^2 t}$, which is used to estimate the volatility σ_u from data on the fluctuations of stock prices as a function of time. By designing a portfolio of uncorrelated stocks, one can reduce the risk associated with the fluctuations of a given stock and predict the portfolio's future value (a prediction used in the pricing of options). One of the main difficulties of this type of predictions is to estimate the volatility (assumed to be time independent) and to take into account the existence of correlations in the stock-market fluctuations.

Gambler's ruin. An important application of random walks and the diffusion equation is known as Gambler's ruin. The goal is to find the long time probability $\mathcal{P}_r(t)$ that a gambler (random walker) starting with an initial capital u_0 gets ruined in a time t. In the ecological context that problem translates into the probability of extinction within a time t of a species with u_0 individuals.

The capital of the gambler fluctuates as a random walk (or diffusion process) with standard deviation σ_u (proportional to the average amount of money a gambled at each time step Δt: $\sigma_u^2 = a^2/4\Delta t$) and drift $-v < 0$ (reflecting the fact that gambling is usually not a zero-sum game: the casino takes its cut of the gambled money). Starting from an initial distribution of capital: $\mathcal{P}(u,0) = \delta(u - u_0)$ the problem is to solve for $\mathcal{P}(u,t)$ satisfying the diffusion equation, Eq.5.14:

$$\frac{\partial \mathcal{P}}{\partial t} = \frac{\sigma_u^2}{2}\frac{\partial^2 \mathcal{P}}{\partial u^2} + v\frac{\partial \mathcal{P}}{\partial u}$$

The condition that a gambler stops playing when ruined implies the boundary condition $\mathcal{P}(0,t) = 0$. The solution of the diffusion equation with this boundary condition is obtained by the method of images introduced in the context of electrostatics (where one had to solve a similar Poisson equation for the electric field of a given charge placed in front of a grounded plane, see section 3.1.5.1). It consists in introducing a random walker (charge) of strength $-\exp(2u_0v/\sigma_u^2)$ at a symmetric position with respect to the absorbing plane, i.e. at $-u_0$. The probability distribution $\mathcal{P}(u,t)$ in the half plane of interest ($u > 0$) is then given by the sum of the probability distributions (Eq.5.13) associated with each walker:

$$\mathcal{P}(u,t) = \frac{1}{\sqrt{2\pi\sigma_u^2 t}}\left[\exp\left(-\frac{(u - u_0 + vt)^2}{2\sigma_u^2 t}\right) - e^{2u_0v/\sigma_u^2}\exp\left(-\frac{(u + u_0 + vt)^2}{2\sigma_u^2 t}\right)\right]$$

One can verify that $\mathcal{P}(0,t) = 0$ for all times. The probability that the gambler's capital is still positive at time t (the "survival" probability) is[1] :

$$Sur(t) = \int_0^\infty \mathcal{P}(u,t)du = \frac{1}{2}\left[1+\text{erf}\left(\frac{u_0-vt}{\sqrt{2\sigma_u^2 t}}\right)\right] - \frac{e^{2u_0 v/\sigma_u^2}}{2}\left[1-\text{erf}\left(\frac{vt+u_0}{\sqrt{2\sigma_u^2 t}}\right)\right]$$

The probability of ruin at precise time t, $\mathcal{P}_r(t)$ satisfies: $\mathcal{P}_r(t)dt = Sur(t) - Sur(t+dt)$, hence[2]:

$$\mathcal{P}_r(t) = -\frac{dSur}{dt} = \frac{u_0}{\sqrt{2\pi\sigma_u^2 t^3}}\exp\left(-\frac{(u_0-vt)^2}{2\sigma_u^2 t}\right)$$

In the limit $t \to \infty$ the probability of ruin (survival) decays exponentially with time $\mathcal{P}_r(t) \sim e^{-v^2 t/2\sigma_u^2}$, i.e. avoiding ruin becomes increasingly unlikely. Surprisingly though, that probability of surviving at time t is larger, the larger the amount a of money bet (i.e. the larger σ_u^2). Notice that even if $v = 0$ (zero sum game) the gambler is still ruined if he/she plays long enough albeit with a probability that decays slower as $\mathcal{P}_r(t) \sim t^{-3/2}$ rather than exponentially. If $-v > 0$ and the drift is away from the absorbing condition at $u = 0$ —a situation which is relevant in an evolutionary/ecological context when a mutant with better fitness arises —the previous result implies that the probability of mutant extinction becomes exceedingly small as time and the mutant population increases.

5.1.3 CENTRAL LIMIT THEOREM

An important theorem in Statistics is the Central Limit Theorem which states that if some random *uncorrelated* variables $\{x_i\}$ are distributed with a probability $\mathcal{P}(x)$, then the mean estimated over N points $\bar{x}_N$:

$$\bar{x}_N = \frac{1}{N}\sum_{i=1}^N x_i \tag{5.20}$$

is (in the limit $N \gg 1$) a random variable, Gaussian distributed with mean $\langle x \rangle$ and standard deviation $\sigma_{x_N} = \sigma_x/\sqrt{N}$. This theorem illuminates the centrality of the Gaussian distribution in Statistics: no matter what is the probability distribution of some measured data (height, weight, speed, pH, wealth, life-span, etc.), the probability distribution of the mean estimated over N uncorrelated data points is Gaussian and the error on that estimated mean σ_{x_N} decreases as $1/\sqrt{N}$.

---

[1] Notice that:

$$\frac{1}{\sqrt{2\pi}\sigma}\int_0^\infty dx e^{-(x-x_0)^2/2\sigma^2} = \frac{1}{2}[1+\text{erf}(x_0/\sqrt{2}\sigma)]$$

[2] Using the fact that: $\text{derf}(z)/dz = 2e^{-z^2}/\sqrt{\pi}$, one obtains (with $z \equiv (u_0 \pm vt)/\sqrt{2}\sigma_u$):

$$\frac{\text{derf}(z)}{dt} = \frac{2}{\sqrt{\pi}}e^{-z^2}\frac{dz}{dt} = \frac{2}{\sqrt{\pi}}e^{-z^2}\left(\frac{\pm v}{\sqrt{2\sigma_u^2 t}} - \frac{u_0 \pm vt}{2\sqrt{2\sigma_u^2 t^3}}\right) = \frac{-u_0 \pm vt}{\sqrt{2\pi\sigma_u^2 t^3}}e^{-z^2}$$

To prove that important theorem, consider the Fourier transform $\Phi(k)$ of the probability distribution $\mathcal{P}(x)$, see Appendix A.4:

$$\Phi(k) = \frac{1}{\sqrt{2\pi}} \int e^{ikx} \mathcal{P}(x) dx \tag{5.21}$$

$\Phi(k)$ is known as the generating function, since all the moments $\langle x^n \rangle$ $(n = 1, 2, \cdots)$ of $\mathcal{P}(x)$ can be derived from it. The first moment (the mean) is:

$$\langle x \rangle = -i \sqrt{2\pi} \frac{\partial \Phi}{\partial k}\Big|_{k=0} = \int x \mathcal{P}(x) dx$$

Similarly the nth moment:

$$\langle x^n \rangle = (-i)^n \sqrt{2\pi} \frac{\partial^n \Phi}{\partial k^n}\Big|_{k=0} = \int x^n \mathcal{P}(x) dx$$

For the estimated mean $\bar{x}_N$ the generating function will be:

$$\Phi_N(k) = \frac{1}{\sqrt{2\pi}} \int \cdots \int e^{ik\bar{x}_N} \mathcal{P}(x_1, x_2, \dots x_N) dx_1 \dots dx_N$$

Expanding the exponential (using $e^\xi = 1 + \xi + \xi^2/2 + \dots$) yields[3]

$$\Phi_N(k) = \frac{1}{\sqrt{2\pi}} \int \cdots \int (1 + \frac{ik}{N} \sum_i x_i - \frac{k^2}{2N^2} \sum_{i,j} x_i x_j + \dots) \mathcal{P}(x_1, x_2, \dots x_N) dx_1 \dots dx_N$$

$$= \frac{1}{\sqrt{2\pi}} [1 + ik \langle x \rangle - \frac{k^2 N(N-1)}{2N^2} \langle x \rangle^2 - \frac{k^2}{2N} \langle x^2 \rangle + \dots]$$

$$= \frac{1}{\sqrt{2\pi}} [1 + ik \langle x \rangle - k^2 \langle x \rangle^2 /2 - k^2 \sigma_x^2/2N + \dots]$$

$$\approx \frac{1}{\sqrt{2\pi}} \exp\left(ik \langle x \rangle - k^2 \sigma_x^2/2N\right)$$

---

[3] Notice that: $\int \cdots \int \sum_i x_i \mathcal{P} dx_1 \dots dx_N \equiv \sum_i \int \cdots \int x_i \mathcal{P} \Pi_n dx_n = N \langle x \rangle$ and:

$$\int \cdots \int \sum_{i,j} x_i x_j \mathcal{P} \Pi_n dx_n = \sum_{i \neq j} \int \cdots \int x_i x_j \mathcal{P} \Pi_n dx_n + \sum_i \int \cdots \int x_i^2 \mathcal{P} \Pi_n dx_n = N(N-1) \langle x \rangle^2 + N \langle x^2 \rangle$$

The distribution $\mathcal{P}_N(\bar{x}_N)$ is then obtained by the inverse Fourier transform of $\Phi_N(k)$:

$$
\begin{aligned}
\mathcal{P}_N(\bar{x}_N) &= \frac{1}{\sqrt{2\pi}} \int dk\, e^{-ik\bar{x}_N} \Phi_N(k) \\
&= \frac{1}{2\pi} \int dk\, e^{-k^2\sigma_x^2/2N - ik(\bar{x}_N - \langle x \rangle)} \\
&= \frac{1}{2\pi} \exp\left[-\frac{(\bar{x}_N - \langle x \rangle)^2}{2\sigma_x^2/N}\right] \int dk \exp\left[-\frac{\sigma_x^2}{2N}\left(k + i\frac{\bar{x}_N - \langle x \rangle}{\sigma_x^2/N}\right)^2\right] \\
&= \frac{1}{\sqrt{2\pi\sigma_x^2/N}} \exp\left[-\frac{(\bar{x}_N - \langle x \rangle)^2}{2\sigma_x^2/N}\right]
\end{aligned}
$$

Hence, as stated above the distribution of the estimated mean $\bar{x}_N$ is Gaussian with mean $\langle x \rangle$ and standard deviation $\sigma_x/\sqrt{N}$. In fact, one can view the random walk as a special case of the Central Limit Theorem. In a random walk each step is a random variable δx_i uncorrelated with the preceding step and which can adopt two values: $\delta x_i = a$ with probability $p = 1/2 + \epsilon$ and $\delta x_i = -a$ with probability $q = 1/2 - \epsilon$. The mean value of that random variable is: $\langle \delta x_i \rangle = a(p - q)$ and its standard deviation is $\sigma_{\delta x}^2 = 4a^2 pq$. By the Central Limit Theorem the distribution of the distance $x = N(\sum_i \delta x_i/N)$ traveled by a random walker after $N = t/\Delta t$ steps is Gaussian with mean $\langle x \rangle = N\langle \delta x_i \rangle = Na(p-q) = (2a\epsilon/\Delta t)t = vt$ and variance $\sigma_x^2 = N^2(\sigma_{\delta x}^2/N) = (a^2/\Delta t)t = 2Dt$ (as derived previously, Eq.5.13).

The Central Limit Theorem is helpful when estimating the probability of events. One usually measures the frequency $f_i = M_i/N$ of a given event and not its probability $\mathcal{P}_i$, defined as the limit of f_i when $N \to \infty$ (see Eq.5.1) which cannot be achieved experimentally. The observation of event i is thus a binary process which outcome "1" (i.e. observed) has probability $\mathcal{P}_i$ and outcome "0" (not observed) has probability $1 - \mathcal{P}_i$. By the central limit theorem, we know that after N trials, the deviation of the frequency f_i (i.e. the measured mean $\bar{x}_N$ of the N trials) from the probability $\mathcal{P}_i$ (the mean when $N \to \infty$) is given by:

$$
\mathcal{P}(f_i) = \frac{1}{\sqrt{2\pi\sigma^2/N}} \exp\left[-\frac{(f_i - \mathcal{P}_i)^2}{2\sigma^2/N}\right] \tag{5.22}
$$

where $\sigma^2 = \mathcal{P}_i(1 - \mathcal{P}_i)$ is the variance of a binary distribution. This is a result that politicians who are fond of polls should learn! If a poll performed on 1000 persons predicts a victory for a candidate with $f_i = 52\%$ of voting intentions, the probability of defeat (i.e. $\mathcal{P}_i < 0.5$) is still a non-negligible $\sim 10\%$ (integrating Eq.5.22 for $-\infty < \mathcal{P}_i < 0.5$ with $\sigma = 0.5$).

5.2 ◆ MISSING INFORMATION, GUESSING AND ENTROPY

The central problem of Statistical Mechanics is how to make a good guess about the probabilistic behavior of many body systems (gases, liquids, solids, glasses, magnetic systems, etc.). The space of possible states is assumed to be known: for example the various spin orientations for magnetic systems or the positions and momenta (the phase space) of the molecules of a gas or a liquid. In equilibrium Statistical Mechanics, as the system doesn't evolve in time the probability of occupation of these various states is assumed to be time independent. The problem then is to determine the occupation probabilities when certain macroscopic (i.e. ensemble) properties are known, such as the number of particles, the energy or the magnetization. Often only the mean values of these properties are actually known, for example the mean energy of a gas or the mean magnetization of an ensemble of spins. How then can one estimate the various states occupation probabilities taking into account all the available information and only that information (i.e. not making uncontrolled assumptions)?

In some ways the problem is similar to that facing a gambler playing a card game: how to assess the strength of one's hand, i.e. estimate the probabilities of some card configurations (states) based only on the knowledge of the cards played without making subjective assumptions (e.g. relying on one's luck). The problem is one of *missing information*: one doesn't know what suit (spade, heart, diamond or club) a card just handed belongs to or what value (1-10, jack, queen or king) in that suit it has. Clearly the more possibilities n the greater the missing information $I(n)$. The missing information (also known as Shannon entropy) associated with the knowledge of a suit is $I(4)$, the missing information associated with the knowledge of the value of the card is $I(13)$. In absence of a-priori data (i.e. if no card has been revealed), the missing information associated with the knowledge of a card $I(52)$ can be supplied in two steps, by first identifying the suit - supplying information in the amount $I(4)$ - and then its value - supplying information in amount $I(13)$. The information thus supplied piecewise is the same as the information supplied by identifying the card directly, hence: $I(52) = I(4) + I(13)$. More generally the Shannon entropy associated with $n \cdot m$ possibilities obeys:

$$I(n \cdot m) = I(n) + I(m) \tag{5.23}$$

This property together with the fact that if there is only one possibility there is no missing information, i.e.: $I(1) = 0$, implies that:

$$I(n) = k \log n \tag{5.24}$$

Where k is an arbitrary constant, which sets the units of I. For example, if the missing information is measured in bits $k = 1/\log 2$. To convince oneself of the validity of Eq.5.24, notice that by virtue of Eq.5.23, $I(a^n) = I(a) + I(a^{n-1}) = nI(a)$, hence by extension:

$$I(n) \equiv I(e^{\log n}) = I(e) \log n \sim \log n$$

Eq.5.24, which specifies the number of bits of information needed to identify a given card (or state, configuration, etc.), is obviously valid only if one has no reason to believe that the card deck is loaded, i.e. if all cards have the same probability of being handed.

Consider another example: one observes N birds sitting on a telephone line, some of which are black and some of which are white. What is the information content of a configuration with m_1 black birds and m_2 white ones? If all the birds are black there is no missing information. If there is a single white bird among the lot ($m_1 = N-1; m_2 = 1$) then we miss the information as to where it is among the N birds and the Shannon entropy is $k \log(N)$. In other words if there are 1000 birds we need about 10 bits of information ($\log_2(1000) \approx 10$) to tell where the white bird is. If $m_2 > 1$ and birds don't interact (e.g. birds of similar color don't bunch) the number of configurations is $\Omega = N!/m_1!m_2!$ (= number of permutations of N distinct birds divided by the permutations among black and white birds which are indistinguishable). With the use of Stirling's formula, Eq.5.10, the missing information is then:

$$I(\Omega) = k \log(\Omega) = k(\log N! - \log m_1! - \log m_2!)$$
$$\approx k((m_1 + m_2) \log N - m_1 \log m_1 - m_2 \log m_2)$$
$$= -Nk(f_1 \log f_1 + f_2 \log f_2)$$

where $f_i = m_i/N$ is the fraction of birds of color i (=black or white). By the definition of probability, Eq.5.1, in the limit $N \to \infty$ that fraction is the probability $\mathcal{P}_i$ of finding a bird of that color.

That example suggests a way to define the Shannon entropy when the outcomes are not equally probable. Let there be $i = 1, ...M$ possible outcomes with probabilities $\mathcal{P}_i$. Consider N successive experiments such that outcome i is observed m_i times ($N = \sum_i m_i$). For a given set $\{m_i\}$ the number of possible configurations is $\Omega_N = N!/\prod_i m_i!$ and the missing information is:

$$I(\Omega_N) = k \log(\Omega_N) = k(\log N! - \sum_i \log m_i!)$$
$$= k\left(\sum_i m_i\right) \log N - k \sum_i m_i \log m_i$$
$$= -Nk \sum_i^M \frac{m_i}{N} \log \frac{m_i}{N}$$

In the limit of large N, $m_i/N \to \mathcal{P}_i$. The Shannon entropy $S(\{\mathcal{P}_i\})$ is defined as:

$$S(\{\mathcal{P}_i\}) = \lim_{N \to \infty} I(\Omega_N)/N = -k \sum_i^M \mathcal{P}_i \log \mathcal{P}_i \qquad (5.25)$$

As expected the entropy is zero when one of the $\mathcal{P}_i$'s equal one. The missing information is maximal when all outcomes are equally probable.

To prove that last statement we have to show that $S(\{\mathcal{P}_i\})$ is maximized under the constraint $\sum_i^M \mathcal{P}_i = 1$ when $\mathcal{P}_i = 1/M$. To find the extremum of a function $f(x)$ under a constraint $g(x) = g_0$ (a constant) one finds the extrema x_y of the auxiliary function

$\Gamma(x) = f(x) + \gamma g(x)$, where γ is known as a Lagrange multiplier and is determined by the implicit equation $g(x_\gamma) = g_0$. In the particular case of maximizing the Shannon entropy $S(\{\mathcal{P}_i\})$ under the constraint that the $\mathcal{P}_i$'s are probabilities that sum up to one, we shall write:

$$\Gamma(\{\mathcal{P}_i\}) = S(\{\mathcal{P}_i\}) + k(\gamma_0 + 1) \sum_i^M \mathcal{P}_i$$

$$= -k \sum_i^M \mathcal{P}_i \log \mathcal{P}_i + k(\gamma_0 + 1) \sum_i^M \mathcal{P}_i \qquad (5.26)$$

where $\gamma \equiv k(\gamma_0 + 1)$ is the Lagrange multiplier. To find the extrema of $\Gamma(\{\mathcal{P}_i\})$ we take its derivative:

$$\frac{\partial \Gamma}{\partial \mathcal{P}_j} = -k \log \mathcal{P}_j - k + k(\gamma_0 + 1) = 0$$

from which we deduce that:

$$\mathcal{P}_j = e^{\gamma_0}$$

and from the constraint $1 = \sum_i^M \mathcal{P}_i = M e^{\gamma_0}$ we deduce that $\mathcal{P}_i = 1/M$. Hence the Shannon entropy is maximized when all outcomes are equally probable.

5.2.1 INFORMATION, SHANNON ENTROPY AND CODING

The Shannon entropy is a central concept in information theory. It sets the minimal number of bits needed to faithfully reconstruct a given message, state, image, etc., without loosing any information. It is essential in all the methods of text and image compression and more generally compressed sensing.

Consider for example the English alphabet. In a typical text, the probabilities of occurrence $\mathcal{P}_i$ of the 26 letters are not equal. If they were, one would need to represent each character by $\log_2 26 = 4.7$ bits. As it happens the Shannon entropy of the English characters is significantly smaller, about 4.2 bits/character (see Fig.5.5). As a result frequently occurring characters in the alphabet (e.g. e,t,a..) can be represented by less bits than rarer characters (such as z, q, x, ...). In a code proposed in 1952 by D.Huffman[87] the number of bits n_i for a given character i increases roughly as $-\log \mathcal{P}_i$, which results in a code that comes very close to the Shannon limit, see Fig.5.5.

To understand how Huffman coding works consider a hypothetical text written with just $M = 10$ letters appearing with the frequencies shown in Fig.5.6. The Shannon entropy of this text is $S_t = -\sum_i^M \mathcal{P}_i \log_2 \mathcal{P}_i = 2.317$ bits, which is considerably smaller than the missing information if the letters were equiprobable: $\log_2 10 = 3.322$ bits.

letter	probability	$-p_i \log_2 p_i$	Huffman code	number of bits, n_i	$n_i * p_i$
e	0,12702	0,378122335	100	3	0,38106
t	0,09056	0,313788791	000	3	0,27168
a	0,08167	0,295159446	1110	4	0,32668
o	0,07507	0,280432971	1101	4	0,30028
i	0,06966	0,267740001	1011	4	0,27864
n	0,06749	0,262480923	1010	4	0,26996
s	0,06327	0,25196231	0111	4	0,25308
h	0,06094	0,24598228	0110	4	0,24376
r	0,05987	0,243193312	0101	4	0,23948
d	0,04253	0,193740113	11111	5	0,21265
l	0,04025	0,186553413	11110	5	0,20125
c	0,02782	0,143766353	01001	5	0,1391
u	0,02758	0,142870846	01000	5	0,1379
m	0,02406	0,129375902	00111	5	0,1203
w	0,02361	0,127599257	00110	5	0,11805
f	0,02228	0,122275023	00101	5	0,1114
g	0,02015	0,113506488	110011	6	0,1209
y	0,01974	0,111782373	110010	6	0,11844
p	0,01929	0,109875899	110001	6	0,11574
b	0,01492	0,090513801	110000	6	0,08952
v	0,00978	0,065290789	001000	6	0,05868
k	0,00772	0,054172656	0010011	7	0,05404
j	0,00153	0,014308947	001001011	9	0,01377
x	0,0015	0,014071233	001001010	9	0,0135
q	0,00095	0,009537796	001001001	9	0,00855
z	0,00074	0,007696138	001001000	9	0,00666
total:	1	4,175799412			4,20507

Figure 5.5 The probabilities of occurrence of the letters of the English alphabet vary by two orders of magnitude. The Shannon entropy of the English alphabet, $S_e = -\sum_i P_i \log_2 P_i = 4.175$ bits. The fourth column represents the Huffman code of the various letters. The average number of bits/character used by this code to represent an English text is on average 4.20 bits, close to the Shannon limit.

Huffman coding starts by ordering the letters according to their frequencies (in Fig.5.6 the most frequent letter "e" with frequency 0.44 is at the top, the rarer one "d" with frequency 0.004 at the bottom of the first column). One then adds the frequencies of the two rarest occurrences to generate a new ordered list of frequencies. The new element in the list is linked to the two elements in the previous list, the link to the rarest of the two being assigned bit "1", the other bit "0" (or vice-versa).

The procedure is repeated $M - 1$ times until the list consists of just one element with frequency one. The code for a given character is then read from right to left as one proceeds up the tree.

For example in Fig.5.6, the code for letter "h" (with frequency 0.014 shown in the grey oval) ends with 000, as this letter (grey oval) is linked with three bit "0" links to the intermediate node with frequency 0.78 and it begins with 0001 as this

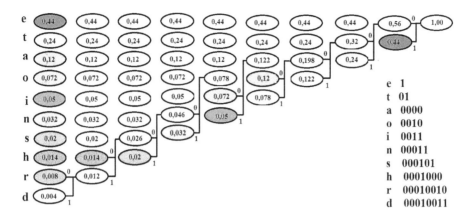

Figure 5.6 Huffman coding. In this example the 10 letters (e,t,a,o,i,n,s,h,r,d) are ranked according to their frequency of appearance (numbers in the colored ovals) in a hypothetical text:the most frequent (e) at the top; the rarest (d) at the bottom. The two rarest frequencies are added and a new list with one less element is obtained and ordered. The procedure is repeated until but one member with frequency one remains. The links between the nodes are assigned bit "1" when linking the least frequent element and "0" otherwise. The code of a given letter is read from right to left as one moves up the tree (from left to right in this example). For example the letter i (pink oval) is linked with bit "1" to a node with frequency 0.122, which is linked to the rightest node by links with bits "1", "0", "0" . Its code is thus: 0011. The mean number of bits required to code this text is 2.362, quite close to the Shannon limit of 2.317.

node is linked to the last node with a bit "1" link followed by three bit "0" links. The Huffman code possesses a crucial property for decoding: it is prefix free, namely no code for one letter is the prefix for the code of another letter. If for example the code for "a" was 01 and the code for "i" was 010, one would not know whether a code 01001100... is to be read as starting with "a" or "i". Such ambiguities are absent in Huffman coding.

5.2.2 ◆ GUESSING THE ODDS FOR FAIR AND LOADED DIE

The role of the missing information, Shannon entropy or just entropy, is not the same in the theory of information and in Statistical Mechanics. In the former, the frequencies of certain symbols in a text, an image, etc. are known and the problem is how to code these symbols in the most efficient way (using as few bits as possible) without losing information. The Shannon entropy sets a lower bound on the possible solutions.

In statistical mechanics, the frequencies of the various equilibrium states of a system of particles (atoms, molecules, spins, etc.) are unknown and the problem is to "guess" them on the basis of data on the ensemble (the number of particles, the mean energy, etc.). Here the entropy is used as a "tool" to make a "guess" on the probability of occupation of these states. Namely one **assumes** that the available data provides all the information there is on the system and one proceeds to maximize the missing information (the entropy) under these conditions. The initial assumption may or may not be correct, but maximizing the entropy has the merit of providing one with a systematic way of making predictions which can be experimentally tested. This approach is an example of those mathematical "miracles that we do not deserve": based on a few known variables, it allows one to deduce the unknown probabilities of many more states (usually in calculus, to solve a problem one needs as many equations as there are unknowns). The logic behind that approach is the "Principle of insufficient reason", introduced as early as 1713 by Bernoulli: in absence of a particular reason to choose one state over another, the only logical choice is that which maximizes our ignorance (any other introduces an unsubstantiated bias). That is why we assign *a-priori* probabilities 1/2 to the observation of the faces of a tossed coin, 1/6 for the faces of a tossed die, etc.

To exemplify the procedure adopted in statistical mechanics let us "guess" the probability of occurrences $\mathcal{P}_i \equiv \mathcal{P}(x_i)$ of the various faces of a die ($x_i = 1, 2, \cdots, 6$) knowing *only* that the mean observed after throwing the die a large number N of times is $\sum_i x_i \mathcal{P}_i = a$ (= 3.5 for an unloaded die). As just argued our best "guess" is based on maximizing the Shannon entropy

$$S(\{\mathcal{P}_i\}) = -k \sum_i \mathcal{P}_i \log \mathcal{P}_i \qquad (5.27)$$

under the constraints:

$$\sum_i \mathcal{P}_i = 1 \text{ and } \sum_i x_i \mathcal{P}_i = a \qquad (5.28)$$

As shown previously, to solve that problem one introduces two Lagrange multipliers $k(\gamma_0 + 1)$ and $-k\alpha$ associated with the two constraints to define the auxiliary function:

$$\Gamma(\{\mathcal{P}_i\}) = -k \sum_i \mathcal{P}_i \log \mathcal{P}_i + k(\gamma_0 + 1) \sum_i \mathcal{P}_i - k\alpha \sum_i x_i \mathcal{P}_i$$

The extrema of Γ are obtained by setting $\partial \Gamma / \partial \mathcal{P}_j = 0$:

$$-k \log \mathcal{P}_j + k\gamma_0 - k\alpha x_j = 0$$

Hence: $\mathcal{P}_i = \exp(\gamma_0 - \alpha x_i)$. The first constraint, $\sum_i \mathcal{P}_i = 1$ implies that:

$$e^{\gamma_0} = 1 / \sum_i e^{-\alpha x_i} \qquad (5.29)$$

The second constraint implies that:

$$a = \sum_i x_i \mathcal{P}_i = \frac{\sum_i x_i e^{-\alpha x_i}}{\sum_i e^{-\alpha x_i}} = -\frac{\partial}{\partial \alpha} \log \sum_{i=1}^{6} e^{-\alpha x_i} = \frac{\partial \gamma_0}{\partial \alpha} \qquad (5.30)$$

Assuming that $\alpha \ll 1$ the exponent can be expanded to yield: $\sum_i e^{-\alpha x_i} = 6 - 21\alpha + 91\alpha^2/2 + O(\alpha^3)$ and after expanding the logarithm to same order, (using $\log(1 + x) = x - x^2/2 + O(x^3)$), one gets:

$$a = -\frac{\partial}{\partial \alpha} \log 6(1 - \frac{7\alpha}{2} + \frac{91\alpha^2}{12}) = -\frac{\partial}{\partial \alpha}(-\frac{7\alpha}{2} + \frac{35\alpha^2}{24}) = 3.5 - 35\alpha/12$$

When $a = 3.5$, $\alpha = 0$ and thus $\mathcal{P}_i = 1/6$, the die is fair: when throwing it all faces are equiprobable. When $\alpha \neq 0$, the die is loaded and we can tune the discrepancy between the probability of appearance of high number faces (≥ 4) vs low number ones (≤ 3) by varying α. For example, when $\alpha = -6/35$ (i.e. $a = 4$) the probabilities of the various faces are:

$$\mathcal{P}_1 = 0.104 \; ; \; \mathcal{P}_2 = 0.123 \; ; \; \mathcal{P}_3 = 0.147 \; ; \; \mathcal{P}_4 = 0.174 \; ; \; \mathcal{P}_5 = 0.206 \; ; \; \mathcal{P}_6 = 0.245$$

These results are our best "guess" when all we know about the die is the mean of its outcomes: the entropy associated to this guess $S_0 = -k\sum_i \mathcal{P}_i \log \mathcal{P}_i$ maximizes our ignorance, i.e. it is larger than the entropy S associated to any other "guess", $\{F_i\}$ satisfying the constraints. To show that consider:

$$S_0 - S = -k \sum_i (\mathcal{P}_i \log \mathcal{P}_i - F_i \log F_i) = k \sum_i (-\mathcal{P}_i \log \mathcal{P}_i + F_i \log \mathcal{P}_i - F_i \log \mathcal{P}_i + F_i \log F_i)$$

$$= k \sum_i (F_i - \mathcal{P}_i) \log \mathcal{P}_i - F_i \log \mathcal{P}_i / F_i = k \sum_i (F_i - \mathcal{P}_i)(\gamma_0 - \alpha x_i) - F_i \log \mathcal{P}_i / F_i$$

$$= k\gamma_0 (\sum_i F_i - \mathcal{P}_i) - k\alpha (\sum_i x_i F_i - x_i \mathcal{P}_i) - k \sum_i F_i \log \mathcal{P}_i / F_i$$

$$= -k \sum_i F_i \log \mathcal{P}_i / F_i \equiv kD(F\|\mathcal{P})$$

Where $D(F\|\mathcal{P})$ is known as the relative entropy or Kullback-Leibler divergence between the best guess probability $\mathcal{P}_i$ and any other guess (or the measured frequency) F_i. Using the fact that $\log x \leq (x-1)$ one can show that the relative entropy is always positive.

$$kD(F\|\mathcal{P}) = S_0 - S = -k \sum_i F_i \log \mathcal{P}_i / F_i \geq k \sum_i F_i (1 - \mathcal{P}_i / F_i) = k \sum_i (F_i - \mathcal{P}_i) = 0$$

$$(5.31)$$

Our best guess $\{\mathcal{P}_i\}$ can be tested by *simulating* a randomly loaded die, namely by rolling a large number (e.g. 20) of unloaded dice and selecting only the outcomes for which the average of these dice is between 3.9 and 4.1. Repeating the experiment many (e.g. 30) times yields the following distribution of frequencies:

$$F_1 = 0.125 \; ; \; F_2 = 0.112 \; ; \; F_3 = 0.128 \; ; \; F_4 = 0.165 \; ; \; F_5 = 0.223 \; ; \; F_6 = 0.247$$

which average is $a = 3.99$. To test if the observed frequencies are compatible with our best "guess", we can use the relative entropy $D(F\|\mathcal{P})$ or the χ-square test (which computes the weighted sum of the square of the difference between the observed frequency and the predicted probability)[4]:

$$\chi^2 = N \sum_{i=1}^{6} \frac{(F_i - \mathcal{P}_i)^2}{\mathcal{P}_i} \approx 2ND(F\|\mathcal{P}) \tag{5.32}$$

If the value of χ^2 is larger than a certain threshold one can reject the null hypothesis (that the data and our best "guess" agree) with a certain degree of confidence (95% is the value typically chosen). For the above results with $N = 600$ trials: $\chi^2 = 5.738$. For a distribution with $n = 6$ possible outcomes and $s = 2$ constraints, see Eq.5.28, the number of degrees of freedom (i.e. independent variables) is $df = n - s = 4$ and the threshold for rejection of the null hypothesis with 95% confidence is $\chi^2_{thresh} = 9.49$. Since the measured value $\chi^2 < \chi^2_{thresh}$, the null hypothesis cannot be rejected and we conclude that the data is consistent with our best "guess". It is easy to check that the data is however inconsistent with a fair die (mean $a = 3.5$; $\mathcal{P}_i = 1/6$) since in this case the value of $\chi^2 \approx 57 > \chi^2_{thresh}$.

Consider now a model for a "real" loaded die, where the face "1" has been loaded with some lead bead to make it more likely to fall face down (and thus increase the probability of face "6" to come up $\mathcal{P}_6 > \mathcal{P}_1$). If the bead has been inserted at the center of the face, the probabilites of observing the four adjacent faces are a-priori equal ($\mathcal{P}_0$). In such case there are only three probabilities to "guess": $\mathcal{P}_1, \mathcal{P}_6$ and $\mathcal{P}_0$. With two constraints ($\mathcal{P}_1 + 4\mathcal{P}_0 + \mathcal{P}_6 = 1$ and $\mathcal{P}_1 + 14\mathcal{P}_0 + 6\mathcal{P}_6 = a > 3.5$) the system is undetermined and there is a continuous manifold of solutions (for example with $a = 4$: $\mathcal{P}_0 = 1/6$ and $\mathcal{P}_1 = 1/15$ and $\mathcal{P}_6 = 4/15$). However, on this manifold there is one solution which maximizes the missing information and is thus our best "guess". One can verify that this solution is:

$$\mathcal{P}_0 = e^{\gamma_0 - 7\alpha/2}; \ \mathcal{P}_1 = e^{\gamma_0 - \alpha}; \ \mathcal{P}_6 = e^{\gamma_0 - 6\alpha}; \ e^{-\gamma_0} = e^{-\alpha} + 4e^{-7\alpha/2} + e^{-6\alpha}$$

A perturbation expansion to first order in α (about $a = 3.5$) yields: $\alpha = 12(3.5 - a)/25$. When $a = 4$, the probabilities are:

$$\mathcal{P}_0 = 0.156963 \ ; \ \mathcal{P}_1 = 0.086143 \ ; \ \mathcal{P}_6 = 0.286005$$

By computing the value $\chi^2 = 41.55$, one can reject with 95% confidence the possibility that the aforementioned frequency data on a randomly loaded die with $a = 4$ were obtained with a "real" loaded die. Since for that die model we assumed that the face "1" has been loaded, the associated missing information (entropy) is smaller than for a randomly loaded die, as can be verified by computing the entropy S for each case or the Kullback-Leibler divergence: $D(\mathcal{P}_{real}\|\mathcal{P}_{random})$.

---

[4]Using the identity $\sum_i (F_i - \mathcal{P}_i) = 0$ and the approximation $\log(1 + x) \approx x - x^2/2$:

$$D(F\|\mathcal{P}) = k \sum_i F_i \log F_i/\mathcal{P}_i = k \sum_i [\mathcal{P}_i + (F_i - \mathcal{P}_i)] \log[1 + (F_i - \mathcal{P}_i)/\mathcal{P}_i] = k \sum_i (F_i - \mathcal{P}_i)^2/2\mathcal{P}_i$$

One can compare our die model to data obtained on a real loaded die. On such a die one measures after 172 throws a mean of $a = 4.5$ with frequencies for the various faces:

$$F_1 = 0.035 \; ; \; F_2 = 0.134 \; ; \; F_3 = 0.117 \; ; \; F_4 = 0.128 \; ; \; F_5 = 0.168 \; ; \; F_6 = 0.418$$

The predictions of the "real" loaded die model with $a = 4.5$ yield $\alpha = -0.4848\cdots$ and probabilities:

$$\mathcal{P}_0 = 0.13054 \; ; \; \mathcal{P}_1 = 0.03885 \; ; \; \mathcal{P}_6 = 0.43866$$

Comparing these probabilities with the observed frequencies of the loaded die yields a value $\chi^2 = 2.38 < \chi^2_{\text{thresh}}$, from which we deduce that our model for a real loaded die is consistent with the data.

Notice however, that had we assumed the die to be loaded differently our best "guess" would have been different. In particular the predictions of the randomly loaded die (i.e. maximization of the missing information under the only constraint that the mean is equal to 4.5) are not consistent with the measurements on a real loaded die and can be rejected. The maximization of the entropy in the case of a loaded die depends very much on the *a-priori* assumptions made, which may not predict correctly the probability of appearance of the various faces. However that approach provides a systematic way of predicting these probabilities which can be tested and falsified: if the frequencies of the various faces are incompatible with the predictions based on our assumptions, these assumptions must be revised. This is an essential characteristics of a scientific theory: it must make falsifiable predictions[5].

As we shall see in the following Nature seems to be fair, it is not loaded unbeknownst to us (with a major caveat discussed in section 5.12). In all the systems studied so far the observed frequencies have been consistent with the probabilities derived from the assumption that what you measure is all there is. In other words, many particle systems at equilibrium are more akin to the unrealistic randomly loaded die described above than to a real one!

Gaussian distribution. As a further example of the power of the information theory approach to probability, we shall show that the Gaussian distribution maximizes the Shannon entropy when all one knows are the mean μ and standard deviation σ^2 of the probability distribution $\mathcal{P}_i = \mathcal{P}(x_i)$. The entropy is:

$$S = -k \text{Tr} \, \mathcal{P} \log \mathcal{P} = -k \sum_i \mathcal{P}_i \log \mathcal{P}_i = -k \int dx \mathcal{P}(x) \log \mathcal{P}(x) \qquad (5.33)$$

where the trace Tr of a function $f(x)$ is the sum (or integral) over all its possible values (the last equality is obtained when $\mathcal{P}(x)$ is a continuous function of x). The

---

[5]Psychoanalysis is not a scientific theory as no observation can falsify its tenets. The theory of evolution on the other hand is a scientific theory: finding human remains in the stomach of a dinosaur (or Homo Sapiens DNA in the stomach of a mosquito trapped in amber millions years ago) would deal a death blow to Darwin's theory.

entropy has to be maximized under the constraints:

$$\text{Tr}\,\mathcal{P} \equiv \sum_i \mathcal{P}_i = 1 \;;\; \text{Tr}\,x\mathcal{P} \equiv \sum_i x_i\mathcal{P}_i = \mu \;;\; \text{Tr}\,x^2\mathcal{P} \equiv \sum_i x_i^2\mathcal{P}_i = \sigma^2 + \mu^2 \quad (5.34)$$

To that purpose we maximize the auxiliary function:

$$\Gamma = -k\text{Tr}\,\mathcal{P}\log\mathcal{P} + k(\gamma_0 + 1)\text{Tr}\,\mathcal{P} - k\gamma_1\text{Tr}\,x\mathcal{P} - k\gamma_2\text{Tr}\,x^2\mathcal{P}$$

Setting $\partial\Gamma/\partial\mathcal{P}_j = 0$, yields:

$$\mathcal{P}_j = e^{\gamma_0 - \gamma_1 x_j - \gamma_2 x_j^2} \quad (5.35)$$

the constraints then imply:

$$\text{Tr}\,\mathcal{P} = 1 \;\rightarrow\; \gamma_0 = -\log\text{Tr}\,e^{-\gamma_1 x - \gamma_2 x^2}$$

$$\text{Tr}\,x\mathcal{P} = \mu \;\rightarrow\; \mu = \sum_i x_i\mathcal{P}_i = \frac{\text{Tr}\,xe^{-\gamma_1 x - \gamma_2 x^2}}{\text{Tr}\,e^{-\gamma_1 x - \gamma_2 x^2}} = \partial\gamma_0/\partial\gamma_1 \quad (5.36)$$

$$\text{Tr}\,x^2\mathcal{P} = \sigma^2 + \mu^2 \;\rightarrow\; \sigma^2 + \mu^2 = \sum_i x_i^2\mathcal{P}_i = \frac{\text{Tr}\,x^2e^{-\gamma_1 x - \gamma_2 x^2}}{\text{Tr}\,e^{-\gamma_1 x - \gamma_2 x^2}} = \partial\gamma_0/\partial\gamma_2$$

Since:

$$e^{-\gamma_0} = \text{Tr}\,e^{-\gamma_1 x - \gamma_2 x^2} = \int_{-\infty}^{\infty} dx\, e^{-\gamma_1 x - \gamma_2 x^2}$$

$$= e^{-\gamma_1^2/4\gamma_2}\int_{-\infty}^{\infty} dx\, e^{-\gamma_2(x+\gamma_1/2\gamma_2)^2} = \sqrt{\pi/\gamma_2}\,e^{\gamma_1^2/4\gamma_2} \quad (5.37)$$

Hence from Eq.5.37:

$$\gamma_0 = -\frac{\gamma_1^2}{4\gamma_2} + \frac{1}{2}\log\gamma_2/\pi$$

$$\mu = \frac{\partial\gamma_0}{\partial\gamma_1} = -\frac{\gamma_1}{2\gamma_2} \quad (5.38)$$

$$\sigma^2 + \mu^2 = \frac{\partial\gamma_0}{\partial\gamma_2} = \frac{\gamma_1^2}{4\gamma_2^2} + \frac{1}{2\gamma_2}$$

from which we deduce that: $\gamma_1 = -\mu/\sigma^2$; $\gamma_2 = 1/2\sigma^2$ and $e^{\gamma_0} = \frac{e^{-\mu^2/2\sigma^2}}{\sqrt{2\pi}\sigma}$ and from Eq.5.35 the probability distribution which maximizes the missing information becomes:

$$\mathcal{P}(x) = \frac{1}{\sqrt{2\pi}\sigma}\exp\left(-\frac{(x-\mu)^2}{2\sigma^2}\right)$$

This is the Gaussian distribution we have encountered many times before.

5.3 PARAMAGNETISM

The procedure we have adopted above to study the probability distributions for a fair or loaded die, namely the maximization of the entropy under constraints, is the procedure adopted for the study of **all** equilibrium processes in Statistical Mechanics[88, 15, 85, 89]. The only difference is the phase-space (the space of possible events) on which the probability is defined: the six faces of the die, the orientations of the spins in a magnetic material, the positions and velocities (or momenta) of the particles in a gas, the local orientation of a polymer chain, etc. In Statistical Mechanics the most common constraint encountered is that the average energy of the system is known or fixed. Therefore to solve for the probability distributions one needs to understand how the energy of the system depends on the phase-space parameters (spin orientation, momenta, distance between particles, etc.). For simplicity we will initially address systems with weakly-interacting components (spins (paramagnets), atoms (ideal gases), monomers (freely jointed chain polymers), etc.). Strictly speaking non-interacting systems cannot exchange energy and reach equilibrium, hence the assumption here is that the interactions are small enough to allow the system to equilibrate without modifying significantly the equilibrium behavior derived by neglecting these interactions. Later on we will address the more difficult problem of systems with non-negligible interactions (ferromagnets, real gases, liquids, etc.).

5.3.1 PARAMAGNETISM AND THE ISING MODEL

A paramagnet is a material, such as iron oxide, which doesn't exhibit permanent magnetization but which can be magnetized in presence of an external magnetic field, see section 4.11.3. It is usually described as an assembly of microscopic magnetic moments (spins) that are disordered by the stochastic blows of their environment. Since the energy of a single spin with magnetic moment $\vec{\mu}_i$ in a magnetic field $\vec{B}$ is $\epsilon_i = -\vec{\mu}_i \cdot \vec{B}$ the total energy E_p of a particular assembly $\{\vec{\mu}_i\}$ of weakly-interacting spins is:

$$E_p(\{\vec{\mu}_i\}) = \sum_{i=1}^{N} \epsilon_i = -\sum_{i=1}^{N} \vec{\mu}_i \cdot \vec{B} \tag{5.39}$$

The phase-space of the spin-orientations has yet to be defined. In the classical Heisenberg model the spins are free to adopt any spatial orientation. In the quantum version the spins can adopt, like the angular momentum, $2S + 1$ discrete values $S_z = -S, \ldots, 0, \ldots, S$). In the simplest quantum version ($S = 1/2$ known as the Ising model) the spins can adopt only two orientations: parallel or anti-parallel to the magnetic field ($\vec{\mu}_i = \pm \mu \hat{B}$), see Fig.5.7. In the following we will solve for the probability distribution of the magnetic moments for an Ising paramagnet. We shall return to the classical Heisenberg model when we examine the behavior of a freely jointed chain polymer[90].

The problem then is to find the probability of occurrence of a spin configuration: $\mathcal{P}(\{\vec{\mu}_i\})$. As argued before our best guess given the mean energy of the system $\langle E \rangle \equiv E$

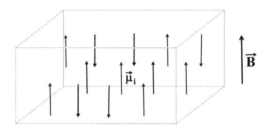

Figure 5.7 In the Ising model of a magnet, the spins with magnetic moment $\vec{\mu}$ can adopt two orientations: parallel $+\mu$ or anti-parallel $-\mu$ to the magnetic field $\vec{B}$. The overall magnetization is given by the sum of the magnetic moments.

is to maximize the entropy:

$$S = -k\text{Tr}\,\mathcal{P}\log\mathcal{P} \equiv -k\sum_{\{\mu_i=\pm\mu\}}\mathcal{P}\log\mathcal{P} \tag{5.40}$$

under the constraints:

$$\text{Tr}\,\mathcal{P} \equiv \sum_{\{\mu_i=\pm\mu\}}\mathcal{P}(\{\vec{\mu}_i\}) = 1\;;\;\langle E_p\rangle = \text{Tr}\,E_p\mathcal{P} \equiv \sum_{\{\mu_i=\pm\mu\}}E_p(\{\vec{\mu}_i\})\mathcal{P}(\{\vec{\mu}_i\}) = E \tag{5.41}$$

where the trace is defined as the sum over all possible spin configurations. As done previously we have to maximize the auxiliary function:

$$\Gamma = -k\text{Tr}\,\mathcal{P}\log\mathcal{P} + k(\gamma_0 + 1)\text{Tr}\,\mathcal{P} - k\beta\text{Tr}\,E_p\mathcal{P}$$

To find the maxima (in fact the extrema) of Γ we set $\partial\Gamma/\partial\mathcal{P}(\{\vec{\mu}_i\}) = 0$ and obtain:

$$\mathcal{P}(\{\vec{\mu}_i\}) = e^{\gamma_0-\beta E_p} = e^{\gamma_0}\exp\left[-\beta\sum_{i=1}^{N}\epsilon_i\right] = e^{\gamma_0}\Pi_{i=1}^{N}e^{\beta\vec{\mu}_i\cdot\vec{B}} \tag{5.42}$$

The constraint $\text{Tr}\,\mathcal{P} = 1$ implies that:

$$e^{-\gamma_0} = \text{Tr}\,e^{-\beta E_p} = \sum_{\{\mu_i=\pm\mu\}}\Pi_{i=1}^{N}e^{\beta\vec{\mu}_i\cdot\vec{B}} \equiv Z \tag{5.43}$$

where Z is known as the partition function. In the Ising model where spins can adopt only two configurations: $\mu_i = \pm\mu$:

$$Z = \sum_{\{\mu_i=\pm\mu\}}\Pi_{i=1}^{N}e^{\beta\vec{\mu}_i\cdot\vec{B}} = (e^{\beta\mu B} + e^{-\beta\mu B})^N = 2^N\cosh^N\beta\mu B \tag{5.44}$$

Notice that since the spins are non-interacting the probability of a given spin μ_i to align parallel or anti-parallel to the magnetic field is independent of the orientations of the other spins:

$$\mathcal{P}_\uparrow = \sum_{\{\mu_{j\neq i}=\pm\mu\}} \mathcal{P}(\{\vec{\mu}_i = \mu\hat{B},\vec{\mu}_j\})$$

$$= e^{\gamma_0}e^{\beta\mu B} \sum_{\{\mu_{j\neq i}=\pm\mu\}} \Pi_j e^{-\beta\epsilon_j} = \frac{e^{\beta\mu B}\sum_{\{\mu_{j\neq i}=\pm\mu\}}\Pi_j e^{-\beta\epsilon_j}}{\text{Tr}\,\Pi_i e^{-\beta\epsilon_i}}$$

$$= \frac{e^{\beta\mu B}}{e^{\beta\mu B}+e^{-\beta\mu B}}$$

$$\text{similarly}: \quad \mathcal{P}_\downarrow = \frac{e^{-\beta\mu B}}{e^{\beta\mu B}+e^{-\beta\mu B}} \tag{5.45}$$

When $\beta = 0$: $\mathcal{P}_\uparrow = \mathcal{P}_\downarrow = 1/2$ and the entropy associated to that spin is : $S = -k\sum_{\mu_i=\pm\mu}\mathcal{P}\log\mathcal{P} = k\log 2$. When $\beta \to \infty$: $\mathcal{P}_\uparrow = 1$, $\mathcal{P}_\downarrow = 0$ and the entropy $S = 0$. The inverse of β thus plays a role similar to the temperature: at low temperatures ($\beta \to \infty$) the spins are aligned whereas at high temperatures ($\beta \to 0$) they are fully disorganized. This relation will be precised in the following.

The partition function plays a central role in Statistical Mechanics, since many relevant variables can be derived from it. For example the mean energy E can be computed from a knowledge of $\mathcal{Z}$:

$$E = \text{Tr}\,E_p\mathcal{P} = \text{Tr}\,E_p e^{\gamma_0-\beta E_p} = -e^{\gamma_0}\frac{\partial}{\partial\beta}\text{Tr}\,e^{-\beta E_p} = -e^{\gamma_0}\frac{\partial}{\partial\beta}e^{-\gamma_0} = \frac{\partial\gamma_0}{\partial\beta}$$

$$= -\frac{\partial}{\partial\beta}\log\mathcal{Z} = -N\mu B\tanh\beta\mu B \tag{5.46}$$

By tuning β we can fix the mean energy of the system (as we could set the loading of a die by tuning α). In particular when $\beta \to \infty$, the energy of the system is minimal: all spins are aligned with the external magnetic field and the total energy is $E = -N\mu B$. On the other hand when $\beta \to 0$ the mean energy of the system $E \to 0$: the spins have equal probability of aligning parallel or anti-parallel to the magnetic field ($\mathcal{P}_\uparrow = \mathcal{P}_\downarrow$). Negative values of β are not physical since they describe a system at equilibrium where high energy states have higher probability of occupancy than low energy ones ($\mathcal{P}_\downarrow > \mathcal{P}_\uparrow$). As previously noticed, the inverse of β acts in fashion similar to our intuitive grasp of temperature: at low temperature (high β) the system has low energy and is ordered, the spins are aligned with the magnetic field, while at high temperatures (low β) the system has high energy and is disordered due to thermal agitation. That intuition can be formalized by writing the entropy S as:

$$S = -k\text{Tr}\,\mathcal{P}\log\mathcal{P} = -k\text{Tr}\,\mathcal{P}(\gamma_0-\beta E_p)$$

$$= -k\gamma_0 + k\beta E = k\log\mathcal{Z} - k\beta\frac{\partial}{\partial\beta}\log\mathcal{Z} \tag{5.47}$$

$$= Nk\log 2\cosh\beta\mu B - Nk\beta\mu B\tanh\beta\mu B$$

where we used Eqs.5.43 and 5.46. As β or B go to zero, the entropy $S \rightarrow Nk\log 2$, i.e. the system is completely disordered and each spin can adopt with equal probability any of two spin states (up or down). On the other hand when β or $B \rightarrow \infty$ ($\log 2\cosh\beta\mu B \rightarrow \beta\mu B$ and $\tanh\beta\mu B \rightarrow 1$) the entropy $S \rightarrow 0$ and the system is fully ordered (all spins are aligned with the magnetic field).

Classically the inverse of the temperature T of a system is defined by the change in entropy when the mean energy of the system is altered. From Eq.5.47 we thus derive:

$$\frac{1}{T} \equiv \frac{\partial S}{\partial E} = k\beta \tag{5.48}$$

The Lagrange multiplier associated with the energy constraint is related to the inverse of the temperature: $\beta = 1/kT$. We shall have more opportunities below to confirm that identification with the classical thermodynamics concept of temperature.

Since βE is dimensionless, kT has units of energy. Therefore temperature could have been measured in units of energy, say Joules (and k would have been dimensionless) but for historical reasons it has its own units and scale. In the Kelvin scale, the temperature difference between boiling and freezing water (at a standard pressure of 1 atmosphere) is $100°K$. The freezing temperature of water is then $273.16°K$. The constant k (also known as Boltzmann constant k_B) sets the change in units from $°K$ to Joules (as one goes from inches to cm). In the MKS system: $k = 1.38 \ 10^{-23}$ J/$°K$ (in the CGS system $k = 1.38 \ 10^{-16}$ erg/$°K$). By definition, see Eq.5.25, the units of k set the units of entropy.

From Eq.5.47, the quantity:

$$\mathcal{F} \equiv kT\gamma_0 = -kT\log\mathcal{Z} = E - TS \tag{5.49}$$

is known as the Helmholtz free energy. Being equivalent to the partition function $\mathcal{Z}$, it plays a similarly important role in the definition of many thermodynamic quantities. Thus having solved for the probability distribution of magnetic moments for an ensemble of non-interacting spins, one can use that solution to compute the value of various observables, such as the mean magnetization:

$$M = \text{Tr} \sum_{i=1}^{N} \mu_i \mathcal{P}(\{\mu_j\}) = \frac{1}{\mathcal{Z}} \sum_{i,\{\mu_j=\pm\mu\}} \mu_i \Pi_j e^{-\beta\epsilon_j}$$

$$= \frac{1}{\mathcal{Z}} \sum_{i,\mu_i=\pm\mu} \mu_i e^{-\beta\epsilon_i} \sum_{\{\mu_{j\neq i}=\pm\mu\}} \Pi_j e^{-\beta\epsilon_j} = \frac{N\mu\sinh\beta\mu B\cosh^{N-1}\beta\mu B}{\cosh^N\beta\mu B}$$

$$= N\mu\tanh\beta\mu B = -kT\frac{\partial\gamma_0}{\partial B} = -(\frac{\partial\mathcal{F}}{\partial B})_T \tag{5.50}$$

The last equation relates the response of the system (here an ensemble of spins) to an external constraint (the magnetic field B). This response (the magnetization M) obtained as the derivative at constant temperature of the Helmholtz free energy $\mathcal{F}$ with respect to the external variable is very general. We shall encounter it in other

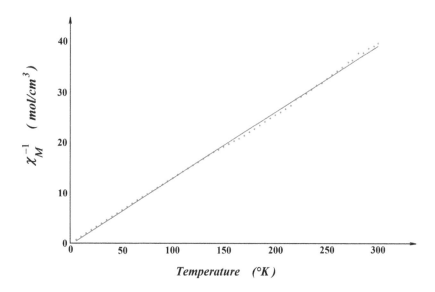

Figure 5.8 Gadolinium (Gd) in Gadolinium oxalate crystallizes in water into a paramagnetic material which susceptibility χ_M is nicely described by the Curie-Law, Eq.5.51 (adapted for the Heisenberg rather than the Ising model of paramagnetism, see Eq.5.80). Data (dots) courtesy of R.Sibille [91]. Continuous line: best linear fit. The measured value of $\chi_M T = N_A \mu^2 / 3k \approx 7.7 \mathrm{erg}^\circ \mathrm{K/G^2 mol} \equiv 7.7 \mathrm{cm^3}^\circ \mathrm{K/mol}$ (CGS units, where $N_A = 6 \ 10^{23}$ atoms/mol is Avogadro's number) allows one to deduce the magnetic moment of a Gadolinium atom $\mu \approx 7.16 \ 10^{-20}$ erg/G. From the relation $\mu = g\mu_B \sqrt{S(S+1)}$ where $g = 2$ is the gyromagnetic ratio of the electron and $\mu_B = 9.274 \ 10^{-21}$ erg/G is the Bohr magneton (the magnetic moment associated to the electron spin, see section 4.6) we derive the Gadolinium total spin state: $S \approx 7/2$ (a result of the presence of 7 unpaired electrons in its $4f$ shell, see section 4.9.1). Due to its high magnetic susceptibility, Gadolinium is often used as a contrast agent in Magnetic Resonance Imaging, see section 4.11.3.1.

contexts (polymer, ideal gas) studied below. The change in the Helmholtz free energy $\delta\mathcal{F}$ stands for the work performed (at constant temperature) on the system to magnetize it $\delta W = -\int M dB$ (or the amount of energy available for isothermal work by the system).

In fact we could have derived Eq.5.50, in a simpler way, by noticing that the mean energy of the system, Eq.5.46 is simply related to the mean magnetization by: $E = -N\mu B \tanh\beta\mu B = \langle \sum_i \vec{\mu}_i \rangle B = -MB$.

From Eq.5.50 the magnetic susceptibility of an Ising paramagnet $\chi_{M,I}$ can be computed:

$$\chi_{M,I} \equiv \left(\frac{\partial M}{\partial B}\right)_{B=0} = \frac{N\mu^2}{kT} \tag{5.51}$$

The behavior of the magnetic susceptibility with temperature, Eq.5.51 is known as the Curie Law and has indeed been observed in paramagnetic materials, see Fig.5.8. For a classical Heisenberg paramagnet (which spins can adopt any orientation in space) a Curie Law is also obtained but with a susceptibility which differs from Eq.5.51 by a factor 1/3 (see section 5.4 below).

Temperature equilibration. It is an everyday experience that if a hot system contacts a cold one for long enough, they equilibrate at a mutual intermediate temperature. In the following we shall show that this behavior can be derived from our information theory approach.

Consider two spin assemblies (two paramagnets) with N_1 and N_2 spins in a magnetic field B with initial average energy E_1, E_2 respectively, see Fig.5.9. Before the two assemblies are put into contact their energies are given by Eq.5.46. Once they are put into contact the average energy E of the combined system is by energy conservation the sum of E_1 and E_2.

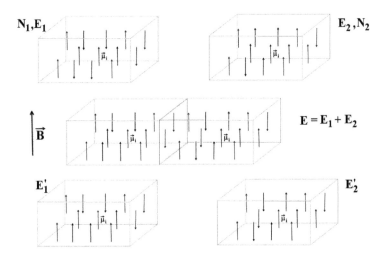

Figure 5.9 Bringing in contact two blocks of weakly interacting spins in an external magnetic field $\vec{B}$ with average internal energies E_1, E_2 (hence temperatures T_1, T_2), results in an equilibration of their temperatures to an intermediate value T such that the mean energy of the blocks in contact becomes $E = E_1 + E_2$. When separated the blocks have new mean energies E_1', E_2' set by their new temperature T.

Finding the probability distribution of spin configurations in the combined system of the N spins in thermal contact ($N = N_1 + N_2$) under the constraints of known average energy E yields, see Eq.5.46:

$$E_1 + E_2 = E = -(N_1 + N_2)\mu B \tanh \beta \mu B = E_1' + E_2'$$

where β is the Lagrange multiplier associated with E. For given initial energies E_1, E_2 and number of spins N_1, N_2, the final mutual temperature T (or its inverse β) will be different from the initial temperatures T_1, T_2 (or their inverse β_1, β_2). We shall show below that in agreement with our daily experience the final temperature is obtained between the two initial temperatures: $T_1 < T < T_2$. Notice that after reaching the temperature T, if the two spin assemblies are separated, they will have the same β (same temperature T) but new average energies $E'_1 = -N_1\mu B\tanh\beta\mu B \neq E_1$ and $E'_2 = -N_2\mu B\tanh\beta\mu B \neq E_2$. Assuming that their initial temperatures: $T_1 > T_2$ (i.e. $\beta_1 < \beta_2$), conservation of energy implies:

$$-(N_1 + N_2)\mu B\tanh\beta\mu B = E = E_1 + E_2$$
$$= -N_1\mu B\tanh\beta_1\mu B - N_2\mu B\tanh\beta_2\mu B$$

Since $\tanh x$ is a monotonously growing function of x:

$$\beta_2 > \beta_1 \quad \rightarrow \quad \tanh\beta_2\mu B > \tanh\beta_1\mu B$$

and therefore:

$$-(N_1 + N_2)\mu B\tanh\beta_2\mu B < -(N_1 + N_2)\mu B\tanh\beta\mu B < -(N_1 + N_2)\mu B\tanh\beta_1\mu B \quad (5.52)$$

which implies $\beta_2 > \beta > \beta_1$ or: $T_1 > T > T_2$. Thus bringing into contact the two spin assemblies results in an equilibration of their initial temperatures at a mutual intermediate value, in agreement with our experience.

The energy (heat) exchanged between the two systems is:

$$\delta Q = E_1 - E'_1 = E'_2 - E_2 \quad (5.53)$$

Heat is a central concept in classical thermodynamics. It has its own units (calories, 1 cal = 4.2 J) even though it is but energy exchanged. We shall now show that the heat exchanged is related to the change in entropy. Using Eq.5.47, and since γ_0 depends on β and B, the change in entropy dS can be written as:

$$dS = -k\,d\gamma_0 + kE\,d\beta + k\beta\,dE \quad (5.54)$$
$$= k(-\frac{\partial\gamma_0}{\partial\beta} + E)d\beta - k\frac{\partial\gamma_0}{\partial B}dB + \frac{dE}{T}$$

By virtue of Eq.5.46 ($E = \partial\gamma_0/\partial\beta$) the first term of the above equation is null. Using Eq.5.50 ($\beta M = \partial\gamma_0/\partial B$) we then get:

$$dE = T\,dS - M\,dB \quad (5.55)$$

At constant magnetic field we see that $\delta Q = dE = T\,dS$. The change in entropy is thus related to the heat exchanged:

$$dS = \frac{\delta Q}{T} \quad (5.56)$$

In fact, Eq.5.55 expresses energy conservation: the change in the mean energy of the system dE is due to injection of heat δQ and to work performed on the system to magnetize it: $\delta W = -MdB$:

$$dE = \delta Q + \delta W \tag{5.57}$$

This equation and similar ones that we will see below sit at the core of the understanding of heat engines, to which we shall return later.

Notice that we can recast Eq.5.55 as a differential equation for $\mathcal{F} = E - TS$, since $dE = d\mathcal{F} + TdS + SdT$:

$$d\mathcal{F} = -SdT - MdB \tag{5.58}$$

from which we again see that the magnetization is the derivative of the free energy with respect to the magnetic field at constant temperature: $M = -(\partial\mathcal{F}/\partial B)_T$. The term $\delta W = -MdB$ is the incremental work in aligning the spins performed by an increase dB in the magnetic field. At constant temperature the change in free energy is due solely to that work.

An equilibration experiment that proceeds slowly enough for the entropy to remain constant (i.e. $dS = 0$) has no heat exchanged and is called an adiabatic process. Magnetic cooling via adiabatic demagnetization is such a process. It is used extensively to reach very low temperatures, often with Gadolinium based compounds due to the high spin of Gd in its ground state (see Fig.5.8). The idea is simple: a magnetic system has its spins aligned in a high magnetic field. It is then isolated (to prevent heat exchange) and the magnetic field is slowly reduced to zero. As a result the mean energy E of the system (which is negative) increases to zero. Setting $\delta Q = TdS = 0$ in Eq.5.55 one has: $dE + MdB = 0$. For small values of the magnetic field B, the energy and magnetization behave as:

$$E = -N\mu B\tanh\beta\mu B \approx -\frac{N\mu^2 B^2}{kT}$$

$$M = N\mu\tanh\beta\mu B \approx \frac{N\mu^2 B}{kT}$$

Therefore at constant entropy:

$$0 = TdS = dE + MdB = \frac{\partial E}{\partial B}dB + \frac{\partial E}{\partial T}dT + MdB$$

$$= -\frac{2N\mu^2 B}{kT}dB + \frac{N\mu^2 B^2}{kT^2}dT + \frac{N\mu^2 B}{kT}dB$$

From which one deduces that : $dT/T = dB/B$ and by integration:

$$T_f = T_i\frac{B_f}{B_i}$$

The smaller the ratio B_f/B_i between the final and initial magnetic fields, the smaller the final temperature T_f. In practice the final magnetic field B_f cannot reach zero. It is set by the residual internal magnetic field generated by the spins themselves (due to the weak interactions that we have neglected all along which tend to align the spins). This adiabatic demagnetization of the spins is often used to cool down to very low temperatures a non-magnetic system to which they are coupled. Using nuclear spins (which produce a weaker residual field due to their very weak coupling) final temperatures below $1\mu°K$ have been reached.

5.3.2 PARAMAGNETISM FROM ENUMERATION OF STATES

Sometimes, one can solve Statistical Mechanics problems by counting the number of configurations with the exact same energy. One then equates that exact energy with the measurable average energy of the system. The success of that approach known as the microcanonical approach (as distinct from the previously exposed canonical one) rests on the fact that the average energy of a system is in principle known with great precision (a result of the central limit theorem and the large number N of particles (atoms, spins, etc.)). For example in the case of paramagnetism, the energy of a system with m spins up and $N - m$ down is:

$$E = -\mu Bm + \mu B(N - m) = -\mu B(2m - N) \tag{5.59}$$

The total number of configurations Ω that possess this energy E is given by the number of ways of ordering m spins up among a total of N spins (as in the example mentioned above of ordering m white birds among N black and white ones):

$$\Omega = \frac{N!}{m!(N - m)!} \tag{5.60}$$

As in the case of an unloaded die, since there is no reason to choose one particular configuration over any other, the probability of any spin configuration is: $\mathcal{P} = 1/\Omega$ and the entropy then becomes:

$$S = k \log \Omega = k[N \log N - m \log m - (N - m) \log(N - m)]$$
$$= -Nk[\frac{m}{N} \log \frac{m}{N} + (1 - \frac{m}{N}) \log(1 - \frac{m}{N})] \tag{5.61}$$

From the definition of temperature, Eq.5.48, we deduce:

$$\frac{1}{T} = \frac{\partial S}{\partial E} = -\frac{1}{2\mu B} \frac{\partial S}{\partial m} = -\frac{k}{2\mu B} \log(N/m - 1) \tag{5.62}$$

From which we derive the frequency of spins up:

$$\frac{m}{N} = \frac{1}{1 + e^{-2\mu B/kT}} = \frac{e^{\mu B/kT}}{e^{\mu B/kT} + e^{-\mu B/kT}} \tag{5.63}$$

the energy of the system E can then be written as :

$$E = -\mu B(2m - N) = -N\mu B \frac{e^{\mu B/kT} - e^{-\mu B/kT}}{e^{\mu B/kT} + e^{-\mu B/kT}} = -N\mu B \tanh \mu B/kT \tag{5.64}$$

which is the result derived using the canonical approach, see Eq.5.46.

5.3.3 SECOND LAW OF THERMODYNAMICS

In the following, we shall use the microcanonical approach to demonstrate the Second Law of Thermodynamics: when many particle systems are brought in contact their entropy always increases. If the entropy decreases locally, it is at the expense of a larger increase somewhere else.

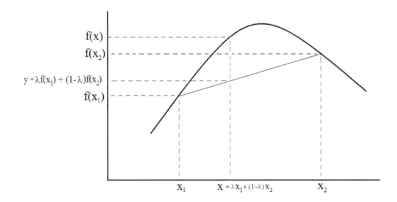

Figure 5.10 For a concave function (one which second derivative satisfies $f''(x) < 0$): $f(\lambda x_1 + (1 - \lambda)x_2) > \lambda f(x_1) + (1 - \lambda)f(x_2)$ as can be easily seen from the above sketch.

Consider the two paramagnetic systems already discussed and shown in Fig.5.9. Let m_i be the number of spins up in the system with N_i spins ($i = 1, 2$). Their entropy before contact is:

$$S_1 + S_2 = -N_1 k[\frac{m_1}{N_1} \log \frac{m_1}{N_1} + (1 - \frac{m_1}{N_1}) \log\left(1 - \frac{m_1}{N_1}\right)]$$

$$- N_2 k[\frac{m_2}{N_2} \log \frac{m_2}{N_2} + (1 - \frac{m_2}{N_2}) \log\left(1 - \frac{m_2}{N_2}\right)]$$

Once the systems are put in contact the energy $E = E_1 + E_2$ being conserved the total number of spins up is : $m_1 + m_2$. The entropy after equilibration is then:

$$S = -(N_1 + N_2)k[\frac{m_1 + m_2}{N_1 + N_2} \log \frac{m_1 + m_2}{N_1 + N_2} + (1 - \frac{m_1 + m_2}{N_1 + N_2}) \log\left(1 - \frac{m_1 + m_2}{N_1 + N_2}\right)]$$

$$\equiv (N_1 + N_2)f(\frac{m_1 + m_2}{N_1 + N_2}) \tag{5.65}$$

Now the entropy per spin $S(x)/N \equiv f(x) = -x \log x - (1 - x) \log(1 - x)$ (with $x = m/N$) is a concave function of x, since its second derivative

$$f'' = -\frac{1}{x} - \frac{1}{1-x} < 0 \qquad \text{for all } x \text{ in } \{0, 1\}$$

For a concave function $f(x)$ the following inequality holds (see Fig.5.10):

$$f(\mathbf{x}) \equiv f(\lambda x_1 + (1 - \lambda)x_2) \geq \lambda f(x_1) + (1 - \lambda)f(x_2) \tag{5.66}$$

Therefore the entropy per spin $S(x)/N$ satisfies:

$$f(\frac{m_1 + m_2}{N_1 + N_2}) = f(\frac{N_1}{N_1 + N_2}\frac{m_1}{N_1} + \frac{N_2}{N_1 + N_2}\frac{m_2}{N_2}) \geq \frac{N_1}{N_1 + N_2}f(\frac{m_1}{N_1}) + \frac{N_2}{N_1 + N_2}f(\frac{m_2}{N_2})$$

This equation can be recast as:

$$S \geq S_1 + S_2 \tag{5.67}$$

The last equation is known as the Second Law of Thermodynamics: in a closed system the entropy never decreases. It may decrease in one part of the system at the expense of a larger increase in another part.

5.4   THE FREELY JOINTED CHAIN MODEL OF A POLYMER

We have already seen the freely jointed chain (FJC) model of a polymer in the context of random walks, see Fig.5.4. We shall now revisit that model with the information theory approach we used to study Ising paramagnetism. In the FJC model a polymer of total contour length L_0 is represented by N straight segments of size b (known as the Kuhn length: $L_0 = Nb$) whose orientations in 3D-space are uncorrelated[90]. Admittedly, this model is a crude simplification of a polymer. A better model would describe a polymer as a flexible string. We shall address such a description later, see section 5.10.3. For the moment let us stick to the simple FJC model. The phase space in which one describes the polymer configurations is thus the space of all possible orientations of its segments: $\vec{b}_i$. Neglecting the interactions between the chain segments (such as the self-avoiding interactions that prevent them from crossing) when the chain is under a tension $\vec{F}$, the energy E_p of a particular configuration $\{\vec{b}_i\}$ is given by:

$$E_p(\{\vec{b}_i\}) = -\sum_i^N \vec{b}_i \cdot \vec{F} \tag{5.68}$$

That equation expresses the fact that pulling on the chain aligns the segments with the direction of the force. Eq.5.68 is identical to the energy of a system of freely rotating dipoles in a magnetic field (the classical Heisenberg model of paramagnetism) mentioned earlier, see Eq.5.39. The problem then is to find the probability of occurrence of a segment configuration: $\mathcal{P}(\{\vec{b}_i\})$. Our best guess given the mean energy of the system E is to maximize the entropy:

$$S = -k\mathrm{Tr}\,\mathcal{P}\log\mathcal{P} \equiv -k\sum_{\{\vec{b}_i\}}\mathcal{P}\log\mathcal{P} \tag{5.69}$$

under the constraints:

$$\mathrm{Tr}\,\mathcal{P} \equiv \sum_{\{\vec{b}_i\}}\mathcal{P}(\{\vec{b}_i\}) = 1 \; ; \; \langle E_p \rangle = \mathrm{Tr}\,E_p\mathcal{P} \equiv \sum_{\{\vec{b}_i\}}E_p(\{\vec{b}_i\})\mathcal{P}(\{\vec{b}_i\}) = E \tag{5.70}$$

where the trace is defined as the sum over all the possible orientations of the segments. As done previously we have to maximize the auxiliary function:

$$\Gamma = -k\mathrm{Tr}\,\mathcal{P}\log\mathcal{P} + k(\gamma_0 + 1)\mathrm{Tr}\,\mathcal{P} - k\beta\mathrm{Tr}\,E_p\mathcal{P}$$

Setting $\partial\Gamma/\partial\mathcal{P} = 0$ yields:

$$\mathcal{P}(\{\vec{b_i}\}) = e^{\gamma_0 - \beta E_p} = e^{\gamma_0 + \beta\sum_i \vec{b_i}\cdot\vec{F}} = e^{\gamma_0}\Pi_i e^{\beta\vec{b_i}\cdot\vec{F}} \tag{5.71}$$

The partition function $\mathcal{Z} = e^{-\gamma_0}$ is then set by the first constraint ($\mathrm{Tr}\,\mathcal{P} = 1$), see Eq.5.43:

$$\mathcal{Z} = \mathrm{Tr}\,e^{-\beta E_p} = \Pi_i \int d\Omega_i e^{\beta\vec{b_i}\cdot\vec{F}} = \left(2\pi\int_0^\pi \sin\theta d\theta e^{\beta b F\cos\theta}\right)^N$$

$$= \left(2\pi\int_{-1}^1 d(\cos\theta)e^{\beta b F\cos\theta}\right)^N = \left(\frac{4\pi\sinh\beta b F}{\beta b F}\right)^N \tag{5.72}$$

where the integration is over all possible orientations : $\int d\Omega = \int_0^{2\pi} d\phi \int_0^\pi \sin\theta d\theta$. The average energy then obeys, see Eq.5.46:

$$E = \frac{\partial\gamma_0}{\partial\beta} = -\frac{\partial}{\partial\beta}\log\mathcal{Z} = -N\frac{\partial}{\partial\beta}[\log(4\pi\sinh\beta b F) - \log(\beta b F)]$$

$$= -NFb\left(\coth\beta Fb - \frac{1}{\beta Fb}\right) = -FL_0\left(\coth\beta Fb - \frac{1}{\beta Fb}\right) \tag{5.73}$$

The entropy of the chain can be computed, as for the Ising model, from Eq.5.71:

$$S = -k\mathrm{Tr}\,\mathcal{P}\log\mathcal{P} = -k\mathrm{Tr}\,\mathcal{P}\log[e^{\gamma_0}\Pi_{i=1}^N e^{\beta\vec{b_i}\cdot\vec{F}}]$$

$$= -k\gamma_0 - k\beta\mathrm{Tr}\,\mathcal{P}\sum_{i=1}^N \vec{b_i}\cdot\vec{F} = k\log\mathcal{Z} + E/T$$

$$= Nk\log\left(\frac{4\pi\sinh\beta b F}{\beta b F}\right) - Nk\beta Fb\left(\coth\beta Fb - \frac{1}{\beta Fb}\right) \tag{5.74}$$

At high temperature ($\beta \to 0$) the entropy saturates: $S \to Nk\log 4\pi$, while at low temperatures ($\beta \to \infty$) the entropy decreases: $S \to -Nk\log\beta b F \to -\infty$. The fact that at low temperatures the entropy is unbounded from below, is a peculiarity of the continuous nature of the phase space of some classical systems (such as the classical Heisenberg model of paramagnetism studied here in which spins can adopt any spatial orientation). Quantum mechanical systems (such as the Ising model studied in section 5.3) have discrete orientations, which set a bound on the entropy difference between the fully disordered and fully ordered states.

The mean extension $L = \hat{F}\cdot\sum_i \vec{b_i}$ is the response of the chain to the force pulling on it and is analogous to the mean magnetization which is the response of the spins to the external magnetic field orienting them. It is given by Eq.5.50 (with the force

F replacing the magnetic field B), namely by the derivative of the free energy $\mathcal{F} = kT\gamma_0 = -kT\log\mathcal{Z}$ with respect to the force at constant temperature:

$$L = -kT\frac{\partial\gamma_0}{\partial F} = \left(\frac{\partial\mathcal{F}}{\partial F}\right)_T = Nb(\coth\beta Fb - \frac{1}{\beta Fb})$$

$$= L_0(\coth\beta Fb - \frac{1}{\beta Fb}) \tag{5.75}$$

As expected from their definitions, the energy $E = FL$. However as can be seen from Fig.5.11 the FJC model is not a good description of a semi-rigid polymer such as DNA at high forces. It does however provide a good approximation to the behavior of polymers at low forces[92], when their relative extension $L/L_0 \ll 1$, i.e when $\beta bF \ll 1$ (for DNA at room temperature with $b = 100$nm, the low force regime implies $F \lesssim 0.04$ pN).

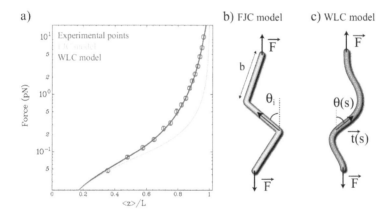

Figure 5.11 (a) The force vs extension behavior of a single DNA molecule (red points) as compared to the predictions of the Freey Jointed Chain (FJC) model of a polymer (b) and the Worm-like Chain (WLC) model (c), see section 5.10.3. Clearly the latter model which describes the polymer as a flexible tube is a much better description of the elastic properties of DNA, than the cruder but simpler FJC model.

In section 5.10.3 we shall study a model (the Worm Like Chain (WLC) model) which describes the polymer as a flexible string and thus takes into account the orientation correlation between successive segments (while still neglecting self-avoiding interactions). The WLC model provides a much better description of the behavior of a single DNA molecule under tension.

As we did for the Ising paramagnet studied in the previous section, we may derive relations between the various thermodynamic functions we have introduced $E, S, \mathcal{F}$, etc.

Using Eq.5.47, and since γ_0 depends on β and F the entropy change is:

$$dS = -kd\gamma_0 + kEd\beta + k\beta dE \tag{5.76}$$

$$= k(-\frac{\partial\gamma_0}{\partial\beta} + E)d\beta - k\frac{\partial\gamma_0}{\partial F}dF + \frac{dE}{T}$$

By virtue of Eq.5.73 the first term of the above equation is null. Using Eq.5.75 we then get:

$$dE = TdS - LdF \tag{5.77}$$

Which we can also write as a differential equation for $\mathcal{F} = E - TS$, since $dE = d\mathcal{F} + TdS + SdT$:

$$d\mathcal{F} = -SdT - LdF \tag{5.78}$$

from which we recover Eq.5.75, namely the relation between the polymer extension and the derivative of the free energy with respect to the force at constant temperature: $L = -(\partial\mathcal{F}/\partial F)_T$. The term $\delta W = -LdF$ is the incremental work in aligning the polymer chain performed by an increase dF in the tension on the polymer (compare with Eq. 5.58).

Since Nb is the total extension L_0 of the chain, in the limit of small forces ($\beta Fb \ll 1$ and using the expansion $\coth x = 1/x + x/3 - \cdots$) Eq.5.75 becomes:

$$L = \frac{L_0 b}{3kT}F \equiv k_{pol}^{-1}F \tag{5.79}$$

where $k_{pol} = 3kT/L_0 b$ is the elasticity of a polymer chain at low forces. As a conventional spring, at low forces the elasticity of a polymer obeys Hooke's law: $F = k_{pol}L$. However in contrast with common materials (e.g. metals) which become softer as they are heated, the elasticity of a polymer increases with temperature. If one heats a rubber band under tension, the rubber band will contract, a property use widely in shrink-wraps. This peculiar behavior of polymers is due to the entropic nature of their stiffness. At high temperature the polymer entropy is larger, the polymer is more compact and is more difficult to stretch[93, 92].

Due to the equivalence between this polymer model and the paramagnetic behavior of freely rotating magnetic dipoles (classical Heisenberg model), we may also deduce from Eq.5.79 the magnetic susceptibility of a Heisenberg paramagnet (by replacing $L_0 b$ by $N\mu^2$ in k_{pol}^{-1} which is the equivalent of $\chi_{M,H}$, see Eq.5.51):

$$\chi_{M,H} = \frac{N\mu^2}{3kT} \tag{5.80}$$

which has been verified experimentally, see Fig.5.8. Notice that appart from the factor $1/3$ this is identical to the magnetic susceptibility of an Ising paramagnet (which can adopt only two orientations). Beyond the formal analogy between very different physical systems which allows us to draw inferences about one system (e.g. Heisenberg paramagnet) from the knowledge of another (e.g. polymer chain), this example demonstrates the unifying power of the information theoretical approach to Statistical Mechanics. We shall have more opportunities to appreciate that power in the examples below.

5.5 ◆ IDEAL GAS

The study of gases (how they expand when heated, how their pressure increases as their volume is reduced at a fixed temperature, etc.) has played a central role in the development of classical thermodynamics. The understanding of the thermodynamic behavior of gases, in particular how work could be extracted from cycles of compression/expansion, was the key ingredient that —through the invention of steam (and later fuel) engines —launched the industrial revolution. Because of its historical importance we shall study quite extensively that system.

The Statistical Mechanics study of a gas follows the procedure we adopted to study a paramagnet or a polymer. First we have to specify the space in which the motion of the gas particle is described. Classically a system of N point particles of mass m in motion is completely determined once the particles' positions and momenta (or velocities) are given. Hence the state of a mono-atomic gas (e.g. noble gases such as He, Ne, etc.) is completely specified by the positions and momenta ($\{\vec{x}_i, \vec{p}_i\}$) of each of its particles. If the gas cannot be modeled as a gas of point particles (for example if it consists of molecules such as O_2, N_2, etc.), internal degrees of freedom (e.g. angular momentum, vibrational state, etc.) have to be specified as well.

The problem is to estimate the probability of finding the gas in a given configuration, $\mathcal{P}(\{\vec{x}_i, \vec{p}_i\})$ under the constraint that its average energy is known. One could also impose a constraint of known mean momentum (or velocity), though by going to a reference frame moving at the mean velocity the mean momentum would then be zero. Since, as we shall see in section 5.5.1 the average energy constraint has zero mean momentum, the mean momentum is not an extra constraint (just as for a fair dice requiring the mean result to be $a = 3.5$ doesn't add a constraint).

We shall initially consider gases, where the interaction potential between particles $V(\vec{x}_i - \vec{x}_j)$ can be neglected in comparison with their kinetic energy: $\epsilon_i = \vec{p}_i^{\,2}/2m$. For such an "ideal" gas the energy E_p of a particular configuration is simply the sum of the kinetic energies of the individual particles:

$$E_p = \sum_i^N \vec{p}_i^{\,2}/2m$$

The statistical properties of such a gas are determined by maximizing the entropy:

$$S = -k\mathrm{Tr}\,\mathcal{P}\log\mathcal{P} \tag{5.81}$$

under the constraints:

$$\mathrm{Tr}\,\mathcal{P} = 1 \; ; \; \langle E_p \rangle = \mathrm{Tr}\,E_p\mathcal{P} = E \tag{5.82}$$

Where the trace is defined as the sum over all possible gas states:

$$\mathrm{Tr}\,\mathcal{P} \equiv \frac{1}{h^{3N}N!} \int \cdots \int d^3x_1 \cdots d^3x_N d^3p_1 \cdots d^3p_N \, \mathcal{P}(\{\vec{x}_i, \vec{p}_i\}) = 1 \tag{5.83}$$

The factor h^{3N} is a normalization factor which takes into account the fact that the precision in a particle's momentum and position is limited by Heisenberg uncertainty principle. That limitation partitions the single particle phase space into cells of volume[6] h^3. The factor $N!$ takes into account the fact that the particles being indistinguishable any permutation of their indices yields the same physical state. To proceed with the estimation of $\mathcal{P}(\{\vec{x}_i, \vec{p}_i\})$, we proceed as usual by maximizing the auxiliary function:

$$\Gamma = -k\mathrm{Tr}\,\mathcal{P}\log\mathcal{P} + k(\gamma_0 + 1)\mathrm{Tr}\,\mathcal{P} - k\beta\mathrm{Tr}\,E_p\mathcal{P}$$

Setting $\partial\Gamma/\partial\mathcal{P} = 0$ yields:

$$\mathcal{P}(\{\vec{x}_i, \vec{p}_i\}) = e^{\gamma_0 - \beta E_p} = e^{\gamma_0}\Pi_{i=1}^N e^{-\beta\vec{p}_i^2/2m} \tag{5.84}$$

The partition function $\mathcal{Z} = e^{-\gamma_0}$ is then set by the first constraint ($\mathrm{Tr}\,\mathcal{P} = 1$), see Eq.5.43:

$$\begin{aligned}
\mathcal{Z} &= \mathrm{Tr}\,e^{-\beta E_p} = \frac{1}{h^{3N}N!}\int\cdots\int d^3x_1\cdots d^3x_N d^3p_1\cdots d^3p_N\,\Pi_i^N e^{-\beta\vec{p}_i^2/2m} \\
&= \frac{V^N}{N!}\left(\frac{1}{h^3}\int d^3p\,e^{-\beta\vec{p}^2/2m}\right)^N = \frac{V^N}{N!}\left(\frac{1}{h}\int_{-\infty}^\infty dp_x\,e^{-\beta p_x^2/2m}\right)^{3N} \\
&= \frac{V^N}{N!}(2\pi m/h^2\beta)^{3N/2} \equiv \frac{V^N}{N!}n_Q(\beta)^N \approx \left(\frac{Ve\,n_Q(\beta)}{N}\right)^N
\end{aligned} \tag{5.85}$$

where we defined the volume $V \equiv \int d^3x$ and used Stirling's formula for large N: $N! \approx (N/e)^N$ and the Gaussian integral: $\int_{-\infty}^\infty dp_x\,e^{-\beta p_x^2/2m} = \sqrt{2\pi m/\beta}$ to derive:

$$n_Q(\beta) \equiv (2\pi mkT/h^2)^{3/2} = 1/\lambda_{DB}^3 \tag{5.86}$$

Where the thermal de-Broglie wavelength λ_{DB} satisfies: $\lambda_{DB} = \langle h/p\rangle/2$:

$$\begin{aligned}
\lambda_{DB} &= \langle h/p\rangle/2 = \frac{h}{2}\left(\frac{\beta}{2\pi m}\right)^{3/2}\int d^3p\,\frac{e^{-\beta\vec{p}^2/2m}}{|\vec{p}|} \\
&= 2\pi h\left(\frac{\beta}{2\pi m}\right)^{3/2}\int_0^\infty pdp\,e^{-\beta p^2/2m} = \pi h\left(\frac{\beta}{2\pi m}\right)^{3/2}\int_0^\infty d(p^2)e^{-\beta p^2/2m} \\
&= \frac{h}{\sqrt{2\pi mkT}}
\end{aligned} \tag{5.87}$$

The thermal de-Broglie wavelength λ_{DB} is associated with the size (i.e. the spatial spread) of the quantum wavepacket describing a particle with momentum p, so that

---

[6]The factor h^3 can be obtained from the quantization of the particle wavenumber (momentum) in a box, see Appendix A.9.1 : $p_i/\hbar = k_i = n_i\pi/L_i$ ($n_i \geq 0$; $i = x, y, z$):

$$\mathrm{Tr}\,\mathcal{P} = \sum_{\{n_i\}}e^{-\beta E(\{n_i\})} = \frac{L_xL_yL_z}{\pi^3}\int_0^\infty dk_x dk_y dk_z e^{-\beta E(\vec{k})} = \frac{V}{h^3}\int_{-\infty}^\infty d^3p\,e^{-\beta E(\vec{p})}$$

$n_Q = \lambda_{DB}^{-3}$ has dimensions of particle density (1/volume). It is the highest particle density allowed for the classical mechanics approximation to hold (i.e. treating particles as possessing definite momentum and position). When the gas density is comparable to or larger than n_Q quantum mechanical effects have to be considered (see section 5.9.3). In practice this only occurs for Helium below about $4°K$ (when $\lambda_{DB} \sim 4\text{Å}$). For other gases, such as Oxygen O_2 with molecular weight $M_{O_2} = N_A m_{O_2} = 32$gr: $\lambda_{DB} = 0.18\text{Å}$, $n_Q = 1.69 \, 10^{26}$ cm^{-3} (at $300°K$) which is much higher than the density of Oxygen gas in standard conditions (see below): $N = N_A = 6 \, 10^{23}$ molecules in a volume of 22.7 l, i.e. $n_{O_2} = 2.67 \, 10^{19}$ cm^{-3}.

The average energy then obeys, see Eq.5.46:

$$E = \frac{\partial \gamma_0}{\partial \beta} = -\frac{\partial}{\partial \beta} \log Z = -N \frac{\partial}{\partial \beta} \log n_Q(\beta) = \frac{3NkT}{2} \tag{5.88}$$

the entropy becomes:

$$S = -k \text{Tr} \, \mathcal{P} \log \mathcal{P} = -k \text{Tr} \, \mathcal{P}(\gamma_0 - \beta E_p) = -k\gamma_0 + k\beta E$$

$$= k \log Z + \frac{3Nk}{2} = Nk \left[\log(V/N) + \frac{5}{2} + \log n_Q(\beta) \right] \tag{5.89}$$

and the Helmholtz free energy is:

$$\mathcal{F} = kT\gamma_0 = -kT \log Z = E - TS = -NkT \left[\log(V/N) + 1 + \log n_Q(\beta) \right] \tag{5.90}$$

In analogy with the magnetization defined as the response of the free energy $\mathcal{F}$ to a change in magnetic field ($M = -(\partial \mathcal{F}/\partial B)_T$, see Eq.5.50), the pressure is defined as the response of the free energy to a change in volume (at constant temperature):

$$P = -\left(\frac{\partial \mathcal{F}}{\partial V} \right)_T = -kT \left(\frac{\partial \gamma_0}{\partial V} \right)_T = NkT/V \tag{5.91}$$

Eq.5.91 is known as the Ideal Gas Law. The pressure which has dimensions of force/area is measured in dynes/cm^2 (CGS system) or N/m^2 (MKS system) also known as Pascals (1 Pa $=$ 1 N/m^2 $=$ 10 dynes/cm^2). Pressure is also sometimes expressed in atmospheres (or bar): 1 bar $\approx 10^5$ Pa, which is roughly the pressure exerted by the atmosphere at sea level. Another unit is the Torr or mmHg which expresses the pressure measured in a Torricelli barometer by a column of mercury of height h mm (1 bar $=$ 760 mmHg). Notice that in the same conditions of volume and temperature different gases with the same number of particles exert the same pressure. Conversely at similar temperature, volume and pressure different gases have the same number of particles (atoms or molecules). Thus in standard conditions of temperature $T = 273.15°K$ ($0°C$), pressure 10^5Pa, and volume $V = 22.711$ liters, the number of particles in a gas is $N_A = 6.022 \, 10^{23}$, known as Avogadro's number. The mass M of a gas with N_A particles (also known as one mole of particles) equals the mass m of a gas particle expressed in grams, e.g. $M = N_A m = 2$g for Hydrogen (H_2) or 28g for Nitrogen (N_2).

When the temperature is fixed Eq.5.91 is known as Boyle's Law after the scientist who discovered it in 1662: PV =const, i.e. at constant temperature the pressure is inversely proportional to the volume of the gas[7]. When the volume is fixed it is known as the Gay-Lussac (or Amontons) law after the scientist who formulated it in 1808: $P \sim T$, the pressure in a gas is proportional to its temperature. When the pressure is fixed it is know as Charles' Law after the scientist who discovered it in 1787: $V \sim T$, at constant pressure the volume of a gas is proportional to its temperature. Charles' Law is interesting since by extrapolating the variation in volume measured at ambient temperatures (say between 0°C and 100°C) to zero volume one can deduce where the zero of temperature is[8]. In 1800, Gay-Lussac estimated it to be −266.6°C a number that was improved in 1848 by Lord Kelvin: −273°C, close to the current definition of the absolute zero of temperature: −273.15°C.

In similitude with the procedure followed for paramagnetic and polymer system we may express the entropy change in terms of the change in internal energy and work. Using Eq.5.89, and since γ_0 depends on β and V:

$$dS = -kd\gamma_0 + kEd\beta + k\beta dE$$
$$= k\left(-\frac{\partial\gamma_0}{\partial\beta} + E\right)d\beta - k\frac{\partial\gamma_0}{\partial V}dV + \frac{dE}{T} \tag{5.92}$$

By virtue of Eq.5.88 the first term of the above equation is null. Using Eq.5.91 we then get:

$$dE = TdS - PdV \tag{5.93}$$

From which we see that the pressure can also be defined by the variation in internal energy as the volume is changed at constant entropy:

$$P = -\left(\frac{\partial E}{\partial V}\right)_S \tag{5.94}$$

We can also write Eq.5.93 as a differential equation for $\mathcal{F}$, since $dE = d\mathcal{F} + TdS + SdT$:

$$d\mathcal{F} = -SdT - PdV \tag{5.95}$$

from which we again see that the pressure is the derivative of the free energy with respect to the volume at constant temperature[9]: $P = -(\partial\mathcal{F}/\partial V)_T$. When compressing a gas ($\delta V < 0$) at constant temperature, the term $\delta W = -PdV > 0$ corresponds to the work performed to increase the free energy of the system. Similarly a decrease in

---

[7] In these conditions: $VdP + PdV = 0$, i.e. $dP/P = -dV/V$ the volume decreases as the pressure is increased. The isothermal compressibility of an ideal gas which measures the relative change in volume as the pressure is increased: $\kappa_T \equiv -(1/V)(\partial V/\partial P)_T$ is thus inversely proportional to its pressure $\kappa_T = 1/P$.

[8] An ensemble of particles cannot remain in gas form down to zero temperature: it transits into liquid phase at a positive temperature when the interparticle interaction competes with the particle's kinetic energy, see section 5.10.2.

[9] Similarly: $S = -(\partial\mathcal{F}/\partial T)_V$ and from the identity $\partial^2\mathcal{F}/\partial T\partial V = \partial^2\mathcal{F}/\partial V\partial T$ one derives the so-called Maxwell relation: $(\partial S/\partial V)_T = (\partial P/\partial T)_V$.

the free energy (at constant temperature) can be coupled to work $\delta W = -PdV < 0$ performed by the expanding gas ($\delta V > 0$) on its surroundings, see section 5.5.3.

By comparing Eq.5.55, Eq.5.77 and Eq.5.93 notice the great similarity between the three different statistical systems that we have just studied: paramagnets, polymers and ideal gases. The magnetic field in the spin system, the tension on the polymer and the volume of the ideal gas are the external control parameters to which the system is responding by changing its magnetization, extension or pressure, see section 5.8.

5.5.1 KINETIC THEORY OF GASES

One of the oldest theory of gases is the so-called kinetic theory of gases which assumed that the probability distribution of the momentum of particles was Gaussian distributed (as a result of multiple collisions and the Central Limit Theorem, see section 5.1.3). This assumption can be derived from the entropy maximization approach used here.

Since we assumed the particles in the gas to be non-interacting, the probability distribution $\mathcal{P}(\{\vec{x}, \vec{p}\})$ for one particle can be deduced from Eqs.5.84 and 5.85 with $N = 1$:

$$\mathcal{P}(\{\vec{x}, \vec{p}\}) = e^{\gamma_0 - \beta E_p} = e^{\gamma_0} e^{-\beta \vec{p}^2/2m}$$

$$= \frac{1}{V n_Q(\beta)} e^{-\beta \vec{p}^2/2m} = \frac{1}{V} \left(\frac{h^2}{2\pi mkT}\right)^{3/2} e^{-\vec{p}^2/2mkT} \tag{5.96}$$

with the normalization: $\text{Tr}\mathcal{P} = (1/h^3) \int d^3x d^3p \; \mathcal{P}(\{\vec{x}, \vec{p}\}) = 1$. The particle is thus distributed uniformly in space with a momentum distribution $\mathcal{P}(\vec{p})$ given by:

$$\mathcal{P}(\vec{p}) = \left(\frac{1}{2\pi mkT}\right)^{3/2} e^{-\vec{p}^2/2mkT} \tag{5.97}$$

Notice that $\mathcal{P}(\vec{p})$ is Gaussian distributed as assumed by Maxwell in 1859 and properly normalized: $\text{Tr}\mathcal{P} = \int d^3p P(\vec{p}) = 1$. The mean momentum $\langle \vec{p} \rangle = 0$ and the variance of the momentum in a given direction (say along the z-axis) is: $\langle p_z^2 \rangle = mkT$. The mean energy of the particle is: $\langle \epsilon \rangle = (p_x^2 + p_y^2 + p_z^2)/2m = 3kT/2$, a result previously derived, see Eq.5.88.

Since $\vec{p} = m\vec{v}$, the velocity distribution $\mathcal{P}(v)$ satisfies:

$$1 = \left(\frac{1}{2\pi mkT}\right)^{3/2} \int d^3p e^{-\beta \vec{p}^2/2m}$$

$$= 4\pi \left(\frac{m}{2\pi kT}\right)^{3/2} \int_0^\infty dv v^2 e^{-mv^2/2kT} \equiv \int_0^\infty \mathcal{P}(v) dv$$

with the velocity distribution:

$$\mathcal{P}(v) = 4\pi \left(\frac{m}{2\pi kT}\right)^{3/2} v^2 e^{-mv^2/2kT} \tag{5.98}$$

This distribution is known as the Maxwell distribution of the velocities of a gas, see Fig.5.12. The most probable velocity v_m (the maximum of the distribution) is obtained by setting: $\partial P/\partial v\big|_{v_m} = 0$, which yields: $v_m = \sqrt{2kT/m}$, while the mean absolute velocity is : $\bar{v} = \int v\mathcal{P}(v)dv = \sqrt{8kT/\pi m} > v_m$. For air (made mostly of nitrogen molecules N_2 ($M_{N_2} = 28$gr) the mean velocity at room temperature (300°K) is $\bar{v} \approx 460$m/s. Finally the velocity variance is: $\langle \vec{v}^2 \rangle = \langle \vec{p}^2 \rangle /m^2 = 3kT/m > \bar{v}^2 > v_m^2$.

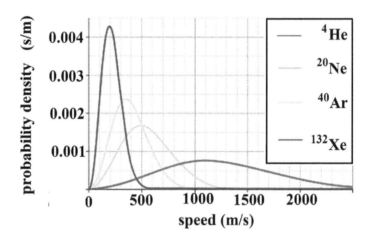

Figure 5.12 Probability distribution of the velocities $\mathcal{P}(v)$ of the atoms of a noble gas. Its units are s/m so that $\int \mathcal{P}(v)dv = 1$.

The flux of gas particles with density $n = N/V$ is: $\vec{J}(v) = n\vec{v}\,\mathcal{P}(v)$. The pressure P_z on a wall parallel to the xy plane is equal to the change in momentum $\Delta p_z = 2mv_z$ of a particle impinging on that wall multiplied by the normal flux J_z, i.e. multiplied by the number of particles colliding with the wall per unit time (those with a velocity $v_z > 0$, i.e. impinging on the wall):

$$P_z = 2mv_zJ_z\big|_{v_z>0} = 2mnv_z^2\mathcal{P}(v)\big|_{v_z>0}$$

The mean pressure P is thus:

$$P = \langle P_z \rangle = 2mn\langle v_z^2 \rangle\big|_{v_z>0} = mn\langle v_z^2 \rangle = \frac{N}{V}\frac{\langle p_z^2 \rangle}{m} = NkT/V$$

which is the Ideal Gas Law, derived previously by maximizing the entropy, Eq.5.91.

If a small hole of area Δa is punched in that wall the number of gas particles lost through that hole per unit time is:

$$\frac{dN}{dt} = -\int_{v_z>0} J_z \Delta a \, dv = -n\Delta a \int_{v_z>0} v_z \mathcal{P}(v) dv$$

$$= -n\Delta a \left(\frac{m}{2\pi kT}\right)^{1/2} \int_0^\infty dv_z \, v_z e^{-mv_z^2/2kT} = -n\Delta a \sqrt{\frac{kT}{2\pi m}} \quad (5.99)$$

Since $n \equiv N/V = P/kT$, the total particle flux effusing through a hole in a container at pressure P is:

$$\Phi_0 \equiv -(1/\Delta a)dN/dt = \frac{P}{\sqrt{2\pi mkT}} \quad (5.100)$$

5.5.2 TEMPERATURE EQUILIBRATION, SPECIFIC HEATS

As we have done for a paramagnetic system let us study the temperature equilibration of an ideal gas. Consider two vessels of volume V_1 and V_2 containing an ideal gas with mean energy E_1 and E_2. According to Eq.5.88, the temperatures T_1, T_2 of these gases obey:

$$E_1 = \frac{3N_1 kT_1}{2} \quad \text{and} \quad E_2 = \frac{3N_2 kT_2}{2}$$

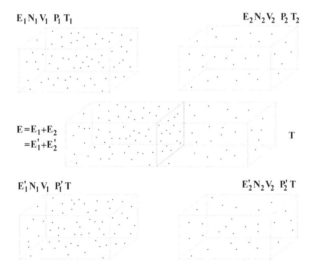

Figure 5.13 Bringing in contact two recipients of volumes V_1, V_2 containing an ideal gas with average internal energies $E_1 = 3N_1 kT_1/2$ and $E_2 = 3N_2 kT_2/2$ results in an equilibration of their temperatures at an intermediate value T such that the mean energy of the blocks in contact becomes $E = E_1 + E_2 = 3(N_1 + N_2)kT/2$. When separated the blocks have new mean energies E_1', E_2' set by their new temperature T. The energy (heat) exchanged is $\delta Q = E_2' - E_2 = 3N_2 k\delta T_2/2 = -3N_1 k\delta T_1/2$ (with $\delta T_i = T - T_i$; $i = 1, 2$).

When the two vessels are brought in contact and let equilibrate, see Fig.5.13, the mean energy of the combined system $E' = E'_1 + E'_2$ will satisfy by Eq.5.88:

$$E' = \frac{3kT(N_1 + N_2)}{2} = E_1 + E_2 \equiv E$$

Clearly if $T_1 > T_2$, the gases will equilibrate at an intermediate temperature T:

$$T_1 > T = \frac{N_1}{N_1 + N_2}T_1 + \frac{N_2}{N_1 + N_2}T_2 > T_2 \qquad (5.101)$$

As the vessels are brought apart the new energy of the gases will be:

$$E'_1 = \frac{3N_1kT}{2} \quad \text{and} \quad E'_2 = \frac{3N_2kT}{2}$$

An amount of energy (heat)

$$\delta Q = E'_2 - E_2 = -(E'_1 - E_1) = 3kN_2(T - T_2)/2 = 3kN_1(T_1 - T)/2$$

has been transferred between the vessels. From Eq.5.89 we deduce that the entropy of an ideal gas can be written as: $S = (3kN/2)\log T + G(N, V)$ where $G(N, V)$ is a function of N and V only (and is not affected by bringing the vessels in contact). Since the entropy is a concave function of T, by Eq.5.66 we see that:

$$(N_1 + N_2)\log T = (N_1 + N_2)\log[\frac{N_1}{N_1 + N_2}T_1 + \frac{N_2}{N_1 + N_2}T_2]$$
$$> N_1 \log T_1 + N_2 \log T_2$$

In other words when the temperatures of the two vessels containing an ideal gas equalize the Second Law of Thermodynamics holds:

$$S > S_1 + S_2$$

One can rewrite the amount of heat exchanged as $\delta Q \equiv \tilde{c}_V N_2 \delta T_2$. The specific heat at constant volume $\tilde{c}_V = 3k/2 = 2.07 \ 10^{-23} \text{J}/°\text{K}$ is then a measure of how much heat can be absorbed by a gas for a temperature increase of $1°\text{C}$. The molar specific heat at constant volume c_V is defined for an Avogadro's number of atoms or molecules $N = N_A = 6.022 \ 10^{23}$ particles/mole:

$$c_V = \tilde{c}_V N_A = \frac{3kN_A}{2} = \frac{3R}{2} \qquad (5.102)$$

where $R \equiv kN_A = 8.3144$ J/mole $°\text{K}$ is known as the gas constant. Since the volume of the vessels is fixed ($dV = 0$), for small temperature changes ($\delta T \ll T$) by virtue of Eq.5.93, we have

$$\delta Q = TdS = dE = 3kN\delta T/2 \qquad (5.103)$$

The equilibration experiment described in Fig.5.13 can also be conducted at constant pressure ($dP = 0$). In that case the heat exchanged is not just due to a change in internal energy: the work performed while changing the volume does also contribute. This can be seen formally by rewriting Eq.5.93 as:

$$TdS = dE + PdV = d(E + PV) - VdP \equiv dH - VdP \qquad (5.104)$$

where $H = E + PV$ is known as the enthalpy. When $dP = 0$ (i.e. for processes occurring at constant pressure) the change in enthalpy defines the heat exchanged, not the change in internal energy:

$$\delta Q = TdS = dH \qquad (5.105)$$

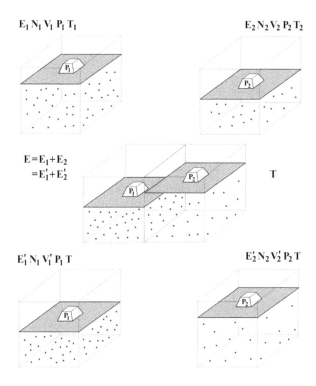

Figure 5.14 Bringing in contact two recipients kept at constant pressures P_1, P_2 and containing an ideal gas with average internal energies $E_1 = 3N_1kT_1/2$ and $E_2 = 3N_2kT_2/2$ (with $T_1 > T_2$) results in an equilibration of their temperatures at an intermediate value $T_1 > T > T_2$ such that the mean energy of the blocks in contact becomes $E = E_1 + E_2 = 3(N_1 + N_2)kT/2$. When separated the blocks have new mean energies E_1', E_2' set by their new temperature T and new volumes $V_1' < V_1, V_2' > V_2$ defined by the ideal gas Law, Eq.5.91: $V_i' = kN_iT/P_i$ (with $i = 1, 2$). The energy (heat) exchanged is equal to the sum of the change in the internal energy of the initially cooler gas $E_2' - E_2$ and the work performed expanding it $P_2(V_2' - V_2)$: $\delta Q = E_2' - E_2 + P_2(V_2' - V_2) = 5kN_2\delta T_2/2 = $ (with $\delta T_2 = T - T_2$).

Consider Fig.5.14, when the two recipients are brought in contact their temperatures equalize and as a result their volume change. The hotter gas cools down and its

volume shrinks while the cooler gas heats up and its volume expands (see Charles'
Law above). Expanding the volume of a gas by δV at constant pressure P necessitates a work: $\delta W = P\delta V = kN\delta T$, hence the amount of heat exchanged at constant
pressure is

$$\delta Q = TdS = dE + PdV = dH = 5kN\delta T/2 \equiv \tilde{c}_P N\delta T$$

where $\tilde{c}_P = 3.45\ 10^{-23} J/°K$ is known as the specific heat at constant pressure. The
molar specific heat at constant pressure c_P is defined for an Avogadro's number of
atoms:

$$c_P = \tilde{c}_P N_A = \frac{5kN_A}{2} = \frac{5R}{2} \tag{5.106}$$

As we shall see below (section 5.6) the specific heats may change if the gas is composed of molecules that can also absorb heat via a change in their rotational (or vibrational) energy, however the difference between the specific heat at constant pressure
and volume is a constant: $\tilde{c}_P - \tilde{c}_V = k$, see Fig.5.15.

Molecular specific heats   (in units of $R = N_A\ k_B = 8.3114$ J/mole °K)

Gas	c_p	c_V	$\gamma = c_p/c_V$
Monatomic			
He	2.50	1.50	1.67
Ar	2.50	1.50	1.67
Ne	2.50	1.53	1.64
Kr	2.50	1.48	1.69
Diatomic			
H_2	3.46	2.45	1.41
N_2	3.50	2.50	1.40
O_2	3.54	2.54	1.40
CO	3.52	2.52	1.40
Polyatomic			
Cl_2	4.17	3.09	1.35
CO_2	4.45	3.43	1.30
SO_2	4.86	3.78	1.29
H_2O	4.26	3.25	1.30
CH_4	4.27	3.26	1.31

Figure 5.15   The molar specific heats at constant volume and pressure of a number of mono-,
di- and poly-atomic gases. Notice that the difference $c_P - c_V \approx kN_A \equiv R = 8.3144$ J/mole °K.

Adiabatic process. Like a paramagnetic system —which can be cooled in a (dissipation free) entropy conserving adiabatic process by slowly decreasing the magnetic field —an ideal gas can be cooled (or warmed) adiabatically, i.e. without heat exchange, by slowly increasing (or decreasing) its volume (or changing its pressure)[10]. In such an adiabatic process while $dS = 0$ the volume, temperature, and pressure of the gas evolve while satisfying Eq.5.91:

$$0 = TdS = dE + PdV = \frac{3Nk}{2}dT + \frac{NkT}{V}dV$$

$$= \tilde{c}_V NdT + (\tilde{c}_P - \tilde{c}_V)\frac{NT}{V}dV$$

which can be rearranged to yield:

$$\frac{dT}{T} = (1 - \frac{\tilde{c}_P}{\tilde{c}_V})\frac{dV}{V} = (1 - \gamma)\frac{dV}{V} \tag{5.107}$$

with the adiabatic index $\gamma \equiv \tilde{c}_P/\tilde{c}_V (= 5/3$ for a monoatomic ideal gas, see Fig.5.15). Integrating this equation yields:

$$TV^{\gamma-1} = Constant \tag{5.108}$$

From the Ideal Gas Law, Eq.5.91, $PV = NkT$. Hence[11]:

$$PV^{\gamma} = Constant \tag{5.109}$$

One can verify that as the system evolves from its initial state T_i, V_i, P_i to its final state T_f, V_f, P_f, the change in internal energy $E_f - E_i = N\tilde{c}_V(T_f - T_i)$ is balanced by the work performed by the system[12]:

$$W_{fi} = \int_{V_i}^{V_f} PdV = \int_{V_i}^{V_f} \frac{NkT}{V}dV = NkT_i V_i^{\gamma-1} \int_{V_i}^{V_f} \frac{dV}{V^{\gamma}}$$

$$= N\tilde{c}_V(\gamma - 1)T_i V_i^{\gamma-1}[\frac{V_f^{-\gamma+1} - V_i^{-\gamma+1}}{1 - \gamma}] = -N\tilde{c}_V(T_f - T_i) \tag{5.110}$$

which is positive if $T_f < T_i$, i.e. if the gas is expanding. The fact that work can be extracted from the expansion of a gas is at the core of all thermal machines.

---

[10]The adiabatic approximation is also valid when processes are so rapid (e.g. sound propagation) that the dissipation can be neglected versus the inertial (kinetic energy) terms. This approximation is sometimes made in the analysis of the quick compression/expansion cycles of fuel engines.

[11]The adiabatic compressibility is: $\kappa_a \equiv -(1/V)(\partial V/\partial P)_S = 1/\gamma P$. It differs from the isothermal compressibility by a factor γ.

[12]Using the specific heat at constant volume to replace: $k = 2\tilde{c}_v/3 = (\gamma - 1)\tilde{c}_v$.

5.5.3 HEAT AND WORK

The essence of all thermal engines is to extract work from a physical system by continuously cycling it through appropriate states. Assuming that the cycling time is slow enough for the system to be instantaneously at equilibrium, one can use the results derived for equilibrium systems (e.g. the Ideal Gas Law) to estimate the work extracted by such a procedure and its efficiency. The theoretical analysis of the efficiency of such engines has been done by S.Carnot in 1824. Without loss of generality he assumed that the thermal engine was working between two temperature reservoirs: a heat source at a high temperature T_h and, heat sink at a low temperature T_l. Work is done while heat is transferred between those reservoirs. An amount δQ_h is injected into the system by the high temperature heat source while an amount δQ_l is lost by the system to the low temperature heat sink. Since during a cycle the total internal energy change is null (the system reverts to its initial state) the work performed by the system is by energy conservation :

$$\delta Q_h - \delta Q_l = \oint \delta Q = \oint dE + \oint P dV = \delta W \qquad (5.111)$$

By loosing heat the source experiences an entropy loss, see Eqs. 5.103 and 5.105: $dS_h = -\delta Q_h/T_h$, while the heat sink gains entropy by the amount $dS_l = \delta Q_l/T_l$. Since by the Second Law of Thermodynamics Eq.5.67 the overall entropy cannot decrease we have:

$$dS_h + dS_l = \delta Q_l/T_l - \delta Q_h/T_h \geq 0 \quad \text{i.e.} \quad \frac{\delta Q_l}{\delta Q_h} \geq \frac{T_l}{T_h} \qquad (5.112)$$

Hence the efficiency η of a thermal engine defined as the ratio of work produced divided by heat injected is limited by:

$$\eta = \frac{\delta W}{\delta Q_h} = 1 - \frac{\delta Q_l}{\delta Q_h} \leq \frac{T_h - T_l}{T_h} \qquad (5.113)$$

To reach a high efficiency the ratio of T_h/T_l should be the largest possible, hence efficient fuel engines (e.g. Diesel motors) have high temperature ratios (or equivalently high compression ratios).

For systems based on cycles of compression/expansion of gases (steam and fuel engines), it is customary to represent the states of the system in a pressure-volume plane (the associated temperature is determined by the Ideal Gas Law, Eq.5.91). For instance consider the following cycle, see Fig.5.16(a): a gas is heated from temperature T_1 to T_2 at constant volume $V_2 = V_1$. It is then left to expand adiabatically up to a volume V_3 (and temperature T_3), at which point it is cooled at constant volume $V_4 = V_3$ to temperature T_4 and compressed adiabatically back to temperature and volume T_1, V_1. The work performed during that cycle is according to Eq.5.110:

$$\delta W = \oint P dV = \int_2^3 \frac{NkT dV}{V} + \int_4^1 \frac{NkT dV}{V}$$
$$= -N\tilde{c}_V(T_3 - T_2) - N\tilde{c}_V(T_1 - T_4) = N\tilde{c}_V(T_2 - T_1) + N\tilde{c}_V(T_4 - T_3)$$
$$= N\tilde{c}_V(T_2 - T_1) + N\tilde{c}_V[\frac{T_1 V_1^{\gamma-1}}{V_4^{\gamma-1}} - \frac{T_2 V_2^{\gamma-1}}{V_3^{\gamma-1}}]$$

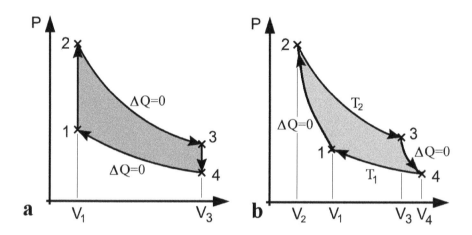

Figure 5.16 (a) A thermal cycle consisting of heating at constant volume V_1 from state 1 to 2, followed by an adiabatic expansion to volume V_3, cooling at constant volume V_3 from state 3 to 4 and final adiabatic compression back to state 1. (b) The Carnot cycle consists of an adiabatic compression from state 1 to 2 followed by an isothermal expansion at temperature T_2 from volume V_2 to V_3, adiabatic cooling from state 3 to 4 and isothermal compression at temperature T_1 back to state 1. The work $\oint P dV$ performed during these cycles is the shaded area.

Using the fact that in the constant volume heating/cooling phases $V_2 = V_1$ and $V_4 = V_3$, we obtain:

$$\delta W = N\tilde{c}_V(T_2 - T_1)[1 - \frac{V_2^{\gamma-1}}{V_3^{\gamma-1}}] = N\tilde{c}_V(T_2 - T_1)[1 - T_3/T_2] \qquad (5.114)$$

That cycle is powered by the heat δQ_{in} injected during the expansion at constant volume V_1:

$$\delta Q_{in} = N\tilde{c}_V(T_2 - T_1) \qquad (5.115)$$

The efficiency η of that cycle is equal to the ratio of the work δW extracted to the heat injected:

$$\eta = \frac{\delta W}{\delta Q_{in}} = \frac{T_2 - T_3}{T_2} \tag{5.116}$$

Since $T_3 > T_4$ (T_4 is the lowest temperature in that cycle) the efficiency of that engine is smaller than Carnot's theoretical limit.

That theoretical limit can be reached with an engine implementing the so-called Carnot cycle. That cycle consists of an adiabatic compression followed by an isothermal expansion (at temperature T_2), an adiabatic cooling and an isothermal compression (at temperature T_1) back to the original state, see Fig.5.16(b). Since by Eq.5.110 the work required during the adiabatic compression phase (between states 1 and 2) is equal to the work produced during the adiabatic cooling phase (between states 3 and 4) the overall work produced by this engine is the difference between the work performed during the isothermal expansion phase (between states 2 and 3) and the one required during the isothermal compression phase (between states 4 and 1):

$$\delta W = \oint P dV = NkT_2 \int_{V_2}^{V_3} \frac{dV}{V} + NkT_1 \int_{V_4}^{V_1} \frac{dV}{V}$$
$$= NkT_2 \log V_3/V_2 - NkT_1 \log V_4/V_1 \tag{5.117}$$

Heat is injected during the isothermal expansion phase and is equal to the work performed during that phase: $\delta Q_{in} = NkT_2 \log V_3/V_2$.

In the adiabatic phases: $(V_4/V_3)^{\gamma-1} = T_2/T_1 = (V_1/V_2)^{\gamma-1}$, see Eq.5.108, hence $V_4/V_1 = V_3/V_2$ and:

$$\eta = \frac{\delta W}{\delta Q_{in}} = \frac{T_2 - T_1}{T_2}$$

which is the theoretical limit established by Carnot, Eq.5.113.

5.6 EQUIPARTITION THEOREM

The ideal gas considered above was mono-atomic and the energy of its atoms was purely kinetic $E_K = \vec{p}^2/2m$. For more complex systems (e.g. molecules) energy may also be stored in rotational and vibrational degrees of freedom. In both cases the energy is a quadratic function of the dynamic variable: the angular momentum $\vec{L}$ in the case of rotation ($E_R = \vec{L}^2/2I$ where I is the moment of inertia) and the displacement $\vec{x}$ in the case of vibration ($E_V = k_s\vec{x}^2/2$, where k_s is the spring or elastic constant). An important theorem of statistical mechanics, known as the equipartition theorem, states that if the energy E_l associated to a certain degree of freedom l (translation, rotation, vibration, ...) is a quadratic function of some dynamical variable ξ_l (linear or angular momentum, displacement, etc.)[13], i.e. $E_l \propto \xi_l^2$, then the mean energy $\langle E_l \rangle$

---

[13] Notice that this is not always true. The potential energy in the gravitational field of a star (or the electric field of a charge) is not a quadratic function of the distance to the center of the star (or the charge).

associated with that degree of freedom is $kT/2$. To prove that theorem let us write the energy E of a single particle as a sum over the energies associated with the variables ξ_l:

$$E(\{\xi_l\}) = \sum_l E_l = \sum_l a_l \xi_l^2/2$$

where the sum is over all degrees of freedom and the a_l's are some constant (e.g. mass, moment of inertia, elastic constant, etc.). The equipartition theorem then states that $\langle E_l \rangle = kT/2$. Using the procedure applied to the study of an ideal gas the probability distribution of the energy E is by Eq.5.84:

$$\mathcal{P}(E(\{\xi_l\})) = e^{\gamma_0 - \beta E} = \frac{\Pi_l e^{-\beta a_l \xi_l^2/2}}{\mathrm{Tr}\Pi_l e^{-\beta a_l \xi_l^2/2}} \tag{5.118}$$

where the last equation ensures that $\mathrm{Tr}\mathcal{P}(\{\xi_l\}) = 1$. The probability distribution of dynamical variable ξ_n is thus:

$$\mathcal{P}(\xi_n) = \frac{e^{-\beta a_n \xi_n^2/2}}{\int d\xi_n e^{-\beta a_n \xi_n^2/2}}$$

which is properly normalized $\mathrm{Tr}\mathcal{P}(\xi_n) = 1$. Therefore the mean energy $\langle E_n \rangle$ associated with that variable is:

$$\langle E_n \rangle = \frac{a_n \langle \xi_n^2 \rangle}{2} = \int d\xi_n \frac{a_n \xi_n^2}{2} \mathcal{P}(\xi_n) = -\frac{\partial}{\partial \beta} \log \int d\xi_n e^{-\beta a_n \xi_n^2/2}$$

$$= -\frac{\partial}{\partial \beta} \log\left(\frac{2\pi}{\beta a_n}\right)^{1/2} = \frac{1}{2\beta} = \frac{kT}{2} \tag{5.119}$$

Hence for a mono-atomic gas which has three translational degrees of freedom with momentum $\vec{p}$ the mean energy for each degrees of freedom is $kT/2$ and the mean overall energy is $E = 3NkT/2$, see Eq.5.88. For a diatomic gas such as N_2, O_2, etc. (for which the first vibrational excited state has an energy usually much larger than kT) there are two additional rotational degrees of freedom (in the plane perpendicular to the axis of the molecule) and therefore the mean overall energy of these diatomic gases is $E = 5NkT/2$, with a subsequent specific heat $\tilde{c}_V = 5k/2$ and adiabatic index $\gamma = c_P/c_V = 7/5$ as indeed observed, see Fig.5.15.

5.7 VARIABLE NUMBER OF PARTICLES

Until now we have assumed that the number of particles (e.g. the atoms in a gas) was fixed and precisely known. In fact the number of particles in a given system is never known precisely: one only knows the average number of particles $\bar{N}$ (for example by measuring their average density). Moreover there are situations in which the average number of particles (just like the mean energy) can vary, for example if the particles participate in a chemical reaction, are exchanged through a semi-permeable membrane or are adsorbed on a surface. The Statistical Mechanics approach of these cases is similar to the previously studied examples, except for the existence of one more constraint on the average number of particles: $\mathrm{Tr}\, n\, \mathcal{P} = \bar{N}$, where the trace is a sum over all possible number n of particles and associated energies $E_p = \sum_i^n \epsilon_i$. For an ideal gas the energy of each particle is: $\epsilon_i = \vec{p}_i^{\,2}/2m + \epsilon_0$, where $\epsilon_0 < 0$ is the ground state (or binding) energy[14] of the atom (molecule), which in the case of exchanging particles cannot be set to zero for all of them. In this so-called Grand Canonical approach, the entropy: $S = -k\mathrm{Tr}\, \mathcal{P}\log\mathcal{P}$ has to be maximized under the three constraints:

$$\mathrm{Tr}\, \mathcal{P} = 1 \quad \langle E_p \rangle = \mathrm{Tr}\, E_p\, \mathcal{P} = E \quad \langle n \rangle = \mathrm{Tr}\, n\, \mathcal{P} = \bar{N}$$

Where the trace is defined as the sum over all possible gas states with n particles and positions/momenta $\{\vec{x}_i, \vec{p}_i\}$ ($i = 1 \cdots, n$). As usual we have to maximize the auxiliary function:

$$\Gamma = -k\mathrm{Tr}\, \mathcal{P}\log\mathcal{P} + k(\gamma_0 + 1)\mathrm{Tr}\, \mathcal{P} - k\beta\mathrm{Tr}\, E_p\mathcal{P} - k\alpha\mathrm{Tr}\, n\mathcal{P}$$

Setting $\partial\Gamma/\partial\mathcal{P} = 0$ yields:

$$\mathcal{P} = e^{\gamma_0} e^{-\beta E_p - \alpha n} \tag{5.120}$$

The condition $\mathrm{Tr}\, \mathcal{P} = 1$ implies as usual, see Eq.5.85, that the partition function is:

$$\mathcal{Z} = e^{-\gamma_0} = \mathrm{Tr}\, e^{-\beta E_p - \alpha n}$$

$$= \sum_{n=0}^{\infty} \frac{e^{-\alpha n}}{h^{3n} n!} \int \cdots \int d^3 x_1 \cdots d^3 x_n d^3 p_1 \cdots d^3 p_n \Pi_i^n e^{-\beta \epsilon_i}$$

$$= \sum_{n=0}^{\infty} \frac{1}{n!} \left(\frac{V}{h^3} \int d^3 p\, e^{-\beta \epsilon - \alpha} \right)^n = \exp\left[\frac{V}{h^3} \int d^3 p\, e^{-\beta \epsilon - \alpha} \right] \tag{5.121}$$

where we used the Taylor expansion of $e^x = \sum_n x^n/n!$, see Appendix A.3.1. For particles with energy: $\epsilon = p^2/2m + \epsilon_0$, integration over the momentum (see section 5.5) yields:

$$-\gamma_0 = \log \mathcal{Z} = V e^{-\alpha - \beta \epsilon_0} n_Q(\beta) \tag{5.122}$$

where:

$$n_Q(\beta) \equiv (2\pi m k T/h^2)^{3/2} = (2\pi m/\beta h^2)^{3/2} = 1/\lambda_{DB}^3 \tag{5.123}$$

---

[14]Notice that the energy of the ground state of bound particles is negative, see for example section 4.1.3, since for the bound state to be stable, its energy must be smaller than the energy of the unbound state, which can be zero if the velocity of the unbound particles is null.

is the maximal particle density before quantum mechanical effects have to be considered, see Eq.5.86. Once the partition function is known, we can compute the constraints. The mean number of particles $\bar{N}$ is given by:

$$\bar{N} = \text{Tr}\, n\, \mathcal{P} = -\frac{\partial \log Z}{\partial \alpha} = V e^{-\alpha - \beta \epsilon_0} n_Q(\beta) \equiv V e^{\beta(\mu - \epsilon_0)} n_Q(\beta) \tag{5.124}$$

Where we defined the chemical potential: $\mu \equiv -kT\alpha$. The units of μ are the same as those of energy, but are often expressed in kJ/mole: 1 kJ/mole = 1.6 10^{-21} J = 0.01 eV. The chemical potential μ is thus related to the particle density $\bar{n} = \bar{N}/V$ or the molar concentration of particles $[c] = 10^3 \bar{n}/N_A$) (where $[c]$ is expressed in moles/l, $\bar{n}$ in particles/cm^3 and N_A = 6.022 10^{23} particles/mole is Avogadro's number):

$$\mu = kT \log(\bar{n}/n_Q) + \epsilon_0 = kT \log[c] + G_0 \tag{5.125}$$

$$\text{where}: \qquad G_0 = \epsilon_0 - kT \log\left(10^3 n_Q/N_A\right) \tag{5.126}$$

is known as the Gibbs free energy of the system, usually quoted in standard conditions ($P = 10^5$ Pa, $T = 273.15°$K, $[c] = 1$ M). At $T = 0$ the chemical potential is equal to the ground state energy of the system $\mu = \epsilon_0 < 0$. As the temperature (and entropy) increases μ decreases (becomes more negative since $\bar{n}/n_Q < 1$).

The mean energy E can be obtained from Eq.5.122:

$$E = -\frac{\partial \log Z}{\partial \beta} = V e^{-\alpha - \beta \epsilon_0} n_Q(\beta)[3kT/2 + \epsilon_0]$$

$$= 3\bar{N}kT/2 + \bar{N}\epsilon_0 \tag{5.127}$$

The pressure P is, see Eq.5.91:

$$P = -kT \frac{\partial \gamma_0}{\partial V} = -kT\gamma_0/V = kT e^{-\alpha - \beta \epsilon_0} n_Q(\beta) = kT\bar{N}/V = kT\bar{n} \tag{5.128}$$

which is the Ideal Gas Law derived in the canonical approach (i.e. when the number of particles N was known precisely, see Eq.5.91).

It is customary to define the so-called Maxwell-Boltzmann distribution:

$$\eta^{MB} \equiv e^{-\beta(\epsilon - \mu)} \tag{5.129}$$

which is related to the energy probability density:

$$E = \frac{V}{h^3} \int d^3 p\, \epsilon\, \eta^{MB} \equiv \frac{V}{h^3} \int d^3 p\, \epsilon\, e^{-\beta(\epsilon - \mu)}$$

$$= V e^{\beta(\mu - \epsilon_0)} n_Q(\beta)[3kT/2 + \epsilon_0] = \bar{N}[3kT/2 + \epsilon_0] \tag{5.130}$$

From the knowledge of the probability distribution $\mathcal{P}$ we can compute the entropy, see Eq.5.89:

$$S = -k\text{Tr}\mathcal{P}\log\mathcal{P} = -k\gamma_0 + k\beta E + k\alpha\bar{N} \tag{5.131}$$

Which can be cast as:

$$TS = -kT\gamma_0 + E - \mu\bar{N} = E + PV - \mu\bar{N} \tag{5.132}$$

The Helmholtz free energy is, see Eq.5.49:

$$\mathcal{F} = E - TS = -PV + \mu\bar{N} \tag{5.133}$$

Since γ_0 depends on V, β, and α, we may rewrite Eq.5.131 as:

$$
\begin{aligned}
dS &= -k(\frac{\partial\gamma_0}{\partial V}dV + \frac{\partial\gamma_0}{\partial\beta}d\beta + \frac{\partial\gamma_0}{\partial\alpha}d\alpha) + kEd\beta + k\beta dE + k\bar{N}d\alpha + k\alpha d\bar{N} \\
&= k(\beta PdV - Ed\beta - \bar{N}d\alpha) + kEd\beta + k\beta dE + k\bar{N}d\alpha + k\alpha d\bar{N} \\
&= k\beta dE + k\beta PdV - k\beta\mu d\bar{N} \tag{5.134}
\end{aligned}
$$

where we used Eq.5.91, Eq.5.127, and Eq.5.124. We can recast this equation as:

$$dE = TdS - PdV + \mu d\bar{N} \tag{5.135}$$

Using the differential form of Eq.5.133: $d\mathcal{F} = dE - TdS - SdT$ and Eq.5.135 yield:

$$d\mathcal{F} = -SdT - PdV + \mu d\bar{N} \tag{5.136}$$

The chemical potential is thus given by the derivative of the free energy with respect to $\bar{N}$ (at constant volume and temperature) $\mu = \partial\mathcal{F}/\partial\bar{N}\big|_{V,T}$. If multiple particle species are involved (as in the chemical reactions studied below), at constant volume and temperature:

$$d\mathcal{F}_{V,T} = \sum_i \mu_i d\bar{N}_i \quad \text{or}: \quad \mu_i = \frac{\partial\mathcal{F}}{\partial\bar{N}_i}\bigg|_{V,T} \tag{5.137}$$

From Eq.5.133, $\mathcal{F} = -PV + \mu\bar{N}$ in its differential form $d\mathcal{F} = -VdP - PdV + \bar{N}d\mu + \mu d\bar{N}$ and Eq.5.136 one derives the Gibbs-Duhem relation:

$$\bar{N}d\mu = -SdT + VdP \tag{5.138}$$

5.7.1   BOILING PRESSURE AND LATENT HEAT OF VAPORIZATION.

The chemical potential plays for the mean particle number the same role as the temperature for the mean energy: when two systems with different chemical potentials μ_1, μ_2 are brought in contact (for example water in its gas or liquid phase) and allowed to exchange their particles they come to an equilibrium such that $\mu_1(P,T) = \mu_2(P,T)$. As the pressure (or temperature) is changed the equilibrium is reached at a different temperature (or pressure): thus due to lower pressure at high altitudes[15], water on the Himalaya boils at a lower temperature than at sea level.

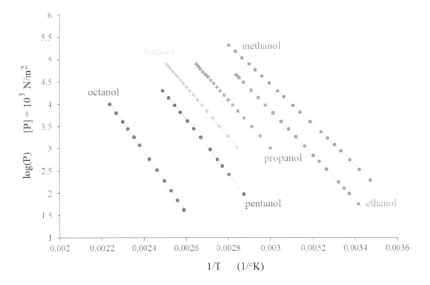

Figure 5.17 Variation of the boiling pressure with temperature for a few alcohols (dots). Notice the nice fit (continuous lines) to the Clausius-Clayperon relation, Eq.5.140. Data from Ambrose and Sprake[94].

Along the co-existence curve, as $d\mu_1 = d\mu_2$ the Gibbs-Duhem relation (Eq.5.138) implies:

$$-(S_1/\bar{N}_1)dT + (V_1/\bar{N}_1)dP = -(S_2/\bar{N}_2)dT + (V_2/\bar{N}_2)dP$$

defining the specific entropy $s_i \equiv S_i/\bar{N}_i$ and specific volume $v_i = V_i/\bar{N}_i$, the equation above can be recast as:

$$\frac{dP}{dT} = \frac{s_2 - s_1}{v_2 - v_1} = \frac{\mathcal{L}}{T\delta v} \qquad (5.139)$$

This equation is known as the Clausius-Clayperon equation. It relates the variation of pressure with temperature along the co-existence line between two phases (e.g. gas and liquid) to the latent heat released at the transition $\mathcal{L} = T\Delta s = T(s_2 - s_1)$ and the change in the specific volumes of the two phases, see below section 5.10.2. For boiling water for example the specific volume of the gas phase is much larger than that of the liquid phase: $v_l \ll v_g = kT/P$ (see above and Eq.5.91) and therefore to a good approximation:

$$\frac{dP}{dT} = \frac{\mathcal{L}P}{kT^2}$$

---

[15]The decrease of pressure with height h is itself a consequence of the equality of the chemical potential of atmospheric slabs at different heights: $\mu(h) = \mu(0)$. In that case $\epsilon_0 = mgh$ is the potential energy of the molecules and from Eq.5.125, one deduces that the air density $\bar{n}(h) = \bar{n}(0)e^{-\beta mgh}$. Hence the pressure $P = kT\bar{n}$ decreases exponentially with height. Thus at 5000m the partial pressure of Oxygen ($P_{O_2} = kT\bar{n}_{O_2}$) is about half its value at sea level, which explains the difficulty of breathing at high altitudes.

which upon integration yields:

$$\log P/P_\infty = -\mathcal{L}(\frac{1}{kT} - \frac{1}{kT_\infty}) \qquad \text{or}: \qquad Pe^{\mathcal{L}/kT} = \text{const.} \qquad (5.140)$$

where P_∞ is the boiling pressure at high temperature (T_∞). Thus at the boiling point of a liquid, the pressure decreases exponentially with the inverse of its boiling temperature, a relation that has been verified for many liquids, see Fig.5.17. At a height of 5000m, as the atmospheric pressure is half its value at sea level, water with a latent heat $\mathcal{L} = 40.66$kJ/mol, boils at $\approx 80°C$.

5.7.2   THE VOLTAGE ACROSS A CELL MEMBRANE.

Consider the case of a semi-permeable membrane, which allows one type of ions (with charge q, say sodium Na^+) to pass through but not another (say Cl^-). If two solutions of $NaCl$ at different concentrations n_1, n_2 are separated by such a membrane the sodium ions will diffuse across the membrane, but not the chloride ions. As a result a charge imbalance is generated which results in an electrical potential across the membrane $V = (\epsilon_0^{(2)} - \epsilon_0^{(1)})/q$ which can be determined from the equality of the chemical potential of the sodium ions: $\mu_1 = \mu_2$ and Eq.5.125 ($\mu = kT \log(\bar{n}/n_Q) + \epsilon_0$):

$$kT \log n_1/n_2 = \epsilon_0^{(2)} - \epsilon_0^{(1)} = qV \qquad (5.141)$$

The Nernst equation, Eq. 5.141, is central to the understanding of cells and neurons in particular. Using membrane inserted proteins, known as ion pumps, cells maintain a steady state characterized by an imbalance in the internal and external concentrations of various ions. The concentration of potassium (K^+) ions inside the cell is $[K^+]_{in} = 140$ mM while it is typically $[K^+]_{out} = 5$ mM outside. Inversely the concentration of sodium (Na^+) ions is $[Na^+]_{in} = 12$mM inside while it is $[Na^+]_{out} = 140$ mM outside the cell. At room temperature $kT/e = 25$ mV, the Nernst potential associated to the potassium ions (charge $q = +e$) is: $V_K = (kT/e)\log[K^+]_{out}/[K^+]_{in} = -84$mV , while that associated with sodium ions (charge $q = +e$) is: $V_{Na} = (kT/e)\log[Na^+]_{out}/[Na^+]_{in} = +60$mV. At steady-state (also known as the rest-state) there is no net current through the membrane: $I = 0$. Since the membrane is much more permeable to potassium ions than to sodium ions, the conductance[16] of potassium g_K is much larger than that of sodium g_{Na} ($\approx g_K/100$) :

$$I = g_K(V_m^0 - V_K) + g_{Na}(V_m^0 - V_{Na}) = 0$$

and the potential V_m^0 across the membrane is closer to the Nernst potential of potassium: $V_m^0 \approx -80$ mV, i.e. the membrane is polarized, see Fig.5.18.

These considerations have led Hodgkin and Huxley to propose in 1952 a mechanism for the transient excitation of a neuron, see Fig.5.18. When the voltage across the neuronal membrane passes a given threshold (about −50mV), sodium (Na^+)

---

[16]The conductance g is the inverse of the resistance R, so that Ohm's law reads: $I = g\Delta V$.

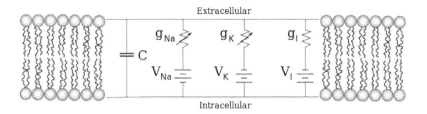

Figure 5.18 The transient excitation of neurons has been explained by Hodgkin and Huxley. They have modeled the neuronal membrane as consisting of parallel voltage sources contributed by the unequal concentration of sodium ($V_{Na} = 60$mV), potassium ($V_K = -84$mV) and other ions (lumped together as leakage voltage, V_l) in series with voltage dependent sodium and potassium conductance $g_{Na}(V_m), g_K(V_m)$, where V_m is the membrane potential. For a membrane of capacitance C the total current I_m through the membrane is given by Ohm's law: $I_m = CdV_m/dt + g_{Na}(V_m)(V_m - V_{Na}) + g_K(V_m)(V_m - V_K) + g_l(V_m - V_l)$. Since the conductivity of potassium is larger than that of any other ion, the steady-state or rest potential of the membrane defined by the condition $I_m = 0$ is: $V_m^0 \approx V_K$. If the membrane is excited (for example via an influx of calcium ions) such that its voltage exceeds a threshold $V_t \sim -50$mV, sodium channels open up (g_{Na} increases) and the membrane depolarizes. As the membrane potential becomes positive these channels close (and potassium ones open) bringing the membrane voltage back to its resting state. The result is a transient activation of the membrane that propagates as a wave up or down neuronal fibers known as axons and dendrites.

channels open up increasing the permeability of the membrane to these ions and depolarizing it. The membrane is driven to the Nernst potential of sodium ions: $V_{Na} = 60$mV. As the membrane voltage becomes positive, sodium channels shut and potassium channels open up: the membrane returns to its steady polarized state while giving rise to a propagating wave (an action potential) down the dendrites or axon of the neural cell due to the sequential activation of voltage sensitive sodium and potassium channels. This wave of ionic current is the signal carried along neurons and processed in our brain, just as the electric current is the signal carried along wires and processed in computers.

5.7.3 CHEMICAL REACTIONS

A chemical reaction is a process that leads to the transformation of one set of chemical substances, the substrates or reagents, into another: the products. Chemical reactions such as fermentation, combustion in fire and the reduction of ores to metals were known since antiquity. In fact one of the goals of Alchemists in the Middle Ages was to transform lead into gold by some chemical reaction, a process that we now know is impossible: lead and gold are elements that cannot be modified by chemical reactions.

That understanding was reached only in the 18th century by Lavoisier as a result of his work on combustion. By carefully weighting lead that had been calcined in a closed vessel filled with air, he showed that the increase in weight of the metal was equal to the loss in weight of the air. He showed that the residual air could support neither combustion nor respiration. Air was thus made of at least two different chemical compounds. He called oxygen the compound reacting with metals when heated (and required for respiration) and determined that it made about 20% of the volume of air. He further showed that oxygen could be released by reducing the heated metal with charcoal. The resulting released air was better at supporting respiration and combustion than regular air. Through further pioneering experiments using only the careful weighting of the substrates and products of chemical reactions, Lavoisier showed that mass was conserved: "rien ne se perd, rien ne se crée, tout se transforme" (nothing is lost, nothing is created, everything is tranformed). Breaking with the classical tradition of the four elements (fire, air, water and earth) he defined an element as any substance that could not be reduced into simpler ones by chemical means. He is thus rightfully seen as the father of modern chemistry and his "Traité élémentaire de Chimie"[95], published in 1789[17], is viewed as the first modern textbook on the subject.

Following up on these ideas, Gay-Lussac and to a fuller extent Dalton revived the atomic theory intuited some 2000 years earlier by Democritus and Lucretius. Gay Lussac showed that the ratio between the volumes of reagent gases and products can be expressed as ratios of whole numbers (their stoichiometry): thus two volumes of hydrogen react with one volume of oxygen to produce one volume of water vapour. This was generalized by Dalton to the assertion (later amply verified) that atoms of different elements combine in simple whole-number ratios to form chemical compounds. In chemical reactions, atoms are thus combined, separated, or rearranged.

Dalton's hypothesis has been confirmed by the QM understanding of the chemical bond as the sharing of valence electrons: a chemical reaction thus involves a redistribution of these shared electrons from the reagents into the products, i.e. the breaking of bonds in the substrates and the formation of new ones in the products[96]. As any QM process this bond reassignment is probabilistic: the larger the concentration of reagents the higher the probability to form products and vice-versa the higher the concentration of products the higher the probability of reforming substrates. There is thus an equilibrium concentration of reagents and products where the formation of the products is balanced by their disappearance (i.e. the reformation of the substrates).

The Grand Canonical approach which accounts for varying number of particles is thus well suited to describe chemical reactions at equilibrium. Consider for example the reaction, $2H_2 + O_2 \rightleftharpoons 2H_2O$. Such an equation can be written as:

$$\sum_i l_i \text{Su}_i = \sum_j m_j \text{Pr}_j \qquad \text{or} \qquad \sum_i b_i B_i = 0 \qquad (5.142)$$

---

[17] He was guillotined 5 years later, after the chief justice declared that "The Republic does not need scientists or chemists"!

where the species, namely the substrates (Su_i, e.g. H_2, O_2) and products (Pr_i, e.g. H_2O) involved in the reaction are labeled by B_i and their stoichiometry by b_i (negative integers $b_i = -l_i < 0$ for substrates and positive integers $b_i = m_i > 0$ for products): $2H_2O - 2H_2 - O_2 \rightleftharpoons 0$. The reaction stoichiometry means that the change in the number of particles of a given species dN_i is proportional to the number of molecules of that species b_i participating in the reaction: $dN_i \propto b_i$.

During a chemical reaction (usually conducted at constant temperature and volume or pressure) the free energy decreases as the system approaches equilibrium. At constant volume, equilibrium is reached when the free energy $\mathcal{F}$ is minimized, namely when $d\mathcal{F}_{T,V} = \sum_i \mu_i d\bar{N}_i = 0$, see Eq.5.137, which implies:

$$0 = \sum_i \mu_i d\bar{N}_i \propto \sum_i \mu_i b_i = \sum_i b_i [kT \log(\bar{n}_i/n_{i,Q}) + \epsilon_0^{(i)}] \qquad (5.143)$$

$$= kT \log\left(\Pi_i [\bar{n}_i e^{\beta \epsilon_0^{(i)}} / n_{i,Q}]^{b_i}\right) \qquad (5.144)$$

Hence when chemical equilibrium is reached the chemical potential of the substrates equals the chemical potential of the products. Eq.5.144 can be recast as:

$$\Pi_i \bar{n}_i^{b_i} = \Pi_i [n_{i,Q} e^{-\beta \epsilon_0^{(i)}}]^{b_i} \equiv K_V^{eq} \qquad (5.145)$$

where K_V^{eq} is the equilibrium constant of the reaction at constant volume. Eq.5.145 is known as the law of mass action. It is a fundamental law of chemistry, which can also be derived by noticing that at equilibrium the rate of disappearance of a substrate dSu_j/dt (or appearance of a product dPr_j/dt) is zero:

$$-\frac{1}{l_j}\frac{dSu_j}{dt} = -k_+ \Pi_i [Su_i]^{l_i} + k_- \Pi_i [Pr_i]^{m_i} = 0$$

where k_+, k_- are the forward and reverse rate of the reaction near equilibrium and $[Su_i], [Pr_i]$ are the concentrations of substrate and products respectively:

$$\frac{\Pi_i [Pr_i]^{m_i}}{\Pi_i [Su_i]^{l_i}} = k_+/k_- = K_V^{eq}$$

Usually one measures the concentration of substrates $[Su_i]$ or products $[Pr_i]$ in molar ($1M = 1$ mole/l) not in particle density: $\bar{n}_i = 10^{-3} N_A [B_i]$. Defining $\Delta b \equiv \sum_i b_i$, the molar equilibrium constant K_M becomes:

$$\frac{\Pi_i [Pr_i]^{m_i}}{\Pi_i [Su_i]^{l_i}} = K_V^{eq} (10^{-3} N_A)^{-\Delta b} = e^{-\Delta G_0/kT} \equiv K_M \qquad (5.146)$$

Where the change in the Gibbs free energy of the reaction: $\Delta G_0 = \sum_i b_i G_0^i$, with $G_0^i = \epsilon_0^{(i)} - \log\left(10^3 n_{i,Q}/N_A\right)$ defined as in Eq.5.126.

Consider for example the ionization of water : $2H_2O \rightleftharpoons H_3O^+ + OH^-$ with a change in binding energy of about $\Delta\epsilon_0 = \sum_i b_i \epsilon_0^{(i)} \sim 1$eV due to the electrostatic potential between the charged ions. Since $n_Q^{H_2O} \sim n_Q^{H_3O^+} \sim n_Q^{OH^-} \sim 0.7\ 10^{26}$cm^{-3}, the equilibrium constant of the reaction is:

$$\frac{[H_3O^+][OH^-]}{[H_2O]^2} \sim e^{-\beta \sum_i b_i \epsilon_0^{(i)}} = e^{-\beta \Delta \epsilon_0} \approx e^{-40} = 4.25 \ 10^{-18}$$

Since for pure water[18] $[H_2O] = 55.5M$ we deduce that : $[H_3O^+][OH^-] \sim 10^{-14}M^2$. As water is electrically neutral: $[H_3O^+] = [OH^-]$, the pH of water defined as $pH = -\log_{10}[H_3O^+] = 7$. Of course this well known result is exponentially dependent on the value of the change in binding energy $\Delta \epsilon_0$ which is not precisely known. In fact it makes more sense to reason backwards and deduce $\Delta \epsilon_0$ from the pH of pure water. If an acid (e.g. HCl) or a base (e.g. NaOH) is added to water they will generate H_3O^+ (or OH^-) ions via the ionization reaction:

$$HCl + H_2O \rightleftharpoons Cl^- + H_3O^+$$
$$NaOH + H_2O \rightleftharpoons HNaOH^+ + OH^-$$

Some of these ions will bind to their counter-ions, OH^- (or H_3O^+) to reform neutral water molecules, such as to preserve the chemical equilibrium, i.e. $[H_3O^+][OH^-] \sim 10^{-14}M^2$. For example if 1mM HCl is added to water it generates about 1mM H_3O^+ ions (since the equilibrium constant of the ionization of HCl, $K_M^{HCl} \gg 1$) which reduces the amount of OH^- ions to 10 pM.

In the above discussion we have considered a reaction conducted at constant Temperature and Volume. A reaction may also be conducted at constant Pressure (and Temperature). In such case we can recast Eq.5.136 as:

$$0 = d(E + PV - TS) - VdP + SdT - \mu d\bar{N} \quad \text{or} \quad dG = -SdT + VdP + \mu d\bar{N}$$

Then if a chemical reaction is performed at constant temperature ($dT = 0$) and pressure ($dP = 0$), it is the change in the Gibbs free energy, $G = E + PV - TS$, which is equal to the change in the chemical energy:

$$dG\Big|_{T,P} = d(E + PV - TS) = \mu d\bar{N} \tag{5.147}$$

Therefore at chemical equilibrium:

$$dG\Big|_{T,P} = \sum_i \mu_i d\bar{N}_i = 0 \quad \text{namely}: \quad \mu_i = (\partial G / \partial N_i)_{T,P}$$

Since the volume is not constant, we have to rewrite Eq.5.145 in terms of the partial pressures of the various species: $P_i = kT\bar{n}_i$:

$$\Pi_i \bar{P}_i^{b_i} = \Pi_i [P_{i,Q} e^{-\beta \epsilon_0^{(i)}}]^{b_i} \equiv K_P^{eq} \tag{5.148}$$

where $P_{i,Q} = kTn_{i,Q}$ and K_P^{eq} is the equilibrium constant of the reaction at constant pressure.

---

[18]The molecular weight of H_2O is 18g and 1 liter of water weights 1 kg, hence the molar concentration of pure water is $[H_2O] = 10^3/18 = 55.5M$.

5.7.4 ADSORPTION

The adsorption[19] of particles on a surface is of practical importance for many engineering and chemical processes. It has been used since World War I in gas masks to adsorb toxic gases on charcoal surfaces. Adsorption is also used in heterogenous catalysis where a catalyst is deposited on a large surface (for example that of a porous matrix) such as to quickly and efficiently capture the reagents, stabilize the activated complex and release the products to the surrounding gas phase. Adsorption is also important in the semiconductor industry where a metal-organic compound is deposited layer by layer on a semiconductor substrate (usually *Si*) in a technique known as Vapour Phase Epitaxy (VPE). Finally adsorption of protein, antigens, viruses, etc. on a surface is key to a family of diagnostic tests known as Enzyme Linked Immunosorbent Assays (ELISA) where the adsorption of an antigen (or a virus) on a surface coated with an appropriate antibody is detected by an enzymatic reaction. Hence, understanding adsorption and the parameters affecting the fractional coverage θ_a (i.e. the percentage of occupied surface) is an important technological question with a wide spectrum of applications.

The first study of adsorption is due to Langmuir who in 1916 modeled the surface as an ensemble of N_s adsorbtion sites S (typically on a solid surface there are about 10^{15} atoms/cm^2 each of which can serve as an adsorption site) onto which molecules (or atoms) from the gas phase A can adsorb via the reaction:

$$A + S \rightleftharpoons AS$$

As in the examples studied above (chemical reaction, semi-permeable membranes) the function controlling the mean number of molecules $\bar{N}_a$ adsorbed on the surface is the chemical potential of the molecules, which at equilibrium has to be equal in both the gas and adsorbed phases $\mu_a = \mu_g$. We have computed above the chemical potential for molecules in the gas phase: $\mu_g = kT \log(\bar{n}_g/n_Q)$ (where $\bar{n}_g$ is the particle density in the gas phase and all energies are measured with respect to the binding energy in the gas phase, i.e. setting $\epsilon_0 = 0$ in Eq.5.125). Let us compute the chemical potential μ_a for molecules adsorbed on the surface. Assume that $N \leq N_s$ molecules are adsorbed with energy $-\epsilon_a$ and $N_s - N$ sites are free . Let $\mathcal{P}(N)$ be the probability of adsorbtion of N molecules with energy $E_p = -N\epsilon_a$. To determine that probability we need to maximize the entropy:

$$S = -k\text{Tr}\,\mathcal{P}\log\mathcal{P} \qquad \text{under the constraints :}$$
$$\text{Tr}\,\mathcal{P} = 1, \quad \langle E_p \rangle = \text{Tr}\,E_p\mathcal{P} = E \qquad \text{and} \quad \langle N \rangle = \text{Tr}\,N\mathcal{P} = \bar{N}_a \qquad (5.149)$$

where the trace is defined over all possible configurations of $N \leq N_s$ adsorbed molecules. As usual maximization of S yields the following equation for the par-

---

[19] Adsorption is the binding of particles to a solid surface (dew is water adsorbed on leaves), whereas absorption is binding to the bulk of a material, such as the absorption of CO_2 by water to yield the weak acid H_2CO_3, which is one of the causes behind the increased acidification of lakes and Oceans.

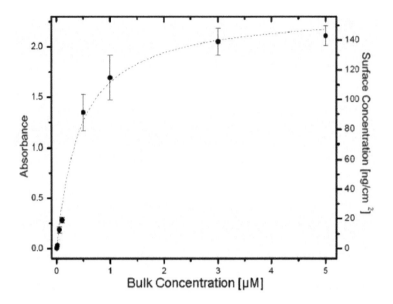

Figure 5.19 Cytochrome C is a small protein (molecular weight: 12 kDalton, $m = 2 \cdot 10^{-20}$g; size: ~ 3 nm) containing a Heme-group that absorbs light at 406nm. Protein adsorption on a flat glass surface can be assessed by measuring the decrease in the intensity of light propagating along the glass surface due to the adsorbed proteins. By varying the concentration $\bar{n}_g$ of Cytochrome C in solution, the equilibrium surface density of adsorbed protein can be measured. The variation of the fractional coverage with bulk concentration follows Eq.5.153 with $c_0 = 10^3 n_0/N_A = 0.4\mu$M (dashed line). Maximal coverage is obtained for a surface density $\bar{N}_a = 7.2 \cdot 10^{12}$ molecules/cm^2 which is roughly what is expected from the size of the molecules assuming close packing on the surface, namely about 10^{13} molecules/cm^2. Reprinted with permission from ref.[97] copyright by The Optical Society.

tition function of adsorbed molecules[20]

$$Z = e^{-\gamma_0} = \mathrm{Tr}\, e^{-\beta E_p - \alpha N} = \sum_{N=0}^{N_s} \frac{N_s!}{N!(N_s - N)!} e^{(-\alpha + \beta\epsilon_a)N}$$

$$= (1 + e^{-\alpha + \beta\epsilon_a})^{N_s} \tag{5.150}$$

Where we used the binomial expansion, Appendix A.3.1, to deduce the last equality. From the knowledge of the partition function Z, we can derive the mean number of adsorbed particles, Eq.5.124:

$$\bar{N}_a = \mathrm{Tr}\, N\, \mathcal{P} = -\frac{\partial \log Z}{\partial \alpha} = \frac{N_s}{1 + e^{-\beta(\mu_a + \epsilon_a)}} = \frac{N_s e^{\beta\mu_a}}{e^{\beta\mu_a} + e^{-\beta\epsilon_a}} \tag{5.151}$$

---

[20]The trace, Tr, is over all states with N adsorbed particles over $N_s \geq N$ possible adsorption sites, of which there are: $N_s!/N!(N_s - N)!$

and the mean energy, Eq.5.88:

$$E = -\frac{\partial}{\partial \beta} \log Z = -\frac{N_s}{1 + e^{-\beta(\mu_a + \epsilon_a)}} \epsilon_a = -\bar{N}_a \epsilon_a \qquad (5.152)$$

The last equality is the expected result: the mean energy of adsorption is equal to the number of adsorbed particles times their adsorption energy gain. At equilibrium $\mu_a = \mu_g$ and thus from Eq.5.125: $e^{\beta \mu_a} = e^{\beta \mu_g} = \bar{n}_g / n_Q = P / P_Q$. Hence the fractional coverage is:

$$\theta_a \equiv \frac{\bar{N}_a}{N_s} = \frac{e^{\beta \mu_a}}{e^{\beta \mu_a} + e^{-\beta \epsilon_a}} = \frac{\bar{n}_g}{\bar{n}_g + n_0} = \frac{P}{P + P_0} \qquad (5.153)$$

with the critical particle density: $n_0 = n_Q e^{-\beta \epsilon_a}$ and pressure: $P_0 = kT n_0$. Fig.5.19 shows the typical adsorption isotherm of a protein (Cytochrome C) adsorbed on glass. The amount of adsorbed protein follows the prediction of the Langmuir model, Eq.5.153 and saturates at a density which is roughly what is expected from their close packing on the surface. For Cytochrome C with mass $m = 2 \ 10^{-20}$g, $n_Q \approx 10^{30}$ cm^{-3}. The measured critical particle density, see Fig.5.19, is $n_0 = 10^{-3} N_A c_0 = 24 \ 10^{13}$ cm^{-3}. Thus we deduce that the adsorption energy: $\epsilon_a = -kT \log n_0 / n_Q \approx 0.9 eV$.

5.8　COMPARISON OF THERMODYNAMIC RELATIONS FOR NON-INTERACTING SYSTEMS

Table 5.1 below summarizes the thermodynamic relations between the various non-interacting systems studied above: magnetic (spin) systems, polymer systems (FJC), ideal gases (with fixed or variable number of particles).

Table 5.1

Summary of thermodynamic relations for non-interacting systems

Variables	Magnetic system	Polymer system	Ideal gas (fixed N)	Ideal gas (variable N)
Phase space	$\vec{\mu}_i = \pm\mu\hat{z}$ (Ising)	$\vec{b}_i$	$\vec{p}_i, \vec{r}_i$	$\vec{p}_i, \vec{r}_i, N$
Energy	$E_p = -\sum_i^N \vec{\mu}_i \cdot \vec{B}_i$	$E_p = -\sum_i^N \vec{b}_i \cdot \vec{F}$	$E_p = \sum_i^N \vec{p}_i^2/2m$	$E_p = \sum_i^N [\vec{p}_i^2/2m + \epsilon_0]$
Constraints	$\langle E_p \rangle = E$	$\langle E_p \rangle = E$	$\langle E_p \rangle = E$	$\langle E_p \rangle = E, \langle N \rangle = \bar{N}$
Partition function $\mathcal{Z}$	$2^N \cosh^N \beta\mu B$	$(4\pi \sinh\beta bF/\beta bF)^N$	$\left(\frac{V}{h^3}\right)^N \frac{(2\pi mkT)^{3N/2}}{N!}$	$\exp\left[(\frac{V}{h^3})(2\pi mkT)^{3/2}e^{-\alpha-\beta\epsilon_0}\right]$
thermodyn. relations	$E = -\partial\log\mathcal{Z}/\partial\beta$	$E = -\partial\log\mathcal{Z}/\partial\beta$	$E = -\partial\log\mathcal{Z}/\partial\beta$	$E = -\partial\log\mathcal{Z}/\partial\beta$ $\bar{N} = -\partial\log\mathcal{Z}/\partial\alpha$; $\alpha = -\beta\mu$
Change in energy	$dE = TdS - MdB$	$dE = TdS - LdF$	$dE = TdS - PdV$	$dE = TdS - PdV + \mu d\bar{N}$
Change in free energy	$d\mathcal{F} = -SdT - MdB$	$d\mathcal{F} = -SdT - LdF$	$d\mathcal{F} = -SdT - PdV$	$d\mathcal{F} = -SdT - PdV + \mu d\bar{N}$
System response	$M = -\left(\frac{\partial\mathcal{F}}{\partial B}\right)_T$	$L = -\left(\frac{\partial\mathcal{F}}{\partial F}\right)_T$	$P = -\left(\frac{\partial\mathcal{F}}{\partial V}\right)_T$	$\mu = \left(\frac{\partial\mathcal{F}}{\partial N}\right)_{V,T}$

5.9 QUANTUM STATISTICS

Quantum Mechanics (QM) enters in essential ways in Statistical Mechanics: it divides the phase space of an ideal gas into cells of volume h^3 and accounts correctly for the $N!$ normalization of identical particles. We have already studied QM systems (paramagnetic spins, photons), but at low temperatures or high densities, the constraints imposed by Pauli's exclusion principle become crucial (even in non-interacting systems): fermions (half integer spin particles) cannot share the same state whereas bosons (integer spin ones) can and do. This has important cosmological and technological implications: the stability of stars and the semiconductor industry are impacted by the quantum statistics of fermions. First though, we shall revisit the radiation of a black-body (already studied in QM, see section 4.1.1) by deriving Planck's law from the constrained maximization of entropy.

5.9.1 PLANCK'S BLACK-BODY RADIATION

As we have seen in section 3.5 warm bodies emit EM radiation, see Fig.3.27. The energy of radiation $E(\nu)$ emitted at frequency ν (or wavenumber k) is a quadratic function of the electric and magnetic fields $(\vec{E}, \vec{B})$, see Eq.3.111.:

$$E(\nu) = \frac{1}{8\pi} \int d^3 r (|\vec{E}(\nu)|^2 + |\vec{B}(\nu)|^2)$$

Hence from the equipartition theorem, see section 5.6, every EM mode of frequency ν present at equilibrium in a cavity (i.e. an oven) at temperature T has mean energy $\langle E(\nu) \rangle = kT$. Integrating over all frequencies leads to the absurd conclusion that the EM radiation inside an oven has infinite energy! This is the so-called Rayleigh-Jeans catastrophe that Planck resolved in 1900, by assuming that the energy of the electromagnetic radiation field was quantized in units (now known as photons) of the frequency ν: $E_n = nh\nu$ (with Planck's constant $h = 6.626 \ 10^{-27}$ erg sec), see section 4.1.1.

We shall now revisit Planck's derivation using the entropy maximization approach used previously to describe ideal gases, spin systems or polymers. Let us therefore find the probability $\mathcal{P}_n$ of observing an EM field with n photons at frequency ν (i.e. energy $E_n = nh\nu$) inside a cavity at temperature T (a similar model was studied by Einstein to derive the specific heat of solids, see Appendix A.9.1). We assume the mean energy $\langle E(\nu) \rangle$ of this EM field to be known, i.e. set by the cavity's temperature. As before we need to maximize the entropy:

$$S = -k \sum_m \mathcal{P}_m \log \mathcal{P}_m$$

under the constraints $\sum_m \mathcal{P}_m = 1$ and $\sum_m E_m \mathcal{P}_m = \langle E(\nu) \rangle$. As usual we maximize the auxiliary function:

$$\Gamma = -k \sum_m \mathcal{P}_m \log \mathcal{P}_m + k(\gamma_0(\nu) + 1) \sum_m \mathcal{P}_m - k\beta \sum_m E_m \mathcal{P}_m$$

Setting $\partial \Gamma / \partial \mathcal{P}_n = 0$ yields:

$$\mathcal{P}_n = e^{\gamma_0(\nu)} e^{-\beta E_n} = \frac{e^{-\beta E_n}}{\sum_n e^{-\beta E_n}}$$

where the last equality ensures that the constraint $\sum_n \mathcal{P}_n = 1$ is satisfied.

$$\gamma_0(\nu) \equiv -\log \mathcal{Z}(\nu) = -\log \left(\sum_{n=0}^{\infty} e^{-\beta E_n} \right) = -\log \left(\sum_{n=0}^{\infty} e^{-n\beta h\nu} \right) = \log\{1 - e^{-\beta h\nu}\} \quad (5.154)$$

and the average energy is then:

$$\langle E(\nu) \rangle = \frac{\partial \gamma_0}{\partial \beta} = -\frac{\partial \log Z}{\partial \beta} = \frac{h\nu}{e^{h\nu/kT} - 1}$$

When $h\nu \ll kT$, one recovers the result expected from the equipartition theorem: $\langle E(\nu) \rangle = kT$, however at large frequencies the average energy of the EM field decays as $\langle E(\nu) \rangle \sim h\nu \exp(-\beta h\nu)$ which when integrated over all frequencies doesn't blow up (see below). This solves the Rayleigh-Jeans catastrophe .

To find the total EM energy inside the cavity we have to integrate the mean energy at a given frequency $\langle E(\nu) \rangle$ over all the frequency range. As already discussed in section 4.1.1, we need to count the number of EM modes inside a cavity of dimensions $L_x \times L_y \times L_z$. Since there are two transverse modes of EM wave (two polarizations) the total energy of the EM field in a cavity can be written as (see also Eq.A.96):

$$E_{tot} = 2 \frac{L_x L_y L_z}{\pi^3} \int_0^{\infty} dk_x dk_y dk_z \langle E(\nu) \rangle = \frac{2V}{\pi^3} \int_0^{\infty} \frac{4\pi}{8} k^2 \langle E(\nu) \rangle dk$$
$$= \frac{8\pi V}{c^3} \int_0^{\infty} \langle E(\nu) \rangle \nu^2 d\nu \equiv V \int_0^{\infty} \rho(\nu) d\nu \quad (5.155)$$

where we used the relation $k = 2\pi\nu/c$. The EM energy density of a black-body at frequency ν, also known as its spectral energy density, is:

$$\rho(\nu) = \frac{8\pi\nu^2 \langle E(\nu) \rangle}{c^3} = \frac{8\pi h}{c^3} \frac{\nu^3}{e^{\beta h\nu} - 1} \quad (5.156)$$

This equation is known as Planck's law for black-body radiation. It is a good approximation to the spectral energy density of the radiation emitted from a hot body (an oven, molten ore, the Sun) and is in agreement over many orders of magnitude with the remnant radiation from the Big-Bang, which turns out to be the best black-body in the Universe (see section 4.1.1.2 for details). One can deduce the temperature of the emitting body from the frequency ν_m (or wavelength λ_m) at which $\rho(\nu)$ (or $\rho(\lambda)$) is maximal:

$$\nu_m \approx 2.85kT/h \qquad \text{or} \qquad \lambda_m \approx hc/5kT \quad (5.157)$$

Planck's law has many applications. For example the discovery by Penzias and Wilson in 1965 of a uniform background radiation emitted around $\nu_m \approx 170\text{GHz}$ was the first evidence of the existence of a remnant $\sim 3°$K radiation from the Big-Bang. From a technological point of view, infra-red cameras capable of visualizing warm-blooded animals such as humans are optimal if their maximal detection efficiency is at a wavelength $\lambda_m \approx 10\mu$m (in the infrared corresponding to the peak of emission of a body at 37°C).

From Planck's law one can compute the total energy density in a hot cavity[21]

$$\bar{\epsilon} = E_{tot}/V = \int_0^\infty \rho(v)dv = \frac{8\pi h}{c^3}(\beta h)^{-4} \int_0^\infty \frac{x^3 dx}{e^x - 1} = \frac{\pi^2 k^4}{15\hbar^3 c^3}T^4 \qquad (5.158)$$

Thus at the center of the Sun, $T \sim 3 \; 10^7$ °K, $\bar{\epsilon}_c = 6 \; 10^9$ J/m³. The mean time t_p for a photon to diffuse via multiple scattering to the surface of the Sun can be estimated from the continuity of energy flux (radiation intensity, see section 4.1.1.1): $\sigma_{SB}T_S^4 = -D\vec{\nabla}\bar{\epsilon}_c \approx D\bar{\epsilon}_c/R_S$, (where $T_S \approx 5770$°K and $R_S \approx 7 \; 10^8$m are the Sun's surface temperature and radius) from which one deduces that: $D \approx \sigma_{SB}T_S^4 R_S/\bar{\epsilon}_c \approx 7 \; 10^6$m²/sec. Hence the time to diffuse out is $t_p = R_S^2/D \approx 2200$ years, see section 5.11.1.1. Compare with the ≈ 8 minutes it takes radiation from the Sun's surface to reach Earth.

The amount of energy radiated (lost) per unit time through a small hole of area ΔS in an oven is given by the integration of over all possible angles of incidence $(d\Omega = \sin\theta d\theta d\phi)$ of the radiation flux component $\vec{J}\cdot\hat{n}$ perpendicular to the hole:

$$-\frac{dE_{tot}}{dt} = \Delta S \int \vec{J}\cdot\hat{n}d\Omega \qquad \text{with}: \qquad \vec{J} = \bar{\epsilon}\,\hat{c}f(\hat{c})$$

where $f(\hat{c})$ is the angular distribution of radiation (with $\hat{c}$ the unit vector in the direction of propagation of the radiation flux, here assumed isotropic: $f(\hat{c}) = 1/4\pi$). Integrating over the angular variable yields:

$$-\frac{dE_{tot}}{dt} = \frac{\bar{\epsilon}c}{4\pi}\Delta S \int \cos\theta d\Omega = \frac{\bar{\epsilon}c}{2}\Delta S \int_0^{\pi/2} \cos\theta \sin\theta d\theta = \frac{\bar{\epsilon}c}{4}\Delta S \equiv I_0\Delta S$$

where the radiation power intensity in W/m² (or erg sec/cm²) is given by Stefan's law:

$$I_0 = \frac{\bar{\epsilon}c}{4} = \pi^2(kT)^4/60\hbar^3 c^2 \equiv \sigma_{SB}T^4 \qquad (5.159)$$

with $\sigma_{SB} = 5.67 \; 10^{-8}$ W m^{-2} °K^{-4}. Hence by measuring the radiation intensity emitted from a hot body one can deduce its temperature. For example the amount of radiation arriving on Earth from the Sun is $I_E = 1.37$ kW/m² (30% of which is reflected by the clouds so that about 1kW/m² is reaching us). Thus from the angle subtended by the Sun ($\theta = 2R_S/R_{SE} = 9,3 \; 10^{-3}$ rad) and the light intensity I_E one can deduce the temperature at the surface of the Sun ($T_S \approx 5770$°K), see section 4.1.1.1.

---

[21] Using: $\int_0^\infty dx x^3/(e^x - 1) = \pi^4/15$.

Radiation Pressure. Just as an ideal gas exerts a pressure on the walls of the vessel containing it, so does the "gas" of photons contained in a cavity at temperature T. To compute this radiation pressure we will use the definition of the pressure, see Eq.5.91: $P = -kT\partial\gamma_0/\partial V$ where γ_0 is obtained by integration over all wavenumbers k (or frequency ν) of the value $\gamma_0(\nu) = \log(1 - e^{-\beta h\nu})$ computed for one frequency (see Eqs. 5.154 and 5.155):

$$
\gamma_0 = \frac{8\pi V}{c^3}\int_0^\infty \log(1 - e^{-\beta h\nu})\nu^2 d\nu = -\frac{8\pi V}{c^3}\int_0^\infty d\nu \frac{\nu^3}{3}\frac{\beta h}{e^{\beta h\nu} - 1}
$$

$$
= -\frac{8\pi V\beta^{-3}h^{-3}}{3c^3}\int_0^\infty \frac{x^3 dx}{e^x - 1} = -\frac{1}{3}\beta\bar{\epsilon}V \tag{5.160}
$$

Therefore the radiation pressure inside the cavity is:

$$
P == -kT\frac{\partial\gamma_0}{\partial V} = \bar{\epsilon}/3 \tag{5.161}
$$

At room temperature ($300°K$) this is a very small force: $P \sim 2 \ 10^{-6}$ N/m^2 = $2 \ 10^{-11}$ atmospheres. Even at the center of the Sun with a temperature of $3 \ 10^7 \ °K$, the radiation pressure at $2 \ 10^9$ atm is still much less than the gravitational pressure $P_g \approx 3 \ 10^{11}$ atm which is counter-balanced by the pressure of the hot gas of ionized particles in the Sun's core ($P_g = nkT/V$ with $n/V \approx 100$ particles per Å^3).

5.9.2 ◆ FERMI-DIRAC STATISTICS

For quantum particles with half integer spins (known as fermions, such as electrons, protons or neutrons), Pauli's exclusion principle, see section 4.9, implies that each quantum state can be occupied by at most one particle[22]. Let us consider a gas of free non-interacting fermions in a box of volume V. If one labels the energy states by their momentum p (with energy ϵ_p) and their degeneracy (the number of states with same energy) by g_p (for free particles in a box, there are two spin states allowed for each energy ϵ_p, hence $g_p = 2$), Pauli's principle implies that the number of particles n_p allowed for each energy level is[23] $n_p = 0, \cdots, g_p$. A state n of the system is then fully specified by the occupation numbers $\{n_p\}$ of the various energy states: its energy is $E_n = \sum_p n_p\epsilon_p$ and its number of particles is $N_n = \sum_p n_p$. The problem therefore is to determine the probability $\mathcal{P}(\{n_p\})$. As usual we have to maximize the entropy:

$$
S = -k\text{Tr}_{\{n_p\}}\mathcal{P}\log\mathcal{P}
$$

---

[22]Particles with same spin have to occupy energy levels with different quantum numbers; if they occupy energy levels with identical quantum numbers they must have opposite spins. A state here is characterized by its spatial and spin quantum numbers and can thus be occupied by at most a single particle (e.g. a single electron).

[23]For example state $n = 2$ in the hydrogen atom has degeneracy $g_p = 8$, since its quantum numbers $l = 0$ or 1; $-l \leq m \leq l$ and an electron can have two spin states for each n, l, m state.

subject to the constraints:

$$\text{Tr}_{\{n_p\}}\mathcal{P} = 1 \quad ; \quad \text{Tr}_{\{n_p\}}E_n\mathcal{P} = E \quad ; \quad \text{Tr}_{\{n_p\}}N_n\mathcal{P} = \bar{N}$$

Maximizing the auxiliary function:

$$\Gamma = S + k(\gamma_0 + 1)\text{Tr}\mathcal{P} - k\beta\text{Tr}E_n\mathcal{P} - k\alpha\text{Tr}N_n\mathcal{P}$$

$$\text{yields:} \qquad \mathcal{P}(\{n_p\}) = e^{\gamma_0}e^{-\beta E_n - \alpha N_n} \tag{5.162}$$

From the first constraint we derive the partition function[24]:

$$Z = e^{-\gamma_0} = \text{Tr}_{\{n_p\}}e^{-\beta E_n - \alpha N_n} = \sum_{\{n_p\}} \frac{g_p!}{n_p!(g_p - n_p)!}e^{-\sum_p(\beta\epsilon_p + \alpha)n_p}$$

$$= \Pi_p \sum_{n=0}^{g_p} \frac{g_p!}{n!(g_p - n)!}e^{-(\beta\epsilon_p + \alpha)n} = \Pi_p[1 + e^{-(\beta\epsilon_p + \alpha)}]^{g_p} \tag{5.163}$$

The factorial term accounts for the number of ways of distributing n_p particles in $g_p \geq n_p$ states. Therefore:

$$\gamma_0 = -\log Z = -\sum_p g_p \log[1 + e^{-(\beta\epsilon_p + \alpha)}]$$

$$= -\frac{V}{h^3}\int d^3p\, g_p \log[1 + e^{-(\beta\epsilon_p + \alpha)}] \tag{5.164}$$

where in case of free particles in a box of volume V we replaced the discrete sum by an integral over all momenta (by mode counting with $\hbar k = p$ as in Eqs. 5.155 and A.96 but with a single (longitudinal, i.e. compressional) oscillation mode). By Eq.5.124 the mean number of particles is:

$$\bar{N} = -\frac{\partial \log Z}{\partial \alpha} = \frac{V}{h^3}\int d^3p \frac{g_p}{e^{\beta(\epsilon_p - \mu)} + 1} \equiv \frac{V}{h^3}\int d^3p\, g_p \eta_p^{FD} \equiv \frac{V}{h^3}\int d^3p\, n_p \tag{5.165}$$

with $\alpha \equiv -\beta\mu$ and where η_p^{FD} is known as the Fermi-Dirac distribution:

$$\eta_p^{FD} \equiv \frac{1}{e^{\beta(\epsilon_p - \mu)} + 1} \tag{5.166}$$

The density of particle with momentum p is then $n_p \equiv g_p\eta_p^{FD}$ and from Eq.5.127, the mean energy is:

$$E = -\frac{\partial \log Z}{\partial \beta} = \frac{V}{h^3}\int d^3p \frac{g_p\epsilon_p}{e^{\beta(\epsilon_p - \mu)} + 1} = \frac{V}{h^3}\int d^3p\, \epsilon_p g_p\eta_p^{FD} \tag{5.167}$$

---

[24]Using the binomial identity: $(a+b)^n = \sum_m(n!/m!(n-m)!)a^m b^{n-m}$, see Appendix A.3.1

At low temperatures ($\beta \to \infty$), the free particle density $n_p = g_p = 2$ when $\epsilon_p < \mu$ and decays rapidly to zero for $\epsilon_p > \mu$. This reflects the fact that at low temperatures fermions (e.g. electrons) fill up all the lowest energy levels up to the maximal energy $\mu \equiv E_F = p_F^2/2m$ known as the Fermi energy As we shall see below for most materials the Fermi energy is way above any reasonable temperature, so that for all practical purposes the low temperature approximation holds. In that limit the density of fermions (the Fermi density) is:

$$n_F = \frac{\bar{N}}{V} = \frac{8\pi}{h^3} \int_0^{p_F} dp \, p^2 = \frac{8\pi p_F^3}{3h^3} \tag{5.168}$$

and the Fermi pressure they exert is:

$$P_F = -kT\frac{\partial \gamma_0}{\partial V} = \frac{2kT}{h^3} \int d^3p \, \log[1 + e^{-\beta(\epsilon_p - \mu)}]$$

$$\approx \frac{8\pi}{h^3} \int_0^{p_F} dp \, p^2(E_F - \epsilon_p) = \frac{8\pi}{15mh^3} p_F^5 = \frac{8\pi h^2}{15m}(\frac{3n_F}{8\pi})^{5/3} \tag{5.169}$$

where we used the fact that in the limit of low temperatures ($p^2/2m = \epsilon_p < E_F = \mu$): $\log[1 + \exp -\beta(\epsilon_p - \mu)] \approx \log[\exp\beta(E_F - \epsilon_p)] = \beta(E_F - \epsilon_p)$.

5.9.2.1   White Dwarfs and Black Holes

These considerations are important when studying the stability of stars. What keeps them from collapsing under their gravitational pull is the pressure exerted by the hot gas inside. However when stars have consumed all their nuclear energy what keeps some from collapsing into a black hole (a star with infinite density) is the Fermi pressure from their electrons (in the case of White Dwarfs) or from their neutrons (in the case of neutron stars where electrons and protons have fused into neutrons). Let us first estimate the gravitational compression. Assuming the density ρ of a star to be constant, its gravitational energy is:

$$E_G = -\int \frac{Gmdm}{r} = -G \int_0^R dr \frac{4\pi\rho r^3}{3} 4\pi\rho r = -\frac{3G}{5}\frac{M^2}{R} \tag{5.170}$$

where $M = 4\pi\rho R^3/3$, R are the star mass and radius respectively and $G = 6.67 \, 10^{-8}$ cm^3 gr^{-1} sec^{-2} is Newton's gravitational constant. The gravitational force is:

$$F_G = -\frac{\partial E_G}{\partial R} = -\frac{3G}{5}\frac{M^2}{R^2} \tag{5.171}$$

and the gravitational compression: $P_G = F_G/4\pi R^2$. In a White Dwarf this compression is balanced by the pressure exerted by the electron gas: $P_G + P_F = 0$. Since the mass of a White Dwarf is essentially contributed by the mass of its protons and neutrons, which in a White Dwarf are present in roughly equal numbers[25] and because

---

[25] The proton and neutron have almost identical masses, $m_n \approx m_p = 1.67 \, 10^{-24}$g = 938 MeV/c^2, while the electron mass $m = 9.1 \, 10^{-28}$g = 0.5 MeV/c^2 is three orders of magnitude smaller.

charge neutrality requires the number of protons to equal the number of electrons: $n_F = \rho/2m_p$:

$$-P_G = \frac{3G}{20\pi}\frac{M^2}{R^4} = P_F = \frac{8\pi h^2}{15m}\left(\frac{3\rho}{16\pi m_p}\right)^{5/3} = \frac{8\pi h^2}{15m}\left(\frac{9M}{64\pi^2 R^3 m_p}\right)^{5/3}$$

from which we derive a relation between the radius of a White Dwarf and its mass:

$$R = \frac{(3/\pi)^{4/3}}{32}\frac{h^2}{Gmm_p^2}(M/m_p)^{-1/3}$$

The radius of a White Dwarf with the mass of the Sun: $M_\odot = 2\ 10^{33}$ gr is thus: $R \approx 0.01 R_\odot \approx 7000$km (about the Earth radius, where $R_\odot \approx 7\ 10^5$ km is the Sun's radius). The density $\rho = 3M_\odot/4\pi R^3$ of a White Dwarf at about one ton/cm^3 is huge: a million time more dense than water!

The first to study this stability of White Dwarfs was S.Chandrasekhar at the age of 20, on his way from India to England to pursue a Ph.D. in Physics at Cambridge! He showed that for relativistic electrons there was an upper mass (known as the Chandrasekhar limit[26] of about $1.4M_\odot$) beyond which the Fermi pressure could not balance the gravitational force. To see how this comes about consider the (high density) extreme relativistic limit where the Fermi energy is much larger than the rest mass of the electron: $E_F \approx p_F c \gg mc^2$. Setting $\epsilon_p = pc$ in Eq.5.169 for the pressure of the Fermi gas leads to a balance between Gravitational and Fermi pressures ($P_F = (3h^3/512\pi)^{1/3}cn_F^{4/3} \propto (M/R^3)^{4/3}$) from which the star's radius but not its mass drops out. This result implies that beyond an upper mass (the Chandrasekhar limit) for which $P_F = P_G$ the Fermi pressure cannot prevent the collapse of the star under its own weight. For this prediction which suggested that the fate of massive stars is to become black holes (stars with infinite densities) he was virulently attacked by Sir A.Eddington (then the leading astrophysicist). Nonetheless he prevailed and earned the Nobel prize for his work some 50 years later (in 1983)! In fact, some stars with a mass M beyond the Chandrasekhar limit may not turn into black holes. For masses $1.4M_\odot \lesssim M \lesssim 3M_\odot$, the electrons and protons fuse into neutrons. Neutrons are also fermions, but due to their larger mass the relativistic argument made by Chandrasekhar becomes valid at higher densities (i.e. for a neutron star mass of about 3 $M_\odot$). For masses beyond the Chandrasekhar limit for neutron stars, i.e. $M \gtrsim 3M_\odot$, no mechanism is known to prevent stars from collapsing under their gravitational pull into black holes.

---

[26]The Chandrasekhar limit $M \approx M_P^3/m_p^2$, where $M_P = (\hbar c/G)^{1/2} = 2.18\ 10^{-8}$ kg is known as the Planck's mass and m_p is the proton mass.

5.9.2.2 ◆ Conductors and Semiconductors

The Fermi-Dirac Statistics of electrons is central to our understanding of conductors and semiconductors and as a result to the development of all the micro-electronics industry. As we have seen in section 4.3.4, in a metal the valence and conduction bands overlap so that some electrons are free to move into the unoccupied states (of momentum $\vec{p}$) of the conduction band. Let N_F be the number of free electrons then by Eq.5.165 (with $g_p = 2$ for the two spin states of the electron) at low temperatures the free electron density n_e is:

$$n_e = \frac{N_F}{V} = \frac{g_p}{h^3} \int_0^{p_F} d^3 p = \frac{4\pi g_p p_F^3}{3h^3} = \frac{8\pi}{3}(2mE_F/h^2)^{3/2}$$

In copper for example the density of free electrons is $n_e = 8.47 \ 10^{22} \text{cm}^{-3}$ and the Fermi energy comes out to be : $E_F = 7\text{eV}$. As explained in section 3.2.1, for copper with an electron mobility $\mu_e = 1.35 \ 10^4 \text{cm}^2/\text{statV·sec}$, this electron density yields a conductivity $\sigma_R = n_e \mu_e e = 5.5 \ 10^{17} \text{sec}^{-1}$. The high temperatures regime is reached when: $kT > E_F$. For a conductor such as copper this occurs above $10^5 \ °\text{K}$, way above the melting temperature. Thus for all practical purposes in a metal we are always in the low temperature regime: $kT < E_F$.

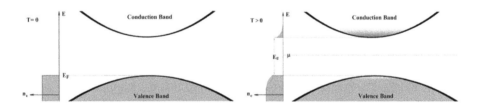

Figure 5.20 In an intrinsic semiconductor at $T = 0°\text{K}$, the valence band is fully occupied, while the conduction band at an energy $E > E_g$ (with a gap energy $E_g \sim 1\text{eV}$) is empty. The Fermi level is at the valence band edge. Since there are no electrons in the conduction band nor holes in the valence band, the material is an insulator. Above $0°\text{K}$, there is a small probability for electrons to thermally hop into the conduction band leaving behind holes in the valence band. Both electrons and holes contribute to the conductance of the semiconductor at non-zero temperatures.

    In an intrinsic semiconductor on the other hand, we have seen in section 4.3.4 that there is a gap of energy E_g of O(1eV) between the valence and conduction bands, see Fig.5.20. At $T = 0°\text{K}$ the valence band is fully occupied by the electrons (there are no holes in it), the Fermi energy level is at the valence band edge. The material is an insulator: electrons cannot flow and conduct electricity as there are no states onto which they can move into upon application of an electric field. Above $0°\text{K}$ there is a small probability (given by the tail of the FD distribution at large β) that electrons can hop into the conduction band. Hence only a small fractions n_e^0 of the electrons are in the conduction band:

$$n_e^0 = \frac{g_p}{h^3} \int d^3 p \, e^{-\beta(\epsilon_p - \mu_0)} \tag{5.172}$$

Where μ_0, the chemical potential of an intrinsic semiconductor has yet to be determined (but corresponds, as we shall see, to an energy around the middle of the band gap, see Fig.5.20). The electrons in the conduction band leave behind them holes in the valence band which also contribute to the conductivity of the material (they provide states with a density n_h^0 which electrons in the valence band can move into). The conductivity of a semiconductor is thus due to the mobility of both electrons and holes (with charge $\mp e$, see section 3.2.1): $\sigma_R = (n_e^0 \mu_e + n_h^0 \mu_h)e$, where μ_e and μ_h are the mobilities of electrons and holes respectively which for crystalline silicon are: $\mu_e = 4.2 \ 10^5$ cm^2/statV·sec; $\mu_h = 1.35 \ 10^5$ cm^2/statV·sec.

Measuring energies from the edge of the valence band, the energy of the electrons in the conduction bands is $\epsilon_p = E_g + \vec{p}^2/2m_c > \mu_0$, where m_c is the effective mass of the electron in the conduction band, see section 4.3.4. Thus the density of electrons in the conduction band is (see also Eq.5.124):

$$n_e^0 = \frac{2e^{-\beta(E_g - \mu_0)}}{h^3} \int_0^\infty dp \, 4\pi p^2 e^{-\beta p^2/2m_c} = 2e^{-\beta(E_g - \mu_0)}(2\pi m_c kT/h^2)^{3/2}$$
$$\equiv n_Q^e e^{-\beta(E_g - \mu_0)} \tag{5.173}$$

Similarly we may consider the density of holes n_h^0 in the valence band. Holes are missing electrons and their probability distribution is thus simply:

$$\eta_h^0 = [1 - \eta_p^{FD}] = [1 - \frac{1}{e^{\beta(\epsilon_p - \mu)} + 1}] \approx e^{\beta(\epsilon_p - \mu_0)} \tag{5.174}$$

where the energy of the holes (with effective mass m_v and charge $+e$) is $\epsilon_p = -\vec{p}^2/2m_v < \mu_0$. The density of holes in the valence band is thus:

$$n_h^0 = \frac{2e^{-\beta\mu_0}}{h^3} \int_{-\infty}^0 dp \, 4\pi p^2 e^{-\beta p^2/2m_v} = 2(2\pi m_v kT/h^2)^{3/2} e^{-\beta\mu_0}$$
$$\equiv n_Q^h e^{-\beta\mu_0} \tag{5.175}$$

Notice that the product $n_e^0 n_h^0 = 4e^{-\beta E_g}(2\pi m_c kT/h^2)^{3/2}(2\pi m_v kT/h^2)^{3/2}$ is a constant independent of the chemical potential (i.e. the number of charge carriers in the material). In fact that result is completely analoguous to the law of mass action, see Eq.5.145 for the reaction: $\emptyset \rightleftharpoons e + h$:

$$n_e^0 n_h^0 = n_Q^e n_Q^h e^{-\beta E_g} \tag{5.176}$$

For an intrinsic semi-conductor (with no impurities injecting charge carriers) $n_e^0 = n_h^0 = n_i$ and the intrinsic density of electrons (and holes) n_i is:

$$n_i = \sqrt{n_Q^e n_Q^h} e^{-\beta E_g/2} = 2(2\pi \sqrt{m_c m_v} kT/h^2)^{3/2} e^{-\beta E_g/2} \approx \frac{2}{\lambda_{DB,e}^3} e^{-\beta E_g/2} \tag{5.177}$$

Where $\lambda_{DB,e}$ is the thermal de-Broglie wavelength of the electron (see Eq.5.87, at 300°K: $\lambda_{DB,e} \approx 44\text{Å}$). In Silicon for example with a gap energy $E_g = 1.1$ eV, the density of charge carriers is $n_i \approx 10^{10}$ cm^{-3} (with m_c, m_v about 10% larger than the electron mass). From the equality $n_i = n_h^0 = n_Q^h e^{-\beta\mu_0}$ we deduce the chemical potential of an intrinsic semiconductor:

$$\mu_0 = \frac{E_g}{2} + \frac{3kT}{4} \log m_v/m_c \sim \frac{E_g}{2}$$

Since the effective masses of the charge carriers (electrons or holes) in the conduction and valence bands are similar: $m_v \sim m_c$, the chemical potential in intrinsic semiconductors is roughly in the middle of the band gap.

5.9.2.3 ◆ Diodes and Photodiodes

As we have seen in the case of water which pH (the density of H_3O^+ ions) could be changed by the addition of a strong acid or base, the density of charge carriers (holes or electrons) in a semiconductor can be affected by the addition of acceptor or donor impurities.

A semiconductor in which donor impurities at concentration n_e^n give-up electrons (negative charges) to the conduction band is known as a **n-type** semiconductor. In such a semiconductor, movement of the electrons (the majority carrier of charge at concentration n_e^n) is the main contributor to the conduction, while movement in the valence band of holes (the minority carrier at concentration $n_h^n \ll n_e^n$) is often negligible.

A semiconductor which is doped with impurities at concentration n_h^p that accept electrons from the valence band, leaving holes (positive charges) in the valence band is known as a **p-type** semiconductor, see section 4.3.4. In such a semiconductor, conduction is mostly determined by the density of holes (the majority carriers at concentration n_h^p) in the valence band and not by the electrons in the conduction band (the minority carrier at concentration $n_e^p \ll n_h^p$).

The chemical potential μ_n of a n-type semiconductor lies close to the bottom of the conduction band so that electrons easily hop from the donor impurity into the conduction band leaving behind fixed holes on the impurities. In that case the density of electrons in the conduction band is set by the impurity concentration $n_e^n > n_i$. From the law of mass-action, Eq.5.176, $n_e^n n_h^n = n_e^0 n_h^0 = n_i^2$. Since the concentration of the minority carriers (holes at concentration n_h^n) is set by the chemical potential μ_n: $n_h^n = n_Q^h e^{-\beta\mu_n}$ and $n_i = n_h^0 = n_Q^h e^{-\beta\mu_0}$, the chemical potential in presence of n-type (donor) impurities becomes:

$$\mu_n = \mu_0 + kT \log n_e^n/n_i > \mu_0 \tag{5.178}$$

Hence as claimed, the chemical potential in an n-type semiconductor lies closer to the conduction band.

Similarly the chemical potential μ_p in a p-type semiconductor lies closer to the top of the valence band. Electrons from the valence band hop onto these impurities (acceptors at concentration n_h^p) leaving holes in the valence band. Since the concentration of the minority carriers (electrons at concentration n_e^p) is set by the chemical potential μ_p: $n_e^p = n_Q^e e^{-\beta(E_g - \mu_p)}$ and $n_e^0 = n_Q^e e^{-\beta(E_g - \mu_0)}$, the law of mass action, Eq.5.176, yields:

$$\mu_p = \mu_0 - kT \log n_h^p / n_i < \mu_0 \qquad (5.179)$$

A pn junction is made when a p-type semiconductor is put in contact with an n-type one. Just as discussed for a semi-permeable membrane, free charges (the ones in the conduction or valence band, not on the impurities) migrate across the junction thus generating an opposing potential V_{pn} until the chemical potential on both side of the junction is balanced. The diffusion of the majority carriers across the junction (free electrons from the n to the p-type and holes from the p to the n-type) results in a diffusion layer (of a fraction of micron in size) on both sides of the junction with a reduced density of charge carriers, hence its name the depletion zone[27]. As a consequence the chemical potential in this diffusion layer is further from both the valence and conduction bands. Just as diffusion of ions across a semi-permeable membrane resulted in an electrical potential difference, see section 5.7.2, the diffusion of charges across a pn-junction creates an electric field directed from the positively charged n-type to the negatively charged p-type and an associated voltage difference V_{pn} that evens-off the chemical potentials on both side of the junction, see Fig.5.21:

$$\mu_n = \mu_0 - eV_{pn} + kT \log n_e^n / n_i = \mu_p = \mu_0 - kT \log n_h^p / n_i$$

So that the voltage across the junction is given by:

$$V_{pn} = (kT/e) \log \frac{n_e^n n_h^p}{n_i^2} \qquad (5.180)$$

For a typical density of impurities in Silicon of about $n_e^n \sim n_h^p \sim 10^{16}$ cm^{-3} the voltage across the junction is: $V_{pn} \sim 0.70V$. This voltage difference (i.e. electric field) prevents holes from flowing into the n-type side (and electrons from flowing into the p-type side). At equilibrium no current flows across the junction: the diffusion current I_{Dh}^0 of holes (the majority charge carriers) from the p to the n-type (see sections 5.1.2 and 5.11.1.1) is equal to the electric-field driven (drift) current I_{Eh}^0 of holes (the minority carriers) from the n to the p-type:

$$I_{Eh}^0 = I_{Dh}^0 \propto n_h^p e^{-\beta eV_{pn}} = n_i^2 / n_e^n = n_h^n \qquad (5.181)$$

---

[27]Charge neutrality implies that the width of the depletion zone on both sides (x_p and x_n) satisfy: $n_e^n x_n = n_h^p x_p$, see Fig.5.21.

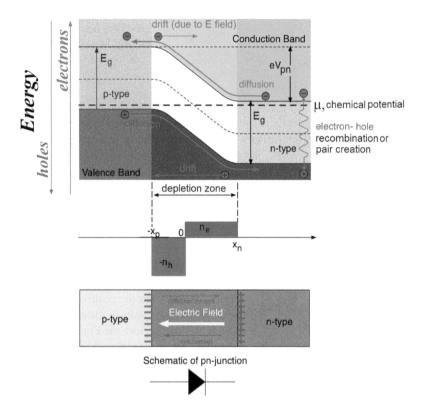

Figure 5.21 A *pn* junction is formed when a semiconductor doped with electron acceptors (*p*) is put in contact with a similar material doped with electron donors (*n*). The electrons in the conduction band (and holes in the valence band) diffuse across the junction such as to equilibrate the chemical potentials on both side of the junction, resulting in an excess of minority charge carriers (i.e. electrons) on the *p*-type side and an excess of holes on the *n*-type side. An electric field is established which counters the movement of electrons from the *n*-type material into the *p*-type one. In this diffusion zone, the chemical potential being far from both the conduction and valence bands, the density of free charges is very small, hence its name: the depletion zone. Current will only flow from the *p*-type into the *n*-type if a reverse electric field is applied on the junction (if the *p*-side is at a higher electrical potential than the *n*-side): a diode is thus established. Notice that minority carriers (e.g. holes in *n*−type semiconductor) can recombine with majority carrier (e.g. electrons) resulting in light emission (as in a Light Emitting Diode (LED)) or energy dissipation. Reproduced with QuantumATK[98] with permission from Synopsis.com.

This drift current I_{Eh}^0 is proportional to the small density of holes (n_h^n, the minority carriers on the n-type side)[28], see Eq.3.40 and Fig.5.21. For a net current to flow through a pn junction, the electric field (i.e. potential V_{pn}) opposing the flow of charges at the junction has to be reduced. If the potential difference across the junction is reduced by a forward bias voltage V (higher on the p-side than on the n-side) the density of holes diffusing from the p-type into the n-type is enhanced by a factor $e^{\beta eV}$ and the diffusion current increases similarly: $I_{Dh} = I_{Dh}^0 e^{\beta eV}$. On the other hand the backward drift current —being proportional to the density of minority carriers (holes) on the n-type side —is not much affected by the forward bias voltage: $I_{Eh} \approx I_{Eh}^0 = I_{Dh}^0$. Thus the overall current (due to the diffusion minus the backward drift currents of holes and electrons) becomes:

$$I_{diode} = I_S(e^{\beta eV} - 1) \tag{5.182}$$

where $I_S = I_{Dh}^0 + I_{De}^0$ is known as the saturation current: it is the leakage current $I = -I_S$ when a reverse bias voltage ($V < 0$) is applied on the junction so that the diffusion current is greatly reduced and the current is only due to the drift current of the minority charges across the junction; typically $I_S \sim 10^{-12}$A.

In summary, in a pn junction current will usually only flow from the p-type into the n-type material upon application of an opposing field, i.e. a higher voltage on the p- than the n-type. Such a device is known as a solid state diode, to differentiate it from the vacuum bulbs used earlier and described in section 3.1.4.3. In an electrical circuit, a pn junction or diode thus plays the role of a tire valve: it allows current to flow in only one direction.

The lifetime of the injected charges (holes moving from the p-type into the n-type and electrons moving in the reverse direction) is set by the recombination time of electrons and holes: $\sim 0.1 - 10 \mu$sec, which is the typical time an electron stays within the conduction band of a p-type material before it spontaneously decays to one of the available states in the valence band). This lifetime sets the typical diffusion length (a few tens of microns) of minority charges in the doped semiconductor (electrons in p-type and holes in n-type). If the momentum of the interacting electron and hole are identical, recombination is usually accompanied by emission of a photon of energy $h\nu \sim E_g$. Otherwise recombination involves coupling to phonon modes (i.e. vibrational modes of the material that absorb the difference in momentum between electron and holes) and dissipation. The former situation is used in Light Emitting Diodes (LED) that besides their small size, robustness and longevity are also very efficient in transforming electric energy into radiation (with less dissipation into heat than other methods, such as the incandescent light bulb). The wavelength of the emitted light is set by the semiconductor gap E_g which depends on the type of materials used: infrared for GaAs, yellow for GaP, green for GaN, blue for SiC.

Conversely, in a photodiode light is transformed into electric energy. Photons absorbed at the pn junction result in electrons being transferred into the conduction

---

[28] Similar arguments apply to the balance between drift and diffusion currents of electrons from the n- to the p-type: $I_{Ee}^0 = I_{De}^0 \propto n_i^2/n_h^p = n_e^p$.

band leaving holes in the valence band. As a result of the electric field across the junction the (negatively charged) electrons are attracted to the n-type side whereas the holes are attracted to the p-type side resulting in a current flowing from the n to the p-type (opposite to the flow in a regular diode). Photodiodes are the basic elements of solar cells, see section 4.1.2. In contrast to regular diodes, photodiodes operate with a reverse voltage on the junction. This reverse voltage results in a larger depletion region and a faster response (due to the smaller capacitance of the junction). This mode of operation is used for light sensitive detection (e.g. cameras). The reverse voltage can in fact be so large that the electric-field across the junction is above the breakdown voltage of the material, as in an avalanche photodiode (APD). In that case the photo-electrons are accelerated (before they are scattered) to such an energy as to generate an avalanche of electrons as in a lightning stroke, see section 3.1.2.1. The photo-electron/hole pair generated by the absorption of a single photon is thus amplified. Avalanche photodiodes are used therefore to study phenomena involving the detection of very low levels of light (single photons). They must be protected against electronic damage due to high currents at intense illumination.

5.9.2.4 ◆ The Field Effect Transistor

The transistor is a semiconductor device[29] that has revolutionized the electronics industry and has allowed for the emergence of information technology, personal computers, the internet, etc. Because of its importance and ubiquity, I believe that understanding its principle of operation should be part of everybody's education.

A transistor is a semiconductor sandwich: it consists of a layer, which conductivity is controlled by a voltage, sandwiched between two layers that provide the appropriate charge carriers (electrons or holes). A transistor has therefore three connections: an input, an output and a control. There are basically two types of transistors: the bipolar transistor used to amplify small signals (such as in a radio receiver) described in Appendix A.9.2 and the Field Effect Transistor (FET) used both as an amplifier and as a toggle switch in digital circuits.

A Metal-Oxide Semiconductor Field Effect Transistor (or MOSFET) such as sketched in Fig.5.22 (and Fig.3.8) consists of a semiconductor channel of one type (here n-type), blocked by a semiconductor of the opposite type. The intensity of the (electron) current flowing in the highly doped n-type semiconductors from the source (S) to the drain (D) is controlled by the availability of electrons in the less doped p-type blocking channel. One can tune the density of these electrons by applying a voltage on the metallic gate (G) which is separated from the p-type channel by a layer of insulating silicon dioxide, i.e. glass. Since the gate and the opposite p-type layer form a capacitor, one can control the density of charges on the p-type side: electrons if the gate voltage is positive; holes if it is negative. This property is used to control the conductance of the channel and turn it on (conducting) or off (insulating).

---

[29]The name transistor (for **trans**fer res**istor**) was given in 1948 by its inventors at Bell Labs - J.Bardeen, W.Shockley and W.Brattain (the 1956 Nobel recipients) - but was popularized 6 years later by the name given to the first radios using that technology.

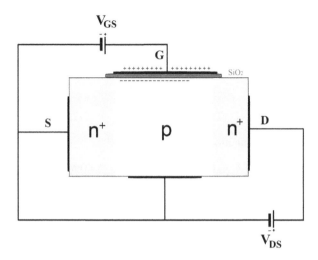

Figure 5.22 A MOSFET transistor consists of a sandwich of p-type squeezed between two layers of n-type semiconductor. A conducting electrode (the gate, G) is insulated from the p-type side by a layer of silicon dioxide (glass), hence in contrast with a bipolar transistor (see Appendix A.9.2) there is no hole current flowing from the gate to the source (S). The voltage on the gate is used to deplete the p-type side of electrons (if the voltage is negative) or of holes (enrich it in electrons if the voltage is positive). One can therefore control (impede or enhance) the flow of electrons from the source (S) to the drain (D) along the narrow (light green) channel close to the insulating layer shown here. Consequently, this device is often used as a current switch.

This is the basis of all the micro-electronic switches used in computers, cell-phones, servers, etc.

5.9.3 BOSE-EINSTEIN STATISTICS

For quantum particles with integer spins known as bosons (for example He^4 atoms[30]), Pauli's exclusion principle doesn't hold and each quantum state can be occupied by any number of particles. Let us consider a gas of non-interacting bosons in a box of volume V. If one labels (as we have done for the Fermi-Dirac statistics) the energy states by their momentum p (with energy $\epsilon_p = p^2/2m$), the number of particles n_p in each energy level is arbitrary (as it was for photons, which are also bosons) $n_p = 0, 1, 2, \cdots$. A state n of the system is fully specified by the occupation

---

[30]The spin state of the nucleus of He^4 composed of two protons and two neutrons is zero. The spin state of its electrons in their ground state is also zero. This is not the case of the He^3 isotope which having a single neutron behaves as a spin=1/2 particle, namely a fermion. He^3 becomes a superfluid only below about 2mK when pairs of atoms couple to form a boson, like electrons in a superconductor.

numbers of its energy levels: $\{n_p\}$. Its energy is $E_n = \sum_p n_p \epsilon_p$ and its number of particles is $N_n = \sum_p n_p$. The problem therefore is to determine the probability of a given configuration: $\mathcal{P}(\{n_p\})$. As usual we have to maximize the entropy:

$$S = -k\mathrm{Tr}_{\{n_p\}}\mathcal{P}\log\mathcal{P}$$

subject to the constraints:

$$\mathrm{Tr}_{\{n_p\}}\mathcal{P} = 1 \quad ; \quad \mathrm{Tr}_{\{n_p\}}E_n\mathcal{P} = E \quad ; \quad \mathrm{Tr}_{\{n_p\}}N_n\mathcal{P} = \bar{N}$$

Maximizing the auxiliary function:

$$\Gamma = S + k(\gamma_0 + 1)\mathrm{Tr}\mathcal{P} - k\beta\mathrm{Tr}E_n\mathcal{P} - k\alpha\mathrm{Tr}N_n\mathcal{P}$$

yields:

$$\mathcal{P}(\{n_p\}) = e^{\gamma_0}e^{-\beta E_n - \alpha N_n} \tag{5.183}$$

From the first constraint we derive the partition function:

$$\mathcal{Z} = e^{-\gamma_0} = \mathrm{Tr}_{\{n_p\}}e^{-\beta E_n - \alpha N_n} = \sum_{\{n_p=0,1,\cdots\}} e^{-\sum_p(\beta\epsilon_p + \alpha)n_p}$$

$$= \Pi_p\sum_{n=0}^{\infty}e^{-(\beta\epsilon_p + \alpha)n} = \Pi_p\frac{1}{1 - e^{-(\beta\epsilon_p + \alpha)}} \tag{5.184}$$

Where we used the formula for the sum of an infinite geometrical series: $\sum_{n=0}^{\infty}q^n = 1/(1-q^n)$ (with $q < 1$). The partition function is then:

$$\gamma_0 = -\log\mathcal{Z} = \sum_p\log[1 - e^{-(\beta\epsilon_p + \alpha)}] = \frac{V}{h^3}\int d^3p\,\log[1 - e^{-(\beta\epsilon_p + \alpha)}] \tag{5.185}$$

By Eq.5.124 the mean number of particles is:

$$\bar{N} = \frac{\partial\gamma_0}{\partial\alpha} = \frac{V}{h^3}\int d^3p\,\frac{1}{e^{\beta(\epsilon_p - \mu)} - 1} \equiv \frac{V}{h^3}\int d^3p\,\eta_p^{BE} \tag{5.186}$$

where the chemical potential μ is defined as before by: $\beta\mu \equiv -\alpha$, see Eq.5.121. The particle density at momentum p is given by the Bose-Einstein distribution (the integrand on the right hand of Eq.5.186):

$$\eta_p^{BE} \equiv \frac{1}{e^{\beta(\epsilon_p - \mu)} - 1} \tag{5.187}$$

For the particle density to be finite at any temperature the term in the exponential must be positive (for all values of ϵ_p), i.e. the chemical potential must be negative: $\mu \le 0$. From Eq.5.127, the mean energy is:

$$E = \frac{\partial\gamma_0}{\partial\beta} = \frac{V}{h^3}\int d^3p\,\frac{\epsilon_p}{e^{\beta(\epsilon_p - \mu)} - 1} = \frac{V}{h^3}\int d^3p\,\epsilon_p\eta_p^{BE} \tag{5.188}$$

At high temperatures $e^{-\beta\mu} \gg 1$, we recover the the Maxwell-Boltzmann distribution for an ideal gas, Eq.5.129: $\eta_p^{MB} = e^{-\beta(\epsilon_p - \mu)}$.

5.9.3.1 Superconductors and Superfluids

As the temperature is lowered, the number of particles N_0 in the ground state $\epsilon_0 = 0$ increases until it becomes a non-negligible fraction of the total number $\bar{N}$, a process known as Bose-Einstein condensation. As a consequence, the QM wavefunction associated with the ground state describes not a microscopic particle but a macroscopic object with weird observable properties.

Thus He^4 below about $2°K$ behaves as a "superfluid", i.e. a fluid with zero viscosity: the entropy in the ground state being null, there can be no heat loss and thus no viscosity. This absence of viscosity has strange consequences: Helium flows in capillaries without resistance, it can creep along the walls of a container (due to the attractive van der Waals force between Helium and the wall's material, see section 3.1.2.2) and spill out without human intervention. This phenomena of Bose-condensation is a rare example of quantum mechanical effects observable at a human scale. A similar phenomenon occurs in conductors, where the coupling of pairs of spin-1/2 electrons creates an effective boson. At low temperature, these so-called Cooper-pairs condense in the ground state to form a macroscopic quantum object: a "superconductor", a state of charge carriers that flow without resistance. Just as for a "superfluid", the entropy of the superconducting condensate being zero, it cannot dissipate heat: it displays zero resistance.

Let us estimate the temperature at which this Bose-condensation occurs. Since the number of particles N_0 in the ground state $\epsilon_0 = 0$ is macroscopic the chemical potential μ at the transition vanishes:

$$N_0 = \frac{1}{e^{-\beta\mu} - 1} \gg 1 \qquad \text{namely}: \qquad \mu \approx -\frac{kT}{N_0} \to 0$$

We can thus determine the temperature T_B ($\beta_B = 1/kT_B$) at which part of the boson gas condenses into the ground state from Eq.5.186 with $\mu = 0$:

$$\bar{n} = \bar{N}/V = \frac{1}{h^3} \int d^3p \frac{1}{e^{\beta_B\epsilon_p} - 1} = \frac{2\pi(2m)^{3/2}}{h^3} \int_0^\infty d\epsilon\, \epsilon^{1/2} \frac{1}{e^{\beta_B\epsilon} - 1}$$

$$= \frac{2\pi(2mkT_B)^{3/2}}{h^3} \int_0^\infty dx \frac{x^{1/2}}{e^x - 1} = \frac{2\pi(2mkT_B)^{3/2}}{h^3} \zeta(3/2)\Gamma(3/2)$$

$$= \zeta(3/2)(2\pi mkT_B/h^2)^{3/2} = \zeta(3/2)/\lambda_{DB,He}^3(T_B) \tag{5.189}$$

Where[31] $\lambda_{DB,He}(T_B)$ is the de-Broglie wavelength for Helium at temperature T_B, Eq.5.87. For Helium with density $\bar{n} = 2.18 \ 10^{22} \ cm^{-3}$, the superfluid transition is predicted to be at $T_B = 3.1°K$. In practice it has been measured at $T_\lambda = 2.2°K$, a small discrepancy due to the fact that He^4 atoms are slightly interacting bosons.

---

[31] Upon change of variable $x \equiv \beta_B\epsilon$; $\zeta(s), \Gamma(s)$ are known respectively as the Riemann zeta-function and gamma function of the variable s: $\zeta(3/2) \approx 2.612$, $\Gamma(3/2) = \sqrt{\pi}/2$.

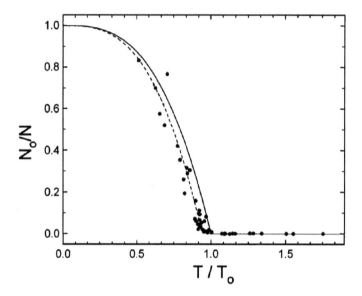

Figure 5.23 Rubidium atoms captured in a magneto-optic trap at very low temperatures behave as bosons with a Bose-condensation temperature $T_B = 0.280\mu°$K. As the temperature is reduced the proportion of condensate increases (dots) as predicted for a gas of bosons trapped in a harmonic potentiel, i.e. as $1 - (T/T_B)^3$ (dashed line, the continuous line is the prediction for $N \to \infty$). Adapted with permission from ref.[99], copyright by the American Physical Society.

Below the transition temperature, i.e. for $T < T_B$, the number of bosons in the ground state N_0 increases at the expense of their number in the excited states: N_e. Since $\mu = 0$, we can use Eq.5.189 to deduce:

$$\frac{\bar{N} - N_0}{V} = \frac{N_e}{V} = \frac{\zeta(3/2)}{\lambda^3_{DB,He}(T)}$$

Dividing that equation by the result of Eq.5.189 yields

$$\frac{N_0}{\bar{N}} = 1 - (T/T_B)^{3/2} \tag{5.190}$$

The ideal boson gas studied above has been achieved experimentally[99] by trapping at a fraction of a $\mu°$K about 40,000 atoms with integer spin in a three dimensional magneto-optic trap (their energy is then $\epsilon_p = \vec{p}^{\,2}/2m + k\vec{r}^{\,2}/2$). A transition to a Bose condensate is observed at $T_B = 0.28\mu$K. For this feat, the authors E.A. Cornell and W.E. Wieman shared the 2001 Nobel prize in Physics with W. Ketterle. Notice that since the atoms are trapped in a harmonic potential and not in a box, the density of the condensates increases below T_B as $1 - (T/T_B)^3$, see Fig.5.23 and not as computed here for atoms in a box of fixed volume, Eq.5.190.

5.10 INTERACTING SYSTEMS

Most of the physical systems of interest (real gases and polymers, ferromagnetic materials, glasses, liquids, etc.) are composed of interacting particles for which the idealizations made previously are not valid. A common feature of these systems is the appearance of phase transitions (in dimensions greater that 1), where the system exhibits a different behavior at higher or lower temperatures. Ferromagnetic systems for example are paramagnetic at high temperatures, while they display a permanent magnetization at low temperatures. Real gases behave as ideal gases at high temperatures (and low densities) but condense into a liquid (high density) phase at low temperatures. Since in the high temperature phase the system often behaves as an ideal one (where interactions can be neglected), perturbations about this ideal solvable case is usually a good approach to study the behavior of the system in the low temperature phase (just below the phase transition). In the following we will apply the maximal entropy formalism to various systems of interacting particles, introducing in each case the approximation used to solve for the probability distribution and studying the behavior near the phase transition (when it exists). Much of the ingeniosity in the study of interacting many particle systems is reflected in the design of realistic yet soluble approximations.

5.10.1 FERROMAGNETISM

A ferromagnet is a material which can be permanently magnetized by an external magnetic field. In such situations one wishes to understand how its magnetization depends on the applied field and below which temperature is it permanently magnetized (at high enough temperatures, thermal fluctuations will overcome the ordering of the internal magnetic field). A ferromagnet is often described as a paramagnet in which a total of N spins (or magnetic moments μ_i) can interact with their neighbors. In most general form the energy of a single spin can be written as:

$$\epsilon_i = -\vec{\mu}_i \cdot \vec{B} - \sum_{j \neq i} J_{ij} \vec{\mu}_i \cdot \vec{\mu}_j \tag{5.191}$$

And for a particular orientation of the N spins the total energy is $\epsilon_p = \sum_i \epsilon_i$ (taking care not to doublecount the interaction between spin i and j). One may consider Ising spins for which $\mu_i = \pm\mu_0$ (parallel or antiparallel to the magnetic field) or Heisenberg spins of fixed amplitude but arbitrary orientation in space. The value of the coupling constants $J_{ij} = J_{ji}$ is often not known precisely and decreases as a function of the distance between the spins. Different approximations are usually made at this point to describe various magnetic systems. In the case of ferromagnets, one usually assumes $J_{ij} = J$ for all i, j or only for j's that are nearest neighbors of i (the assumption we will make in the following). In the case of anti-ferromagnetic systems (for which nearest neighbor spins tend to align in opposite directions) one assumes $J_{ij} = J < 0$ for j's that are nearest neighbors of i. In the case of spin glasses, which exhibit a frozen disordered state at low temperatures, one often draws the J_{ij} from a random distribution (for all i, j or just nearest neighbors).

In the following we shall study the case of an Ising ferromagnet with nearest neighbor interactions. The goal as always is to determine the probability $\mathcal{P}(\{\mu_i\})$ of a given spin configuration $\{\mu_i\}$ $(i = 1, \cdots, N)$. As usual we have to maximize the entropy:

$$S = -k\mathcal{P}\log\mathcal{P} \quad \text{with constraints}: \sum_{\{\mu_i\}} \mathcal{P} = 1 \quad \text{and} \quad \sum_{\{\mu_i\}} \epsilon_p\mathcal{P} = E \qquad (5.192)$$

The result is (see Eq.5.42):

$$\mathcal{P}(\{\vec{\mu}_i\}) = e^{\gamma_0}\Pi_i^N e^{-\beta\epsilon_i} \qquad (5.193)$$

and the partition function is:

$$e^{-\gamma_0} = \mathcal{Z}_N = \sum_{\{\mu_i\}} \exp\left[\beta\sum_i \mu_i B + J\sum_{<ij>}\mu_i\mu_j\right] \qquad (5.194)$$

where the notation $\sum_{<ij>}$ implies summation on the z nearest neighbors only (on a cubic lattice for example: $z = 6$). Exact solutions exist only in one and two dimensions. In 3D the simplest approach is known as the mean field approximation. It consists in approximating the interaction terms by the mean value of the magnetic moment $\langle\mu_i\rangle = m$ (which is determined at the end from the computation of the free energy see Eq.5.50) while neglecting second and higher orders of the fluctuations: $\delta\mu_i = \mu_i - m \ll m$. The interaction sum becomes (taking care not to count the same interaction twice)

$$\sum_{<ij>}\mu_i\mu_j = \sum_{<ij>}(m + \delta\mu_i)(m + \delta\mu_j) = \frac{1}{2}[z\sum_i m^2 + 2z\sum_i m\delta\mu_i + O(\delta\mu_i^2)]$$

$$\approx \frac{1}{2}[z\sum_i m^2 + 2z\sum_i m(\mu_i - m) = zm\sum_i \mu_i - Nzm^2/2 \qquad (5.195)$$

With that approximation the partition function becomes:

$$\mathcal{Z}^{MF} = \sum_{\{\mu_i=\pm\mu_0\}} \exp\left[\beta\sum_i (B + Jzm)\mu_i - \beta NJzm^2/2\right]$$

$$= e^{-\beta NJzm^2/2} \, 2^N \cosh^N \beta\mu_0(B + Jzm) \qquad (5.196)$$

The Helmholtz free energy (see Eq.5.49) becomes:

$$\mathcal{F} = -kT\log\mathcal{Z}^{MF} = NJzm^2/2 - NkT\log 2 - NkT\log\cosh\beta\mu_0(B + Jzm) \quad (5.197)$$

From the expression of $\mathcal{F}$ we can deduce the mean magnetization per site m, see Eq.5.50:

$$m = \frac{M}{N} = -\frac{1}{N}\frac{\partial\mathcal{F}}{\partial B} = \mu_0\tanh\beta\mu_0(B + Jzm) \qquad (5.198)$$

This equation —which can also be obtained by minimizing $\mathcal{F}$ with respect to m —is a self-consistent equation for the mean magnetization per site[32]. It can be solved graphically, see Fig.5.24 by looking for the intersection of the line $y = ax$ with the curve $y = \tanh(x+b)$:

$$m/\mu_0 \equiv y = ax \equiv Tx/T_c \quad \text{with}: \quad x = \beta\mu_0 mJz \equiv T_c m/\mu_0 T$$
$$y = \tanh(x+b) \quad \text{with}: \quad b \equiv \beta\mu_0 B \quad (5.199)$$

where the para to ferro-magnetic transition temperature: $T_c \equiv \mu_0^2 Jz/k$.

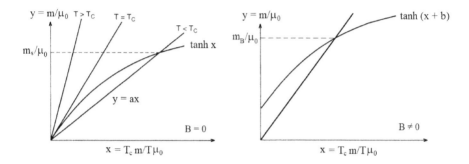

Figure 5.24 The solutions $y \equiv m/\mu_0 = \tanh\beta\mu_0(B + Jzmx) = \tanh(x+b)$ (with $x \equiv \beta\mu_0 mJz \equiv T_c m/\mu_0 T$; $b \equiv \beta\mu_0 B$) can be obtained graphically from the intersection betwen the line $y = ax$ ($a \equiv T/T_c$) and the curve $y = \tanh(x+b)$ (with $b = 0$ or not). When $B = 0$ there is a non-zero solution $m_s \neq 0$ only when $a < 1$, i.e. for $T < T_c$. When $B \neq 0$ there is always a single solution with $m_B \neq 0$.

In absence of magnetic field there are two phases: a high temperature paramagnetic phase $m = 0$ (no intersection for $m \neq 0$) when $a > 1$, i.e. $T > T_c$ and a low temperature ferromagnetic phase when $a = T/T_c < 1$, i.e. $T < T_c$, as there is then a non-zero spontaneous magnetization m_s. In presence of a magnetic field $B \neq 0$: there is always a solution $m_B \neq 0$, the system is magnetized.

From the partition function Z^{MF} we can compute the mean energy of the system in the mean field approximation, using Eq.5.198:

$$E^{MF} = -\frac{\partial}{\partial\beta}\log Z^{MF} = NJzm^2/2 - N\mu_0(B + Jzm)\tanh\beta\mu_0(B + Jzm)$$
$$= NJzm^2/2 - N(B + Jzm)m = -NmB - NJzm^2/2 \quad (5.200)$$

---

[32] Notice the similarity of Eq.5.198 with the equation derived for the magnetization of a paramagnetic material, Eq.5.50.

From the partition function Z^{MF} we can also compute the mean entropy of the system in the mean field approximation. Using the relation derived from Eq.5.198:

$$\beta\mu_0(B + Jzm) = \tanh^{-1} m/\mu_0 \equiv \frac{1}{2}\log\frac{(1+y)}{(1-y)}$$

yields the following relation for the mean field entropy[33]

$$TS^{MF} = kT\log Z^{MF} + E^{MF} = -Nm(B + Jzm) + NkT\log 2\cosh\beta\mu_0(B + Jzm)$$

$$= -NkT[\frac{y}{2}\log\frac{1+y}{1-y} + \frac{1}{2}\log\frac{(1+y)(1-y)}{4}]$$

$$= -NkT[\frac{1+y}{2}\log\frac{1+y}{2} + \frac{1-y}{2}\log\frac{1-y}{2}]$$

$$= -NkT[P_\uparrow \log P_\uparrow + P_\downarrow \log P_\downarrow] \qquad (5.201)$$

where $P_\uparrow, P_\downarrow$ are the mean field probabilities of a spin being aligned parallel or antiparallel to the magnetic field, see Eq.5.45. Above T_c and in absence of magnetic field $m = 0$ and thus $S^{MF} = Nk\log 2$, as for a paramagnetic material. Below T_c, as the spins get ordered the entropy is smaller, see Appendix A.9.3.1.

5.10.2 REAL GASES

The particles in a real gas (and even more so in a liquid) do interact with each other. First at short range, the electron clouds surrounding the nuclei of atoms repel each other: the atoms behave as if they were hard billiard balls. Then there are long range attractive interactions due to van der Waals forces: dipole-dipole interactions (either permanent or instantaneous, induced or not) that are always present, see section 3.1.2.2. As a result of these complex interactions, represented by a potential $U(\vec{x}_{ij})$ (where $\vec{x}_{ij} = \vec{x}_i - \vec{x}_j$ is the distance between two particles) the energy of a particular configuration of N particles in a gas with positions and moments $\{\vec{x}_i, \vec{p}_i\}$ is:

$$E_p = \sum_i \frac{\vec{p}_i^2}{2m} + \sum_{i,j\neq i} U(\vec{x}_i - \vec{x}_j) \qquad (5.202)$$

Proceeding as for an ideal gas, see Eq.5.85, yields the partition function:

$$Z = \text{Tr}\, e^{-\beta E_p} = \frac{1}{h^{3N}N!}\int d^3x_1 \cdots d^3x_N e^{-\beta\sum U(\vec{x}_i-\vec{x}_j)}\int d^3p_1 \cdots d^3p_N \Pi_i^N e^{-\beta\vec{p}_i^2/2m}$$

$$= \frac{(2\pi mkT/h^2)^{3N/2}}{N!}\int d^3x_1 \cdots d^3x_N e^{-\beta\sum_{i,j\neq i} U(\vec{x}_i-\vec{x}_j)} \qquad (5.203)$$

The difficulty in evaluating the partition function lies in computing the spatial integral. One approximation due to van der Waals consists in accounting for the hard

---

[33] Using the identity: $\cosh x = 1/\sqrt{1 - \tanh^2 x} = [(1 - \tanh x)(1 + \tanh x)]^{-1/2}$.

core (i.e. billiard ball) repulsive interaction by reducing the integration volume to $V - Nv_0$ (where v_0 is the effective volume occupied by a single particle). Then, if the attractive interaction potential is small with respect to the thermal energy, i.e. $|\beta U(\vec{r})| \ll 1$ the exponential can be expanded to yield:

$$\int d^3x_1 \cdots d^3x_N e^{-\beta \sum_{i,j\neq i} U(\vec{x}_i - \vec{x}_j)} \approx \int d^3x_1 \cdots d^3x_N (1 - \beta \sum_{i,j\neq i} U(\vec{x}_i - \vec{x}_j))$$

$$= (V - Nv_0)^N - \beta(V - Nv_0)^{N-2} \sum_{i,j\neq i} \int d^3x_i d^3x_j U(\vec{x}_i - \vec{x}_j)$$

$$= (V - Nv_0)^N - \beta(V - Nv_0)^{N-1} \frac{N(N-1)}{2} \int d^3r U(r)$$

$$\approx (V - Nv_0)^N (1 + \frac{\beta N^2 a}{V}) = (V - Nv_0)^N e^{\beta N^2 a/V} \tag{5.204}$$

where the factor $N(N-1)/2$ counts the number of interacting particle pairs and $U(r)$ is an attractive potential (down to the hard core) that depends only on the distance between the two particles $r = |\vec{x}_i - \vec{x}_j|$. The constant $a \equiv -\int d^3r U(r)/2 > 0$ (a is positive since the potential $U(r)$ is attractive, i.e. negative). The partition function is thus:

$$Z = \frac{(2\pi mkT/h^2)^{3N/2}}{N!} (V - Nv_0)^N e^{\beta N^2 a/V} = \frac{n_Q^N}{N!} (V - Nv_0)^N e^{\beta N^2 a/V} \tag{5.205}$$

The average energy is:

$$E = -\frac{\partial}{\partial \beta} \log Z = \frac{3N}{2} kT - \frac{N^2 a}{V} \tag{5.206}$$

The molar ($N = N_A$) specific heat of a real gas $c_V = \partial E/\partial T = 3N_A k/2$ is thus the same as that of an ideal gas. With $v = V/N$ standing for the specific volume, the Helmholtz free energy $\mathcal{F}$ is:

$$\mathcal{F} = -kT \log Z = -NkT \log n_Q - NkT - NkT \log(v - v_0) - \frac{Na}{v}$$

$$\equiv Nf(v,T) \tag{5.207}$$

From the relation $\mathcal{F} = E - TS$, see section 5.90, the entropy of a real gas becomes:

$$S = k \log Z + k\beta E = Nk[\log n_Q + 5/2 + \log(v - v_0)] \tag{5.208}$$

And from the definition of the pressure, see Eq.5.91, the pressure is:

$$P = -\frac{\partial \mathcal{F}}{\partial V} = -\frac{1}{N} \frac{\partial \mathcal{F}}{\partial v} = \frac{kT}{v - v_0} - \frac{a}{v^2} \tag{5.209}$$

From which we derive the van der Waals equation of state for a real gas:

$$(P + a/v^2)(v - v_0) = RT \tag{5.210}$$

Where the specific volumes v are measured in l/mol (hence the replacement of k by $R = N_A k = 8,314$ J/mol°K). For Oxygen the typical values are[34]: $a^{O_2} = 1.382 \, 10^5$ Pa(l/mol)^2, $v_0^{O_2} = 0.0318$ l/mol.

With the definitions:

$$v_c = 3v_0 \qquad \text{for } O_2 \ : \ v_c^{O_2} \approx 0.0954 \text{ l/mol}$$

$$P_c = a/27v_0^2 \qquad \text{for } O_2 \ : \ P_c^{O_2} \approx 50 \text{ bar}$$

$$RT_c = 8a/27v_0 \qquad \text{for } O_2 \ : \ T_c^{O_2} \approx 155 \text{ °K}$$

the pressure, volume, and temperature can be rescaled as dimensionless variables: $P_R = P/P_c$; $v_R = v/v_c$ and $T_R = T/T_c$. In those dimensionless (reduced) variables the van der Waals equation, Eq.5.210, can be written in a universal form valid in principle for all gases and liquids[100]:

$$(P_R + \frac{3}{v_R^2})(v_R - \frac{1}{3}) = \frac{8T_R}{3} \qquad (5.211)$$

The dimensionless constant $z_c = P_c v_c / RT_c = 3/8 = 0.375$ should be the same for all gaseous compounds, as indeed it is though with a slightly smaller value of $z_c \approx 0.30$. As a consequence for O_2 the measured specific volume of O_2 at the critical point is $v_c^{meas} = 0.078$ l/mol and not as deduced from $v_c = 3v_0 = 0.0954$ l/mol).

For any given reduced pressure and temperature, the reduced volume v_R obeys a third order equation which can have either one or three real solutions. At high temperatures ($T_R > 1$) it has one solution which corresponds to a single gas phase. At low temperatures ($T_R < 1$) Eq.5.211 has three solutions, one is unstable (the reduced compressibility is negative: $\kappa_R = -V_R^{-1} \partial V_R / \partial P_R < 0$, i.e. the volume increases as the pressure is increased) and the two others are stable: they correspond to coexisting liquid and gas phases with respective reduced volume $v_{R,1}, v_{R,2}$ (with $v_{R,1} < v_{R,2}$). At any given temperature T (or T_R), that coexistence is stable at a single value of the (reduced) pressure ($P_{R,1} = P_{R,2} \equiv P_R$) for which the chemical potential of the two phases are equal ($\mu_1 = \mu_2$). As the control parameters are pressure and temperature, the chemical potential is given by the derivative of the Gibb's free energy:

$$\mathcal{G} = \mathcal{F} + PV \equiv Nf(v,T) + NPv$$

with respect to the particle number: $\mu = \partial \mathcal{G}/\partial N = f(v,T) + Pv$, see Eq. 5.147. Equality of the chemical potentials implies that:

$$f(v_{R,1}, T_R) + P_R v_{R,1} = f(v_{R,2}, T_R) + P_R v_{R,2}$$

---

[34] For water, the measured values at the critical point are: $T = T_c = 647°$K, $P = P_c = 220$ bar , $v_c^{meas} = 0.055$l/mol. From the values of T_c and P_c one deduces: $a^W = 5.536 \, 10^5$ Pa(l/mol)^2, $v_0^W = 0.0305$ l/mol $< v_c^{meas}$. Notice however that the value of v_0^W is larger than the specific volume of water at 300°K deduced from its density (1 g/cm^3): $v = 0.018$ l/mol, which using Eq.5.210 would result for water at 300°K in a non-physical negative pressure. Due to specific interactions (hydrogen bonds) between its molecules, water at room temperatures (i.e. far from T_c) cannot be described by Eq.5.210 .

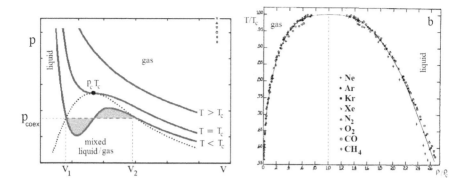

Figure 5.25 a) variation of the pressure vs. specific molar volume $v = V/N_A$ (where N_A is Avogado's number, i.e. the number of particles in a mole of gas) at different temperatures for a van der Waals gas. Below the critical temperature T_c there exist a regime of coexistence between a liquid and a gas phase. In this mixed phase the equilibrium pressure is determined from the Maxwell construction, i.e. the pressure is such that the two shaded areas are equal. (b) Measurement of the density of the liquid and gas normalized by the critical density $\rho_c = 1/v_c$ vs. the reduced temperature $T_R = T/T_c$. Notice that the reduced data measured for different gases fall on the same curve as predicted by the van der Waals theory. Reprinted from ref.[100] with the permission of AIP publishing.

which can be recast as:

$$P_R = -\frac{f(v_{R,2}, T_R) - f(v_{R,1}, T_R)}{v_{R,2} - v_{R,1}} \tag{5.212}$$

Since $P_R = -\partial f(v, T)/\partial v$ Eq.5.212 imply that:

$$\int_{v_{R,1}}^{v_{R,2}} P_R dv = P_R(v_{R,2} - v_{R,1})$$

Graphically this corresponds to the so-called Maxwell construction shown in Fig.5.25: at coexistence of the two phases the hashed areas are equal. The fact that Eq.5.211 is universal implies that the reduced volumes of the gas and liquid phases of different compounds should be identical, as they are indeed observed to be, see Fig.5.25 (b). For Oxygen at coexistence between the liquid and gas phase at $130°K$ ($T_R = 0.8387$) the specific volume of the liquid phase is: $v_{R,1} = 0.555$(i.e. $v_l^{O_2} = v_{R,1}v_c^{meas} = 0.0433$ l/mol) and the gas phase: $v_{R,2} = 5.173$ (i.e. $v_g^{O_2} = v_{R,1}v_c^{meas} = 0.403$ l/mol). The pressure at co-existence is $P_R = 0.35$ (i.e. $P^{O_2} = 17.5$ bar).

5.10.2.1   The Joule Effect and Refrigeration

Anybody who has released the gas in an air-spray has experienced the subsequent cooling of the container. This effect is known as the Joule effect. It expresses the decrease in the temperature of a gas when it is allowed to expand freely from an initial volume V_1 to a final volume $V_2 > V_1$. Since no work is performed on (or by) the gas as it expands its internal energy E is unchanged during this rapid and irreversible process. From Eq.5.206, we deduce that:

$$E = \frac{3N}{2}kT_1 - \frac{N^2 a}{V_1} = \frac{3N}{2}kT_2 - \frac{N^2 a}{V_2} \tag{5.213}$$

In terms of the specific volumes ($v_i \equiv V_i/N$; $i = 1, 2$) the final temperature T_2 can be expressed as:

$$T_2 = T_1 - \frac{2a}{3k}(\frac{1}{v_1} - \frac{1}{v_2}) \tag{5.214}$$

The larger the difference between v_2 and v_1 the lower the temperature after expansion. In particular if the molecules are initially in a liquid phase (as the coolant in a fridge) from which they expand into a gas phase, the decrease in temperature can be very significant. The Joule effect is used in many refrigeration systems. Following a period of expansion and cooling (used to chill the internal volume of a fridge), the gas is compressed while in contact with the external environment (which acts as a thermal bath at constant temperature) and usually liquified again and the procedure repeated.

Notice that there is no Joule effect for an ideal gas ($a = 0$): its free expansion occurs at constant temperature. It is however different from isothermal expansion. In free expansion, the entropy of the expanding gas (real or ideal) increases, while that of the environment stays constant: the process is rapid and irreversible (i.e. the total entropy has increased irreversibly since from the Second Law of Thermodynamics, section 5.3.3, the total entropy can never decrease).

In an isothermal expansion on the other hand, the gas expands by pushing against a piston slowly enough to be considered at equilibrium at all times. The increase in its entropy $\Delta S_{gas} = Nk\log(V_2/V_1)$ is balanced by a reduction in the entropy of the environment on which work $\Delta W = \int PdV = NkT\log(V_2/V_1)$ has been performed by the piston ($\Delta S_{env} = -\Delta W/T$, see section 5.5.3). Hence the overall entropy (system + environment) is unchanged. This is the hallmark of a reversible process.

5.10.3   DNA AS A MODEL POLYMER

The freely jointed chain model for polymers that we studied earlier is not a good description of the behavior of a semi-rigid polymer such as DNA, see Fig.5.11. The assumption that a polymer chain can be modeled as a chain of freely jointed segments doesn't take into account the bending rigidity of the polymer which results in a correlation of the orientation $\hat{t}_i$ of successive segments, see Fig.5.26. To take into account that bending rigidity, the polymer can be modeled as a chain with N segments of size a which are much smaller than the distance over which the orientations of chain segments become uncorrelated, a distance known as the persistence length of the chain[90]. Its energy becomes:

$$E_p\{\hat{t}_i\} = -a\sum_i \hat{t}_i \cdot \vec{F} - B\sum_i \hat{t}_i \cdot \hat{t}_{i+1} \tag{5.215}$$

The first term on the right describes the effect of tension on the polymer (which tends to align it with the direction of the force $\vec{F}$), the second term accounts for the polymer rigidity which results in a tendency of successive fragments to align (B is known as the bending modulus). The solution of that model known as the Worm Like Chain (WLC) model[101] of a polymer proceeds by deriving a recursion relation between the partition function Z_N for a chain of N segments and one of $N-1$ segments. By a perturbation expansion in the segment size ($a \to 0$) a differential equation is obtained for Z_N which solution allows one to determine the behavior of Z as a function of the length of the polymer $L = Na$ and the tension F exerted on it[101, 75]. Proceeding

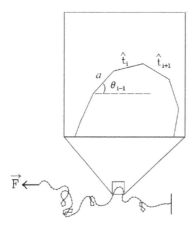

Figure 5.26 Discretization of a continuous polymer by a set of small segments of size a and orientation $\hat{t}_i$.

as previously done for the freely jointed chain model of polymers yields the partition function Z_N as:

$$Z_N(\hat{t}_{N+1}) = \text{Tr } e^{-\beta E_p} = A \int d^2 t_1 \cdots d^2 t_N \exp\left\{\beta \sum_i [a\hat{t}_i \cdot \vec{F} + B\hat{t}_i \cdot \hat{t}_{i+1}]\right\}$$

$$= A \int d^2 t_N Z_{N-1}(\hat{t}_N) \exp\{[\beta aF\cos\theta + \beta B\hat{t}_N \cdot \hat{t}_{N+1}]\} \tag{5.216}$$

Where A is a normalization constant, θ the angle between the tangent vector $\hat{t}_N$ and the direction of the force (which defines the z-axis) and the integration is over all the possible orientations of the tangent vector $\hat{t}_N$, namely the surface of the unit sphere (i.e. the solid angle $d\Omega = \sin\theta d\theta d\phi$) . We shall use the relation:

$$\beta B \hat{t}_N \cdot \hat{t}_{N+1} = -\beta B(\hat{t}_{N+1} - \hat{t}_N)^2/2 + \beta B \equiv -\vec{u}^2/2 + \beta B$$

In the rigid polymer limit $\beta B \to \infty$, the main contribution to the integral in Eq.5.216 arises from tangent vectors that are almost aligned: $\hat{t}_N \approx \hat{t}_{N+1}$. The difference $\hat{t}_{N+1} - \hat{t}_N$ is a vector in a 2D plane tangent to the unit sphere[35]. In other words in that limit the vector $\vec{u} \equiv \sqrt{\beta B}(\hat{t}_{N+1} - \hat{t}_N)$ is a vector defined over the whole 2D plane ($d^2 t \to d^2 u/\beta B$). Defining the persistence length ξ by $\xi/a \equiv \beta B$, we may expand Eq.5.216 as:

$$\mathcal{Z}_N(\hat{t}_{N+1}) = \frac{Ae^{\xi/a}}{\xi/a} \int d^2 u \mathcal{Z}_{N-1}(\hat{t}_{N+1} - \vec{u}\sqrt{a/\xi}) e^{\beta aF\cos\theta} e^{-\vec{u}^2/2}$$

$$\approx \int d^2 u [\mathcal{Z}_{N-1} + \frac{\vec{u}^2}{2\xi/a} \mathcal{L}^2 \mathcal{Z}_{N-1}][1 + \beta aF\cos\theta] e^{-\vec{u}^2/2}$$

$$\approx \mathcal{Z}_{N-1}(\hat{t}_{N+1}) + \frac{a}{2\xi} \mathcal{L}^2 \mathcal{Z}_{N-1} + \beta aF\cos\theta \mathcal{Z}_{N-1} \qquad (5.217)$$

where the normalization constant is set as $Ae^{\xi/a}a/\xi = 1/2\pi$, so that in the limit $a \to 0$, $\mathcal{Z}_{N-1} \to \mathcal{Z}_N$. The terms inside the integral are expanded to first order in a. The operator $\mathcal{L}^2$ is the angular momentum operator[36]. It arises from the Taylor expansion of $\mathcal{Z}_{N-1}$ about $\hat{t}_{N+1}$ (the first term in the expansion, a term in $\vec{u}$ cancels out when integrated). Let us define the distance along the chain of segment j as $s = ja$. In the limit $a \to 0$, the term $(\mathcal{Z}_N - \mathcal{Z}_{N-1})/a \to \partial\mathcal{Z}(s)/\partial s$. Thus one obtains the following Schrödinger-like equation for $\mathcal{Z}$:

$$\xi \frac{\partial\mathcal{Z}}{\partial s} = \frac{1}{2}\mathcal{L}^2\mathcal{Z} + \alpha\cos\theta\mathcal{Z} \qquad (5.218)$$

where $\alpha \equiv \beta F\xi$. Since the right hand side of Eq.5.218 is independent of s we seek a solution: $\mathcal{Z}(s,\phi,\theta) = \mathcal{Z}(\phi,\theta)e^{-gs/\xi}$, where the eigenvalue g satisfies the equation:

$$\frac{1}{2}\mathcal{L}^2\mathcal{Z} + \alpha\cos\theta\mathcal{Z} = -g\mathcal{Z} \qquad (5.219)$$

The above equation is identical to Schrödinger's equation for a rotating diatomic molecule, see section 4.10.2, whose dipole moment $\vec{\mu}$ interacts with an electric field $\vec{E}$ (to generate a potential energy $V = -\vec{\mu} \cdot \vec{E} \equiv \alpha\cos\theta$, see section 4.11.2). It can be solved by the same methods as the QM problem. For a very long chain $L \gg \xi$ the behavior of the partition function $\mathcal{Z}(L)$ is dominated by the eigenstate with the smallest eigenvalue g_0 (the ground state of the equivalent QM problem[37]):

$$\mathcal{Z}(L,\phi,\theta) = \mathcal{Z}(\phi,\theta)e^{-g_0 L/\xi}$$

---

[35]Think of integrating over an area on Earth where the position of the sun at noon ($\sim \hat{t}_N, \hat{t}_{N+1}$) is about the same. The difference $\hat{t}_{N+1} - \hat{t}_N$ is a vector in a flat plane tangent to the Earth surface.

[36]$\mathcal{L}^2 = \{\frac{1}{\sin\theta}\frac{\partial}{\partial\theta}(\sin\theta\frac{\partial}{\partial\theta}) + \frac{1}{\sin^2\theta}\frac{\partial^2}{\partial\phi^2}\}$, see section4.7

[37]The contribution of the eigenstates with eigenvalues $g_n > g_0$ scale as $\exp(-(g_n - g_0)L/\xi)$ and gets negligibly small as $L \gg \xi$.

The extension of the chain at a given tension is then given by Eq.5.75:

$$l = -\left(\frac{\partial F}{\partial F}\right)_T = kT\frac{\partial \log \mathcal{Z}}{\partial F} = -\frac{kTL}{\xi}\frac{\partial g_0}{\partial F} \tag{5.220}$$

At zero force ($\alpha = 0$, zero field in the QM problem) the eigenstates of Eq.5.219 are the spherical harmonics: $Y_{lm}(\theta,\phi)$ and its eigenvalues are $g_l = l(l+1)/2$, see section 4.10.2. At low forces $\alpha = \beta F\xi \ll 1$, second order perturbation theory around the exact $\alpha = 0$ solution (see section 4.11) yields[38]: $g_0 = g_0^{(2)} = -\alpha^2/3$ and therefore:

$$\frac{l}{L} = \frac{2F\xi}{3kT} \tag{5.221}$$

which is the result obtained for the Freely Jointed Chain (FJC) model, section 5.4, provided we identify the Kuhn length to be twice the persistence length: $b = 2\xi$. At larger forces the lowest eigenvalue of Eq.5.219 has to be solved numerically. Nonetheless a pretty good approximate formula valid over the whole range of forces and extensions (with a relative maximal error of $\sim 10\%$) has been proposed by Marko and Siggia[101]:

$$\frac{F\xi}{kT} = \frac{l}{L} - \frac{1}{4} + \frac{1}{4(1-l/L)^2} \tag{5.222}$$

The relative extension l/L as a function of force computed from Eq.5.220 differs significantly from the FJC model at high force but is in excellent agreement with the measurements performed on single DNA molecules, see Fig.5.11. The fit of these data to the predictions of the WLC model provides the most accurate method for estimating the persistence length of DNA.

The understanding of the elastic properties of DNA (both under tension and torsion[102]) has laid the foundation for the single molecule study of its interactions with a large panel of DNA binding proteins (histones, transcription factors, etc.) and enzymatic motor proteins[75] (DNA-polymerases, helicases, topoisomerases, etc.). This has revolutionized biophysics and opened a new vista on the *in-vitro* study of biological processes.

---

[38]To second order in α the shift $g_0^{(2)}$ in the lowest eigenvalue $g_0 = 0$ is given by Eq.4.107: $g_0^{(2)} = |\langle 0,0|\alpha\cos\theta|1,0\rangle|^2/((g_0 - g_1))$. Since $\langle 0,0|\alpha\cos\theta|1,0\rangle = \int d\phi d\cos\theta\, Y_{10}(\theta,\phi)\alpha\cos\theta Y_{00}(\theta,\phi) = \alpha/\sqrt{3}$, we obtain $g_0^{(2)} = -\alpha^2/3$, see section 4.11.2.

5.11   OUT OF EQUILIBRIUM SYSTEMS

The behavior of out of equilibrium systems presents the most difficult and unsolved problems in Science. Systems that are far from equilibrium (e.g. the weather) are notoriously difficult to model and predict. Their outcome are often extremely sensitive to initial conditions and microscopic details, hence the difficulty in predicting the probability of earthquakes, hurricanes or stock-market crashes. In some sense these systems are "loaded": unknown microscopic details affect the probability of macroscopic events. In the context of non-linear systems, this is popularly known as the "butterfly effect": the flapping of a butterfly wings in one place may affect the probability of a storm in a different place months or years later. Notice however that we don't know how to estimate that effect: it maybe large, it maybe small... no one knows. As argued all along, one should not be surprised that we cannot make predictions (even probabilistic ones) on the world surrounding us, but rather be surprised by the fact that we **can** often make accurate predictions at the very small or very large scale and on systems at equilibrium where Nature seems to be "fair".

5.11.1   NEAR EQUILIBRIUM TRANSPORT PROPERTIES

Notwithstanding the above general remarks, much progress can be made in the understanding of systems near equilibrium, i.e. systems which mean macroscopic properties (energy, momenta, magnetic field, etc.) are not too different from their value at equilibrium, even if they vary both in space and time. In such situations the equilibrium results (e.g. probability distributions) can be used to extrapolate the statistical and dynamical properties close to equilibrium.

Quick estimates of various transport properties can be obtained using the approach and results from the kinetic theory of gases, see section 5.5.1. These transport properties relate the spatial variation in one quantity (velocity, temperature, particle density, etc.) to some force or flux. For example the diffusion constant D connects between the gradient in the particle density and the particle flux: $\vec{J} = -D\vec{\nabla}n$, while the thermal conductivity κ connects the temperature gradient with the thermal flux: $\vec{J}_Q = -\kappa\vec{\nabla}T$. Finally the viscosity η of a fluid relates the shear in flow velocity $v_x(z)$ to the force per unit area (the drag) exerted by the flow on the surface shearing it: $F_x/A = -\eta\partial v_x/\partial z$.

To derive these relations from equilibrium statistical mechanics, I will adopt a semi-quantitative approach[89] that is intuitive though not rigorous, see Fig.5.27. The spatial variation of the quantity in question (velocity, temperature, particle density) is discretized in slabs of thickness $l_{m,z}$ (where $l_{m,z} = l_m\cos\theta$ is the projection of the mean free path of the particles l_m along the direction of variation, the z-axis). Within one slab the results obtained at equilibrium are assumed to hold, i.e. the particle density n, the temperature T, the mean magnitude of the velocity $\bar{v}$ are assumed to be uniform and given by their equilibrium values. Based on this assumption one then computes the force/flux at a mid plane. This semi-quantitative approach, which will be detailed below, provides estimates of various transport coefficients (D, κ, η) that display the correct functional dependence on the thermodynamic parameters (temperature, pressure, particle density).

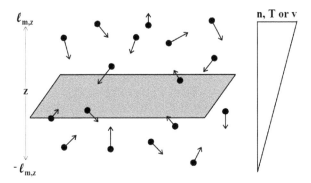

Figure 5.27 Estimation of transport properties. Gradients of particle number n, temperature T or velocity v_x are assumed to exist along the z-axis. By estimating the transfer across the shaded area of particle number, heat or momentum within a mean-free path slice $l_{m,z}$, one can obtain an estimate of the particle diffusion D, heat conductivity κ or viscosity η. For details see text.

5.11.1.1 Diffusion

Let us for example revisit the diffusion of particles already studied in section 5.1.2. Consider a gas of particles displaying a gradient of particles density (or chemical potential): $\partial n/\partial z \neq 0$. The particle flux through the plane at position z, $J_z(z)$, is the result of the imbalance between the mean flux of particles at $z + l_{m,z}$, half of which moving downwards with velocity v_z^- and the mean flux of particles at $z - l_{m,z}$, half of which moving upwards with velocity v_z^+ ($\langle v_z^+ \rangle = -\langle v_z^- \rangle = \bar{v}\langle\cos\theta\rangle$):

$$J_z = \langle n(z + l_{m,z})v_z^- + n(z - l_{m,z})v_z^+ \rangle/2$$
$$\approx \frac{1}{2}n(z)(\langle v_z^- \rangle + \langle v_z^+ \rangle) + \frac{1}{2}\langle l_{m,z}v_z^- - l_{m,z}v_z^+ \rangle \partial n/\partial z$$
$$= -\langle l_{m,z}v_z^+ \rangle \partial n/\partial z \approx -l_m\bar{v}\,\partial n/\partial z \int_0^1 \cos^2\theta d\cos\theta$$
$$= -\frac{l_m\bar{v}}{3}\,\partial n/\partial z \equiv -D\partial n/\partial z \tag{5.223}$$

where $\bar{v}$ is the mean velocity of a gaz particle, see Eq.5.98 and the integration is over the half sphere ($0 < \theta < \pi/2$, the upper plane where v_z^+ is defined). Generalizing to an arbitrary gradient we can write: $\vec{J} = -D\vec{\nabla}n$, where D is the coefficient of self-diffusion of a gas (it is different from the diffusion of a particle in a viscous medium, see below).

Since the mean free path of hard-sphere particles with diameter d and collision cross section $\sigma = \pi d^2$, satisfies: $\sqrt{2}nl_m\sigma = 1$, the constant of self-diffusion is seen to be given by:

$$D = \bar{v}/3 \sqrt{2} n\sigma \sim T^{3/2}/P$$

where we used the fact[39] that $\bar{v} \approx v_{rms} = \sqrt{3kT/m}$ and $P = nkT$, see section 5.5.1. Notice that D depends on both the temperature and the pressure of the gas. The density of an ideal gas in standard conditions ($T = 300°$K, $P = 1$ atm.), is $n_A = 2.7 \ 10^{25}$ m^{-3}. For a gas such as Nitrogen (molecular mass: $M = mN_A = 28$gr) with molecular diameter $d \approx 3$ Å, the mean free path is: $l_m \approx 100$ nm and the self-diffusion constant is : $D = l_m\bar{v}/3 \approx 1.63 \ 10^{-5}$m^2/sec (with $\bar{v} \approx v_{rms} = 500$ m/sec, see section 5.5.1), in agreement with the observed value ($D \approx 1.8 \ 10^{-5}$m^2/sec).

Particle (mass) conservation implies that:

$$\frac{\partial n}{\partial t} + \vec{\nabla} \cdot \vec{J} = 0 \tag{5.224}$$

Since $\vec{J} = -D\vec{\nabla}n$, the evolution of the particle density obeys the diffusion equation (compare to Eq.5.16 when the advection velocity is zero):

$$\frac{\partial n}{\partial t} = D\nabla^2 n$$

When particles are entrained by a flow $\vec{u}$, the particle flux is: $\vec{J} = n\vec{u} - D\vec{\nabla}n$ and mass conservation implies:

$$\frac{\partial n}{\partial t} + \vec{\nabla} \cdot (n\vec{u}) = D\nabla^2 n$$

5.11.1.2 Heat Conduction

Consider now the flux of heat $\delta Q = n\tilde{c}_p\delta T$ (see section 5.5.2) in a gas at constant pressure generated by a temperature gradient: $\partial T/\partial z \neq 0$. Proceeding as previously the heat flux (at constant pressure) through a plane at z results from an imbalance between the heat transported from the upper half space to the lower one and the heat being transported in the reverse direction :

$$J_{Q,z} = \frac{n}{2}\tilde{c}_p \langle T(z+l_{m,z})v_z^- + T(z-l_{m,z})v_z^+ \rangle = -n\tilde{c}_p \langle l_{m,z}v_z^+ \rangle \ \partial T/\partial z$$

$$= -\frac{n\tilde{c}_p l_m\bar{v}}{3}\partial T/\partial z = n\tilde{c}_p D\partial T/\partial z \equiv -\kappa\partial T/\partial z \tag{5.225}$$

and generally: $\vec{J}_Q = -\kappa\vec{\nabla}T$ with:

$$\kappa = n\tilde{c}_p D = \bar{v}\tilde{c}_p/3 \sqrt{2}\sigma$$

---

[39]For the collision of a moving particle with particles at rest (e.g. electron diffusing from impurities within a typical time τ), the mean free path obeys: $v_{rms}\tau \equiv l_m = 1/n\sigma$. For a gas of moving particles one has to consider the relative velocity of two colliding particles $v_{rel} = \sqrt{\langle(\vec{v}_1 - \vec{v}_2)^2\rangle} = \sqrt{2\langle v^2\rangle} = \sqrt{2}v_{rms}$. The volume swept by a particle between collisions is thus: $\sigma v_{rel}\tau = 1/n$ and therefore the mean free path between collisions obey: $\bar{v}\tau \approx v_{rms}\tau \equiv l_m = 1/\sqrt{2}n\sigma$.

Since for a diatomic gas such as Nitrogen $\tilde{c}_p = 7k/2 = 4.83\ 10^{-23}$ J/°K, see Fig.5.15, we derive a value for the heat conductivity of: $\kappa \approx 18.5\ 10^{-3}$ W/°Km, not far from the actual measurement ($\kappa^{meas} = 26\ 10^{-3}$ W/°Km). Notice that κ depends only on the temperature of the gas (via $\bar{v}$), not its pressure. That is different from the self-diffusion constant D which depends on both temperature and particle density (i.e. pressure).

The conduction of heat is not the only determinant in the time required to equilibrate the temperature in a given system. The heat capacity of that system is an important factor too. Energy conservation implies that:

$$\frac{\partial \delta Q}{\partial t} + \vec{\nabla} \cdot \vec{J}_Q = 0 \tag{5.226}$$

Since $\delta Q = n\tilde{c}_p \delta T$ (with $\delta T = T - T_0$ where T_0 is the temperature fixed at one boundary) one notices that the temperature T obeys the diffusion equation:

$$\frac{\partial T}{\partial t} = D_T \nabla^2 T$$

with $D_T = \kappa/n\tilde{c}_p$. A temperature perturbation localized over a region of size ϵ will decay within a typical time $\tau = \epsilon^2/D_T$. Notice that for an ideal gas the particle and temperature diffusion constants are identical: $D = D_T = l_m\bar{v}/3$. For air in standard conditions the thermal diffusivity is thus: $D_T \approx 10^{-5}$m²/sec.

5.11.1.3 Viscosity

Finally consider the estimation of the viscosity η of an ideal gas. The force per unit area F_x/A (the shear force or drag) exerted on a given slab of gas at position z when subjected to a shear $\partial v_x/\partial z > 0$ is equal to the momentum transfer mv_x across the plane at z:

$$F_x/A = \frac{n}{2} \langle mv_x(z+l_{m,z})v_z^- + mv_x(z-l_{m,z})v_z^+ \rangle$$

$$= -mn\langle l_{m,z}v_z^+ \rangle \partial v_x/\partial z = -\frac{mnl_m\bar{v}}{3}\partial v_x/\partial z$$

$$= -\frac{m\bar{v}}{3\sqrt{2}\sigma}\partial v_x/\partial z \equiv -\eta \partial v_x/\partial z \tag{5.227}$$

The shear force F_x is directed against the flow just above the z–plane (where the faster moving flow is dragged by the slower moving flow below) and along the flow just below it (as the slower moving flow is entrained by the faster moving one above), a reflection of the equality of action and reaction. For Nitrogen gas, the estimated value of the viscosity: $\eta = m\bar{v}/3\sqrt{2}\sigma \approx 1.8\ 10^{-5}$Pa·sec agrees with the observed value ($\eta^{meas} = 1.75\ 10^{-5}$Pa·sec). In CGS the unit of viscosity is known as the Poise: 1 Poise $= 0.1$ Pa·sec. Notice that the viscosity depends only on the temperature of the gas (via $\bar{v}$) and not on its pressure: the viscosity of a very dilute gas is the same as that of a highly compressed one. For fluids, the viscosity is two orders of magnitude higher, but its estimation is controversial and complicated by the interactions between fluid particles. For water the measured viscosity at 20°C is $0.89\ 10^{-3}$Pa·sec $=$ $8.9\ 10^{-3}$Poise. It is strongly dependent on temperature and in contrast to gases which viscosity increases with temperature, the viscosity of water decreases by a factor ~ 4 at 100°C.

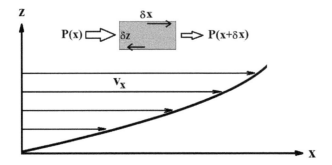

Figure 5.28 Force balance on a fluid element (shaded area) subjected to a pressure gradient along the x-axis and to a shear stress due to the variation of the fluid velocity v_x along the z-axis (due for example to the presence of a wall at $z = 0$.

5.11.2 HYDRODYNAMICS: THE NAVIER-STOKES EQUATIONS

The above considerations and estimates set the foundation for the vast field of hydrodynamics. Consider the flow of an incompressible fluid such as water which has fixed density $\rho = mn$. To determine the acceleration of a fluid element of volume δV, we have to estimate the forces exerted on it. For simplicity let us assume that the fluid is moving along the x-axis under the action of a pressure difference resulting in a force:

$$F_{p,x} = \delta y \delta z [P(x) - P(x + \delta x)] = -\delta V \frac{\partial P}{\partial x}$$

If in addition to a pressure gradient the fluid is also under a shear stress, due for example to the presence of walls on which the fluid velocity is null, the shear force $F_{s,x}$ exerted at a distance z from a wall by the shear along x is given, see Fig.5.28, by the sum of the forces pulling on the top surface of the fluid element and that opposing it on the bottom one:

$$F_{s,x} = \delta x \delta y \eta \left(\frac{\partial v_x}{\partial z} \Big|_{z+\delta z} - \frac{\partial v_x}{\partial z} \Big|_{z} \right) = \delta V \eta \frac{\partial^2 v_x}{\partial z^2}$$

The sum of the pressure and shear forces results in acceleration of a fluid element of mass $\delta m = \rho \delta V$ along the x-direction:

$$\rho \frac{dv_x}{dt} = -\frac{\partial P}{\partial x} + \eta \frac{\partial^2 v_x}{\partial z^2}$$

This equation can be generalized to arbitrary velocities, shear and pressure gradients by replacing v_x by $\vec{v}$, the spatial first derivative by the gradient $\vec{\nabla}$ and the second derivatives by the Laplacian: ∇^2. Since $\vec{v}(t, \vec{r})$ is a function of both space and time the differential:

$$\frac{d\vec{v}}{dt} = \frac{\partial \vec{v}}{\partial t} + (\frac{\partial x}{\partial t}\frac{\partial}{\partial x} + \frac{\partial y}{\partial t}\frac{\partial}{\partial y} + \frac{\partial z}{\partial t}\frac{\partial}{\partial z})\vec{v} = \frac{\partial \vec{v}}{\partial t} + (\frac{d\vec{r}}{dt} \cdot \vec{\nabla})\vec{v} = \frac{\partial \vec{v}}{\partial t} + (\vec{v} \cdot \vec{\nabla})\vec{v}$$

Since $d\vec{r}/dt = \vec{v}$. The general force balance on a fluid element therefore obeys the so-called Navier-Stokes equation:

$$\rho \frac{d\vec{v}}{dt} = \rho \frac{\partial \vec{v}}{\partial t} + \rho(\vec{v} \cdot \vec{\nabla})\vec{v} = -\vec{\nabla}P + \eta \nabla^2 \vec{v} \qquad (5.228)$$

This equation is the foundation of the vast field of fluid dynamics[40]. Although solutions exist in some simple cases (e.g. laminar pipe flow), because of its non-linear character (due to the nonlinear velocity term: $(\vec{v} \cdot \vec{\nabla})\vec{v} \propto v^2$) there are no general solutions to Eq.5.228. Worse, it is not known if depending on initial and boundary conditions a solution always exists and if all solutions are smooth, i.e. if some don't blow up in finite time. Consequently proving the existence and smoothness of solutions to the Navier-Stokes equation is one of the major unsolved problems of Mathematics: there is a prize of $1 Million for whoever provides a proof or a counter-example, i.e. a blowing-up solution. Nonetheless, this equation is used in all numerical simulations of aerodynamic (or hydrodynamic) flows around airplanes (or boats). It is essential in computing the aerodynamic lift on the wings of a plane or the viscous drag on a boat. It is also central in simulations of the weather where predicting the movement of clouds, hurricanes, etc. is a major issue.

5.11.3 DISSIPATION-FLUCTUATION THEOREM

Since the 1980's various methods (such as atomic force microscopy and optical, magnetic and acoustical tweezers) have emerged that allow for the visualization and manipulation of single atoms and molecules[75]. These methods use a microscopic sensor (cantilever or bead) to exert a force and/or monitor the position/extension of

---

[40]For example the sound velocity c_s can be derived from Eq.5.228 when the dissipation term (last term on the right) is negligible versus the inertial term: $\rho dv/dt$. In absence of dissipation sound propagation is an adiabatic process. Since the flux $\vec{J} = \rho \vec{c}_s$ is constant: $\rho dc_s/dt = -c_s d\rho/dt = -dP/dx = -dP/c_s dt$ from which one derives $c_s^2 = dP/d\rho = \gamma P/\rho = 1/\kappa_a \rho$, where κ_a is the adiabatic compressibility, see footnote 11. In standard conditions ($P = 10^5$ N/m^2, $T = 0°$C) $\rho = 1.22$ kg/m^3 and the sound velocity in air is $c_s = 340$ m/sec.

the molecule. The extent of the noise associated to the exerted force or measured position (i.e. the fluctuations of the sensor) is proportional as we shall show, to the friction (dissipation) experienced by the sensor. The smaller the dissipation, the smaller the fluctuations, the more precise is the measurement. To demonstrate that point we shall revisit the Brownian motion[103] of a small particle of mass m and radius r immersed in a viscous fluid of viscosity η and subjected to an external force F_{ex} and to the random shocks of the surrounding fluid F_{ran}. The equation of motion of that particle along any axis follows Newton's law in a viscous medium, also known as Langevin's equation:

$$m\frac{dv}{dt} = -\gamma v + F_{ex} + F_{ran} \tag{5.229}$$

where γ is the friction on the particle due to the viscous fluid[41]. As the dimension of the viscosity η is pressure·time for γv to have dimension of force: $\gamma \sim \eta r$ (the exact result for a spherical particle (due to Stokes) is: $\gamma = 6\pi\eta r$). Since the shocks on the particle from the surrounding fluid molecules (the Langevin force) are uncorrelated: $\langle F_{ran}(t) \cdot F_{ran}(0) \rangle = g\delta(t)$, where g sets the strength of the force fluctuations on the particle (which are proportional to the friction γ as we shall show below). To solve for the velocity it helps to take the Fourier transform of Eq.5.229:

$$im\omega\tilde{v} + \gamma\tilde{v} = \tilde{F}_{ex}(\omega) + \tilde{F}_{ran}(\omega)$$

Let us assume for now that $F_{ex} = 0$, then: $\tilde{v} = \tilde{F}_{ran}(\omega)/(im\omega + \gamma)$. For uncorrelated shocks, the Langevin force spectrum is: $\langle |\tilde{F}_{ran}(\omega)|^2 \rangle = g$, where g the spectral density is expressed in units of N^2/Hz. Using Parseval's theorem, see Appendix A.4, we can deduce the velocity fluctuations:

$$\langle v^2 \rangle = \int dt \langle v(t)^2 \rangle = \frac{1}{2\pi} \int_{-\infty}^{\infty} d\omega \frac{\langle |\tilde{F}_{ran}(\omega)|^2 \rangle}{m^2\omega^2 + \gamma^2} = \frac{g}{2\pi\gamma^2} \int_{-\infty}^{\infty} \frac{d\omega}{(m\omega/\gamma)^2 + 1}$$

$$= \frac{g}{2\pi m\gamma} \int_{-\infty}^{\infty} \frac{dx}{x^2 + 1} = \frac{g}{2m\gamma} \tag{5.230}$$

From the equipartition theorem: $\langle v^2 \rangle = kT/m$, we deduce that: $g = 2kT\gamma$. This relation is known as the dissipation-fluctuation theorem: it relates the strength of the force fluctuations g on the micro-bead (sensor) to the friction (dissipation) γ. The more one averages over the environmental shocks, i.e. the smaller the low-pass frequency window δf, the smaller the variance of the Langevin force:

$$\langle \delta F^2 \rangle_{\delta f} = \frac{1}{2\pi} \int_{-2\pi\delta f}^{2\pi\delta f} \langle |\tilde{F}_{ran}(\omega)|^2 \rangle \, d\omega = 4kT\gamma\delta f$$

---

[41] All motions on Earth are subject to a viscous drag γv which is why Aristotle (as many philosophers since) believed that motion required a mover: the viscous drag had to be countered by an external force F_{ex}. The random force F_{ran} due to shocks from the surrounding molecules was intuited by Lucretius more than 2000 years ago, see section 5.1.1.

A similar relation relates the voltage fluctuations in an RC circuit to the dissipation (the resistance R): $\langle \delta V^2 \rangle_{\delta f} = 4kTR\delta f$. These considerations are crucial to estimate the signal to noise ratio in a number of systems (for example in single molecule manipulation studies or in a radio receiver with frequency bandwidth $2\delta f$, see section 3.3.2.4).

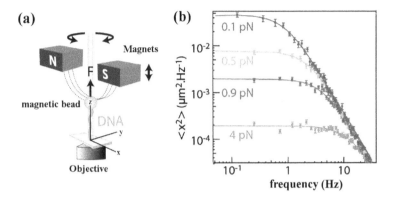

(a) Magnets N F S magnetic bead DNA Objective

(b) 0.1 pN, 0.5 pN, 0.9 pN, 4 pN, $\langle x^2 \rangle$ ($\mu m^2 \cdot Hz^{-1}$), frequency (Hz)

Figure 5.29 (a) Schematics of a magnetic trap for single DNA manipulations. A magnetic bead (used both as an actuator to pull on the molecule and a sensor to measure its extension) is tethered to a surface by a single DNA molecule of extension L and pulled by the magnetic field of a pair of small magnets. The force F exerted on the bead is proportional to the magnetic field gradient. As in a pendulum the tangential restoring force is $F_x = -(F/l)x \equiv -\kappa x$, where x is the displacement of the bead from its equilibrium position. (b) The spectrum $\langle |x(\omega)|^2 \rangle$ of the bead's fluctuations as a function of the force on the bead. At low frequencies ($\omega < \omega_0 = \kappa/\gamma$) the fluctuations are independent of ω and increase at low forces (low κ) whereas at high frequencies they decay as ω^{-2}, as for a freely diffusing particle.

Suppose now that the particle is trapped in a potential (as in all single molecule manipulation methods) and thus subjected to a restoring force $F_{ex} = -\kappa x$ (where κ is the elastic stiffness of the trap), see Fig.5.29. The equation of motion for its fluctuations along x may then be written as:

$$m\frac{d^2x}{dt^2} + \gamma\frac{dx}{dt} + \kappa x = F_{ran}$$

The solution obtained in Fourier space is written as:

$$\langle |\tilde{x}(\omega)|^2 \rangle = \frac{2kT\gamma}{|-m\omega^2 + i\gamma\omega + \kappa|^2}$$

If the dissipation term is much larger than the inertial term (as is often the case), i.e. if $\gamma > \sqrt{m\kappa}$, we may neglect the inertial term:

$$\langle |\tilde{x}(\omega)|^2 \rangle = \frac{2kT\gamma}{\gamma^2\omega^2 + \kappa^2}$$

The spectrum of fluctuations is known as a Lorentzian. At short times (or high frequencies $\omega \gg \omega_0 = \kappa/\gamma$) the behavior is diffusive: $\langle|\tilde{x}(\omega)|^2\rangle = 2kT/(\gamma\omega^2)$. On times $t \ll \omega_0^{-1}$ the particle diffuses over a distance $\langle x^2\rangle = 2(kT/\gamma)t = 2Dt$ where the diffusion coefficient (compare with Eq.5.13):

$$D = kT/\gamma \qquad\qquad (5.231)$$

This relation between diffusion and dissipation was first derived by Einstein in 1905 and many considered it then as the definite proof[42] of the existence of atoms[103]. At low frequencies: $\langle|\tilde{x}(\omega)|^2\rangle = 2kT\gamma/\kappa^2$. When averaged over a frequency window $|\delta f| < \omega_0/2\pi$ the extent of the particle fluctuations

$$\langle x^2\rangle = \frac{4kT\gamma\delta f}{\kappa^2}$$

increase with the dissipation[43]. These considerations are essential in the choice and design of the appropriate sensor for single molecule manipulations and in the analysis of the data coming out of these experiments. The smaller the dissipation γ, the larger the trap stiffness κ and the longer the averaging time (the smaller δf), the smaller the sensor fluctuations $\langle x^2\rangle$ and the more precise the measurements.

The set-up shown in Fig.5.29, has been used to study the elastic properties of a DNA chain (see section 5.10.3). By measuring the transverse fluctuations $\langle x^2\rangle$ of a magnetic bead tethered to a surface by a DNA molecule of extension l and subjected to the traction F of a magnetic field, the trap elasticity $\kappa = F/l$ could be estimated and thus the force F deduced (from the measurement of l). Thus the WLC model of polymer could be tested and validated as an accurate description of the behavior of double-stranded DNA. This validation has opened the way to a detailed study of DNA associated molecular motors (DNA polymerases, helicases, topoisomerases, etc.) using their alteration of the molecule's extension as a proxy for their activity[75].

Einstein relation, Eq.5.231, gave rise for a few years to a peculiar paradox related to DNA segregation during cell division. DNA is a very long molecule (the total DNA length in a human cell is about 3m long, split among 23 pairs of chromosomes). Prior to cell division all chromosomes are duplicated and then segregated along the division plane of the cell. Since DNA in the nucleus is highly entangled (think of a dish of spaghetti), an estimate of the time required to dis-entangle the molecule is many orders of magnitude larger than the cell division time[92, 93]. Indeed for a DNA molecule to segregate from such a pack, constrained as it is by

---

[42] Einstein's relation connected apparently disparate phenomena: diffusion, temperature, the specific heat of gases and Avogadro's number (via $k = R/N_A$)

[43] Notice however that the overall fluctuations in x are independent of the dissipation and obey the equipartition theorem:

$$\langle x^2\rangle = \frac{1}{2\pi}\int_{-\infty}^{\infty} d\omega \frac{2kT\gamma}{\gamma^2\omega^2 + \kappa^2} = kT/\kappa$$

the other molecules, it will have to slither along its whole length l while experiencing a drag $\gamma \sim \eta l$ (with $\eta \approx 0.01$ Poise). From Eq.5.231, diffusion via reptation for a $l = 1$cm DNA would take a time $t_{rep} = l^2/2D = l^2\gamma/2kT = \eta l^3/2kT \sim 10^4$ years(!), much longer than the typical division time of about one day (for a human cell). This conundrum was solved with the discovery by J.Wang in 1971 of topoisomerases, a class of enzymes that actively pass one DNA molecule through another, thus allowing them to dis-entangle like "phantom" chains (non-avoiding random walks). In such case, modeling DNA molecules as random coils with radius of gyration: $R_g = \sqrt{lb}$ (where $b = 100$nm is the DNA Kuhn length), see section 5.1.1, allows for an estimate of their dis-entanglement time

$$t_{phan} = R_g^2/2D = R_g^2\gamma/2kT = 6\pi\eta R_g^3/2kT = 3\pi\eta(lb)^{3/2}/kT \sim 1 \text{ day}$$

From a technological viewpoint, notice that polymer entanglement and van der Waals attraction, see section 5.10.2, is the cause of the stickiness of some glues. These often consists of long polymer chains that by being entangled and attracted to the surfaces they bind ensure their adhesion[92].

5.12 BIOLOGY

Biological systems are the quintessential out-of-equilibrium systems: life dissipates energy and generates entropy. Equilibrium is death. Yet in contrast with physical non-equilibrium systems (weather, earthquakes, etc.), living systems are predictable and robust to changes in initial or boundary conditions. Organisms develop in a very orderly and reproducible way: the fertilized egg of a fly gives rise to a fly, that of a human to a baby. Bacteria respond reproducibly to changes in their source of food, osmotic pressure, temperature, etc. In that sense biological systems resemble man-made machines (those engineered out-of-equilibrium systems) and have indeed been compared to them for a long time, using analogies to clocks and automatons when these were fashionable or to computers when these were invented.

There is however a fundamental difference between biological systems (be they molecular motors, cells, organisms, insect-societies) and engineered systems. The former have arisen from tinkering by evolution, rather than being optimally designed for some goal. Biological systems are often closer to Rube Goldberg contraptions than well designed engineered solutions. As coined by J.Monod, biological systems are a result of Chance and Necessity[104]: random mutations that affect function and physico-chemical constraints that select among these mutations the one best fit to reproduce. Since the space of possible mutations is much larger than the population of competing individuals, the selected solution is often not the optimal one.

This has two important consequences. First, biological systems are history dependent (contingent). As formulated by T.Dobzhansky: "Nothing in Biology makes sense except in the light of evolution"[105]. This fact invalidates statistical mechanics approaches to most biological systems: we are the lucky winners of the evolutionary roulette and have no way to replay that game. Second, extremal principles often do not work in Biology, in contrast with physics where they are the driving force of the dynamics (e.g. minimization of energy or action, maximization of entropy). Except for being better at reproduction than the competition, we do not know what principles (if any) underlie the complexity of life-forms. This is in contrast with engineering solutions often designed to minimize energy consumption while providing for more resilient, faster, or stronger machines.

In Biology, the unity of Science is reflected not in the miraculous explanatory power of Mathematics, but in the no-less amazing unity of building blocks and designs among all life-forms[105, 106, 107]. They all seem to share a common ancestor, use the same DNA, RNA, genetic code and amino-acids to build cells and organisms using very similar designs. To quote J.Monod again: "what is true for the bacteria is true for the elephant". Once a design has been selected by evolution, it is usually kept and at best tinkered with, rarely dropped to be replaced by a radically new design, see Fig.5.30. Understanding biological systems is therefore a problem in reverse engineering: understanding how a given function is actually realized, not how it can be intelligently (i.e. optimally) designed.

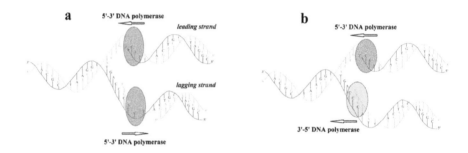

Figure 5.30   (a) DNA is a double-strand molecule, each strand running anti-parallel to its opposite complementary strand. The replication of DNA is performed by an enzyme, a 5'-3' DNA polymerase which synthesizes a strand complementary to the existing one from its 5' (phosphate) end to its 3' (hydroxyl) end. Replication thus proceed continuously on the leading 3'-5' strand and discontinuously on the lagging 5'-3' strand. The replication of the lagging strand is a complex process involving many enzymes and is consequently more prone to replication errors than replication of the leading strand. (b) A conceptually simpler replication mechanism (not chosen by evolution) involving the continuous synthesis of the complementary strands by two different DNA polymerases: the regular 5'-3' DNA polymerase and a hypothetical 3'-5' DNA polymerase.

These considerations notwithstanding, biological systems can be understood and described by physico-chemical concepts[108]. As we have seen DNA is an ideal Worm Like Chain polymer (see section 5.10.3), neuronal conduction can be understood in terms of Nernst law, Eq.5.141 and biochemical reactions are not different from chemical reactions. The description of the dynamics of biological systems (e.g. developmental networks) as non-linear dynamical systems (described by appropriate

Ordinary or Partial Differential Equations (ODEs or PDEs)) is by now well established and often quite successful in generating testable predictions.

The information theory approach described here has also been used to model the outcome of complex biological systems such as proteins and neural networks. Proteins (and neural networks) are not random polymers (or networks): they have been selected by evolution to perform a certain task. In a protein of size N performing a certain function (DNA replication, signaling, enzymatic catalysis, etc.) the frequency of occupancy (or non-occupancy) of a given site i by one of the 20 possible amino-acids is not uniform. A mutation (e.g. replacing one amino-acid by an other) can completely alter the protein function and could have been selected against or accompanied by a compensatory mutation elsewhere in the protein that would have restored the function. In that sense the protein "die" is loaded: certain amino-acids are preferred at certain sites. Similarly in a neural network of size N, neurons do not fire at random. The connections between neurons have been selected by evolution to ensure that their activity is coordinated to produce a desired outcome given a certain input. Here again the neuronal "die" has been loaded: certain activities and connections are preferred at certain nodes of the network.

As in our approach of loaded dice (see section 5.2.2), the current approach to these biological systems is to use entropy maximization to best fit a model to the observed data, for example the frequency $F_o(\sigma_i)$ of observing a given amino-acid (or vacancy) σ_i at a certain position i in a protein family performing the same function in different organisms (or the frequency of the firing state σ_i of neuron (node) i in a neural network). The best-fit model can then be used to make testable predictions on the structure or function of new proteins or the outcome of a neural network to novel inputs.

The goal therefore is to find the probability distribution $\mathcal{P}_m(\{\sigma_i\})$ which maximizes the entropy $S = -\mathcal{P}_m \log \mathcal{P}_m$ under some observable constraints, such as the mean firing rate (or amino-acid) at site i:

$$\sum_{\{\sigma_i\}} \sigma_i F_o(\{\sigma_i\}) \equiv \langle \sigma_i \rangle_o$$

and the correlation between firing rates (amino-acids) between two sites ($i < j$):

$$\sum_{\{\sigma_i\}} \sigma_i \sigma_j F_o(\{\sigma_i\}) \equiv \langle \sigma_i \sigma_j \rangle_o$$

Obviously higher moments can also be considered, but that unduly multiplies the number of fit parameters. As we have seen before (see section 5.10.1) the probability distribution is given by:

$$\mathcal{P}_m(\sigma_i) = \frac{1}{Z} \exp\{\sum_i \lambda_i \sigma_i + \sum_{i<j} \lambda_{ij} \sigma_i \sigma_j\}$$

where λ_i and λ_{ij} are Lagrange multipliers associated with the observables which have to be best fitted to the data (as we have done for loaded dice), namely they have to minimize the Kullback-Leibler divergence between the observed frequencies and modeled probabilities, Eq.5.31:

$$D(F_o\|\mathcal{P}_m) \equiv -\sum_{\{\sigma_i\}} F_o(\{\sigma_i\})\log\mathcal{P}_m(\{\sigma_i\})/F_o(\{\sigma_i\}) \qquad (5.232)$$

which implies:

$$\frac{\partial D(F_o\|\mathcal{P}_m)}{\partial \lambda_l} = \sum_{\{\sigma_i\}} (\mathcal{P}_m - F_o)\sigma_l = \langle\sigma_l\rangle_m - \langle\sigma_l\rangle_o = 0 \qquad (5.233)$$

$$\frac{\partial D(F_o\|\mathcal{P}_m)}{\partial \lambda_{kl}} = \sum_{\{\sigma_i\}} (\mathcal{P}_m - F_o)\sigma_k\sigma_l = \langle\sigma_k\sigma_l\rangle_m - \langle\sigma_k\sigma_l\rangle_o = 0 \qquad (5.234)$$

Finding a minima of $D(F_o\|\mathcal{P}_m)$ can be done by a steepest descent algorithm (which however may get stuck in a local minimum) or by Monte Carlo methods (see appendix A.9.4). Due to the large number of fit parameters, this is often a time consuming computation, which success and accuracy depends to a large extent on the size (and absence of bias) of the sample used to compute the frequencies $F_o(\{\sigma_i\})$. With a fast growing database of protein folds and full structures (tens of thousands), this maximal entropy approach begins to make sense and to yield testable predictions.

However as pointed out by Jonas and Kording in a provocative paper[109] even if a maximal entropy approach can simulate the data coming out of a microprocessor performing a certain task (playing a given game) and can make testable predictions it fails in providing an understanding of the logic of the network. A similar problem plagues the understanding of deep learning networks, which are neural networks made of layers of neuronal (integrate and fire) nodes that feed-forward to the next layer. These networks can be trained to produce desired outputs for given inputs (by using back-error propagation to best-fit the couplings λ_{ij} between neurons from adjacent layers). While these networks are very successful at learning a specific task (e.g. translating a text, recognizing a face, winning at Chess or Go, etc.), we do not understand how the network does it. If we cannot understand or reverse-engineer man-made networks with a clear logic (microprocessors) or architecture (deep learning networks), how can we expect to be successful at understanding and reverse-engineering a neural network that is the outcome of tinkering by evolution?

A Appendix

A.1 SOME PHYSICAL CONSTANTS

name	symbol	value (CGS)	value (MKS)	comments
Gravitational constant	G	$6.67 \ 10^{-8}$ dynes cm^2/g^2	$6.67 \ 10^{-11}$ N m^2/kg^2	gravitational acceleration on Earth: $g = 9.8$m/s^2
velocity of light	c	$2.997 \ 10^{10}$ cm/s	$2.997 \ 10^{8}$ m/s	
Planck's constant	h $\hbar$	$6.626 \ 10^{-27}$ erg s $1.054 \ 10^{-27}$ erg s	$6.626 \ 10^{-34}$ J s $1.054 \ 10^{-34}$ J s	$4.135 \ 10^{-15}$ eV s $\hbar \equiv h/2\pi = 6.583 \ 10^{-16}$ eV s
Boltzmann constant	k_B or k	$1.38 \ 10^{-16}$ erg/°K	$1.38 \ 10^{-23}$ J/°K	at 20°C ($T = 293$°K): $k_B T = 4 \ 10^{-14}$erg $= 4 \ 10^{-21}$ J $= 25$ meV
Stefan-Boltzmann constant	σ_{SB}	$5.6704 \ 10^{-5}$ erg cm^{-2}s^{-1}°K^{-4}	$5.6704 \ 10^{-8}$ J m^{-2}s^{-1}°K^{-4}	$\sigma_{SB} \equiv \pi^2 k_B^4/60\hbar^3 c^2$
Avogadro's number	N_A	$6.022 \ 10^{23}$ particles		no. particles in 1 mole
electron/proton charge	e	$4.8 \ 10^{-10}$ statC	$1.6 \ 10^{-19}$ C	fine structure constant: $\alpha \equiv e^2/\hbar c \approx 1/137$
electron mass	m_e	$9.109 \ 10^{-28}$ g	$9.109 \ 10^{-31}$ kg	0.51 MeV/c^2
proton mass	m_p	$1.672 \ 10^{-24}$ g	$1.672 \ 10^{-27}$ kg	938 MeV/c^2
Bohr's radius	r_0	$0.53 \ 10^{-8}$ cm	$0.53 \ 10^{-10}$ m	$\hbar^2/me^2 = 0.53$ Å
Bohr magneton	μ_B	$9.284 \ 10^{-21}$ erg/G	$9.284 \ 10^{-24}$ J/T	$\mu_B \equiv \frac{e\hbar}{2m_e c} = 5.788 \ 10^{-5}$ eV/T

DOI: 10.1201/9781003218999-A

name	symbol	value (CGS)	value (MKS)	comments
Sun's radius	R_S	$6.96\ 10^{10}$ cm	$6.96\ 10^8$ m	also denoted $R_\odot$
Sun Mass	M_S	$1.988\ 10^{33}$ gr	$1.988\ 10^{30}$ kg	also denoted $M_\odot$
Sun-Earth distance	R_{SE}	$1.496\ 10^{13}$ cm	$1.496\ 10^{11}$ m	
Earth radius	R_E	$6.371\ 10^8$ cm	$6.371\ 10^6$ m	
Earth mass	M_E	$5.972\ 10^{27}$ g	$5.972\ 10^{24}$ kg	
Thermal de-Broglie wavelength (electron)	$\lambda_{DB,e}$	$\approx 44\ 10^{-8}$cm	$\approx 44\ 10^{-10}$m	$\lambda_{DB,e} = h/\sqrt{2\pi m_e kT}$ ≈ 44Å (at : $T = 300°$K)
Thermal de-Broglie wavelength (proton)	$\lambda_{DB,p}$	$\approx 1\ 10^{-8}$cm	$\approx 1\ 10^{-10}$m	$\lambda_{DB,p} = h/\sqrt{2\pi m_p kT}$ ≈ 1Å (at : $T = 300°$K)

A.2   LINEAR ALGEBRA

In the following, I shall recall a few concepts from Linear Algebra of relevance to the material presented here. The solution of a linear equation $Ax = b$ (where A is a matrix and x, b are vectors) can be obtained by successive eliminations of the unknowns, a procedure known as Gauss-Jordan elimination and exemplified below:

$$
\begin{array}{llcll}
(a) & 2x_1 + 2x_2 - x_3 = 3 & \rightarrow & (a') & 2x_1 + 2x_2 - x_3 = 3 \\
(b) & -2x_1 + x_2 - x_3 = 1 & \text{add (a) : } (b') & & 3x_2 - 2x_3 = 4 \\
(c) & 2x_1 - x_2 + 2x_3 = 0 & \text{subtract (a) : } (c') & & -3x_2 + 3x_3 = -3
\end{array}
$$

$$
\begin{array}{lcll}
& \rightarrow & (a'') & 2x_1 + 2x_2 - x_3 = 3 \\
& \rightarrow & (b'') & 3x_2 - 2x_3 = 4 \\
\text{add (b') : } & & (c'') & x_3 = 1
\end{array}
$$

which can be summarized in a series of subtractions of successive rows in the $(n + 1) \times n$ matrix formed by A and the vector b as a new column:

$$
\begin{pmatrix} 2 & 2 & -1 & 3 \\ -2 & 1 & -1 & 1 \\ 2 & -1 & 2 & 0 \end{pmatrix} \rightarrow \begin{pmatrix} 2 & 2 & -1 & 3 \\ 0 & 3 & -2 & 4 \\ 0 & -3 & 3 & -3 \end{pmatrix} \rightarrow \begin{pmatrix} 2 & 2 & -1 & 3 \\ 0 & 3 & -2 & 4 \\ 0 & 0 & 1 & 1 \end{pmatrix}
$$

Once the nth unknown is determined (here $x_3 = 1$) the others can be found by substitution of the successively determined variables (here $x_2 = 2$ and then $x_1 = 0$).

Determining the inverse of a matrix A is equivalent to solving n sets (for the n columns of A^{-1}) of n equations in n unknowns (the values of the components of a given column of $\mathbf{A}^{-1}$). The inverse of a matrix $\mathbf{A}$ can thus be determined by the Gauss-Jordan elimination method on the $2n \times n$ matrix formed by the combination of $\mathbf{A}$ and $\mathbf{I}$. For example in the previous example:

$$\begin{pmatrix} 2 & 2 & -1 & 1 & 0 & 0 \\ -2 & 1 & -1 & 0 & 1 & 0 \\ 2 & -1 & 2 & 0 & 0 & 1 \end{pmatrix} \to \begin{pmatrix} 2 & 2 & -1 & 1 & 0 & 0 \\ 0 & 3 & -2 & 1 & 1 & 0 \\ 0 & -3 & 3 & -1 & 0 & 1 \end{pmatrix} \to \begin{pmatrix} 2 & 2 & -1 & 1 & 0 & 0 \\ 0 & 3 & -2 & 1 & 1 & 0 \\ 0 & 0 & 1 & 0 & 1 & 1 \end{pmatrix}$$

One then proceeds backwards from the nth row by eliminating the off diagonal elements:

$$\begin{pmatrix} 2 & 2 & -1 & 1 & 0 & 0 \\ 0 & 3 & -2 & 1 & 1 & 0 \\ 0 & 0 & 1 & 0 & 1 & 1 \end{pmatrix} \to \begin{pmatrix} 2 & 2 & 0 & 1 & 1 & 1 \\ 0 & 3 & 0 & 1 & 3 & 2 \\ 0 & 0 & 1 & 0 & 1 & 1 \end{pmatrix} \to \begin{pmatrix} 2 & 0 & 0 & 1/3 & -1 & -1/3 \\ 0 & 3 & 0 & 1 & 3 & 2 \\ 0 & 0 & 1 & 0 & 1 & 1 \end{pmatrix}$$

The product of the diagonal elements in the left half of the above matrix yields the determinant of $\mathbf{A}$ (i.e. $\det\mathbf{A} \equiv |\mathbf{A}| = 6$). Finally dividing by the diagonal values yields:

$$\begin{pmatrix} 1 & 0 & 0 & 1/6 & -1/2 & -1/6 \\ 0 & 1 & 0 & 1/3 & 1 & 2/3 \\ 0 & 0 & 1 & 0 & 1 & 1 \end{pmatrix}$$

The right half of this matrix is the inverse of $\mathbf{A}$:

$$\mathbf{A}^{-1}\mathbf{A} = \begin{pmatrix} 1/6 & -1/2 & -1/6 \\ 1/3 & 1 & 2/3 \\ 0 & 1 & 1 \end{pmatrix}\begin{pmatrix} 2 & 2 & -1 \\ -2 & 1 & -1 \\ 2 & -1 & 2 \end{pmatrix} = \begin{pmatrix} 1 & 0 & 0 \\ 0 & 1 & 0 \\ 0 & 0 & 1 \end{pmatrix}$$

The determinant of $\mathbf{A}$ can also be computed from the Levi-Civita equation

$$|\mathbf{A}| = \sum_{i_1,i_2,\dots,i_n=1}^{n} \epsilon_{i_1 \dots i_n} A_{1,i_1} A_{2,i_2} \dots A_{n,i_n} \tag{A.1}$$

where $\epsilon_{i_1 \dots i_n} = 1$ if the permutation of the indices is even, -1 if it is odd and 0 if two indices are equal. An alternative is to compute the determinant via the recursion relation:

$$|\mathbf{A}| = \sum_{i=1}^{n} (-1)^{i+1} A_{1,i} |\mathbf{C_i}|$$

Where $\mathbf{C_i}$ is a the $(n-1) \times (n-1)$ cofactor matrix built from $\mathbf{A}$ by removing the terms in row 1 and column i. Thus for a 3×3 matrix:

$$\begin{vmatrix} a & b & c \\ d & e & f \\ g & h & i \end{vmatrix} = a \begin{vmatrix} e & f \\ h & i \end{vmatrix} - b \begin{vmatrix} d & f \\ g & i \end{vmatrix} + c \begin{vmatrix} d & e \\ g & h \end{vmatrix}$$

$$= a(ei - hf) - b(di - gf) + c(dh - ge)$$

Some useful properties of the determinant are: $|\mathbf{AB}| = |\mathbf{A}||\mathbf{B}|$ and thus $|\mathbf{A}^{-1}\mathbf{A}| = |\mathbf{A}^{-1}||\mathbf{A}| = 1$; $|\mathbf{A}^T| = |\mathbf{A}|$ (where $\mathbf{A}^T$ is the transpose of $\mathbf{A}$, see below); $|c\mathbf{A}| = c^n|\mathbf{A}|$ (where c is a number).

A.2.1 EIGENVALUES AND EIGENVECTORS.

A central problem in Quantum Mechanics is finding solutions to the following equation:

$$\mathbf{A}v = \lambda v \qquad \text{or} \qquad (\mathbf{A} - \lambda\mathbf{I})v = 0 \tag{A.2}$$

The values of λ for which non-trivial solutions (i.e. $v \neq 0$) exist are known as the eigenvalues of $\mathbf{A}$ and the vectors v are its eigenvectors. Non-trivial solutions exist only if $\det(\mathbf{A} - \lambda\mathbf{I}) = 0$. For an $n \times n$ matrix, finding its eigenvalues implies finding the n zeros of a polynomial of order λ^n. In Quantum Mechanics $\mathbf{A}$ is a Hermitian matrix (one satisfying $A_{ji} = A_{ij}^*$) and its n eigenvalues are real numbers (the results of measurements), which can always be found numerically. Once the eigenvalues $\lambda^{(i)}$ have been found, the eigenvectors $v^{(i)}$ can be determined from Eq.A.2 by setting the value of one of the components (say $v_1^{(i)} = 1$) and solving the resulting $n - 1$ equations in $n - 1$ unknown (the other components of $v^{(i)}$) as described above. The eigenvectors are then normalized, i.e. divided by their norm:

$$|v^{(i)}| = \sqrt{\sum_j v_j^{(i),*} v_j^{(i)}}$$

The $n \times n$ matrix $\mathbf{S}$ which columns are the normalized eigenvectors of $\mathbf{A}$, diagonalizes $\mathbf{A}$:

$$\mathbf{AS} = \mathbf{\Lambda S} \qquad \text{namely} \qquad \mathbf{S}^T\mathbf{AS} = \mathbf{\Lambda} \tag{A.3}$$

where $\mathbf{\Lambda}$ is a matrix whose diagonal elements are the eigenvalues of $\mathbf{A}$. Since the eigenvectors are normalized the inverse of $\mathbf{S}$ is its transpose: the matrix which rows are the complex conjugate of the eigenvectors ($S_{ij}^T = S_{ji}^*$):

$$\mathbf{S}^{-1}\mathbf{S} = \mathbf{S}^T\mathbf{S} = \mathbf{I}$$

For example the eigenvalues of the matrix:

$$\mathbf{A} = \begin{pmatrix} 1 & 1 & 0 \\ 1 & 2 & 1 \\ 0 & 1 & 1 \end{pmatrix}$$

satisfy the equation: $(1 - \lambda)[(2 - \lambda)(1 - \lambda) - 2] = (1 - \lambda)\lambda(\lambda - 3) = 0$ which solutions are: $\lambda = 0, 1, 3$. The normalized eigenvector $v^{(1)}$ associated with eigenvalue $\lambda_1 = 0$ is: $v^{(1)} = (1, -1, 1)/\sqrt{3}$; the eigenvector associated with $\lambda_2 = 1$ is: $v^{(2)} = (1, 0, -1)/\sqrt{2}$;

the eigenvector associated with $\lambda_3 = 3$ is: $v^{(3)} = (1,2,1)/\sqrt{6}$. The diagonalizing matrix $\mathbf{S}$ is thus:

$$\mathbf{S} = \frac{1}{\sqrt{6}}\begin{pmatrix} \sqrt{2} & \sqrt{3} & 1 \\ -\sqrt{2} & 0 & 2 \\ \sqrt{2} & -\sqrt{3} & 1 \end{pmatrix}$$

Obviously by construction: $\mathbf{S}^{-1} = \mathbf{S}^T$. One can check that $\mathbf{A}$ is diagonalized by $\mathbf{S}$ to yield a matrix $\mathbf{\Lambda}$ which values on the diagonal are the eigenvalues of $\mathbf{A}$.:

$$\mathbf{S}^T\mathbf{A}\mathbf{S} = \frac{1}{6}\begin{pmatrix} \sqrt{2} & -\sqrt{2} & \sqrt{2} \\ \sqrt{3} & 0 & -\sqrt{3} \\ 1 & 2 & 1 \end{pmatrix}\begin{pmatrix} 1 & 1 & 0 \\ 1 & 2 & 1 \\ 0 & 1 & 1 \end{pmatrix}\begin{pmatrix} \sqrt{2} & \sqrt{3} & 1 \\ -\sqrt{2} & 0 & 2 \\ \sqrt{2} & -\sqrt{3} & 1 \end{pmatrix} = \begin{pmatrix} 0 & 0 & 0 \\ 0 & 1 & 0 \\ 0 & 0 & 3 \end{pmatrix} = \mathbf{\Lambda}$$

A.3 VECTOR CALCULUS

Vector calculus is to calculus what vectors are to scalars in algebra. In calculus, the basic component is a scalar function, e.g. $f(x)$, on which various operations (differentiation and integration) are defined. The basic components of vector calculus are vector fields: vectors which are functions of the coordinates, e.g. $\vec{A}(\vec{r}) = (A_x(\vec{r}), A_y(\vec{r}), A_z(\vec{r}))$. In vector calculus the differentiation (or nabla) operator: $\vec{\nabla}$ can act on a scalar function to generate a vector field:

$$\vec{\nabla}\phi \equiv (\partial_x\phi, \partial_y\phi, \partial_z\phi)$$

This operation known as the gradient measures the rate and direction of change in a scalar field. The nabla operator can also act on a vector field to generate a scalar function (like the dot product of vectors in algebra):

$$\vec{\nabla}\cdot\vec{A} = \partial_x A_x + \partial_y A_y + \partial_z A_z$$

This operation known as the divergence measures the magnitude of a source or sink at a given point in a vector field. Finally the differentiation operator can act on a vector field to generate an other vector field (like the cross product of vectors in algebra):

$$\vec{\nabla}\times\vec{A} = (\partial_y A_z - \partial_z A_y)\hat{x} - (\partial_x A_z - \partial_z A_x)\hat{y} + (\partial_x A_y - \partial_y A_x)\hat{z} = \begin{vmatrix} \hat{x} & \hat{y} & \hat{z} \\ \partial_x & \partial_y & \partial_z \\ A_x & A_y & A_z \end{vmatrix}$$

This operation known as the curl or rotational measures the tendency to rotate about a point in a vector field. In calculus one has rules for the operation of the differentiation operator on the sum and product of two functions, e.g. $\partial_x(fg) = (\partial_x f)g + f(\partial_x g)$. Similarly in vector calculus one has rules for the action of the nabla operator on sum and products of scalar functions or vector fields. The following identities are often used:

- $\vec{A} \times \vec{B} = -\vec{B} \times \vec{A}$
- $(\vec{A} + \vec{B}) \cdot \vec{C} = \vec{A} \cdot \vec{C} + \vec{B} \cdot \vec{C}$
- $(\vec{A} + \vec{B}) \times \vec{C} = \vec{A} \times \vec{C} + \vec{B} \times \vec{C}$
- $\vec{A} \cdot (\vec{B} \times \vec{C}) = \vec{C} \cdot (\vec{A} \times \vec{B}) = \vec{B} \cdot (\vec{C} \times \vec{A})$
- $\vec{A} \times (\vec{B} \times \vec{C}) = (\vec{A} \cdot \vec{C})\vec{B} - (\vec{A} \cdot \vec{B})\vec{C}$

- $\vec{\nabla} \times (\vec{\nabla}\phi) = 0$
- $\vec{\nabla}(\psi + \phi) = \vec{\nabla}\psi + \vec{\nabla}\phi$
- $\vec{\nabla}(\psi\phi) = \phi\vec{\nabla}\psi + \psi\vec{\nabla}\phi$
- $\vec{\nabla} \cdot (\psi\vec{A}) = \psi\vec{\nabla} \cdot \vec{A} + \vec{A} \cdot \vec{\nabla}\psi$
- $\vec{\nabla} \times (\psi\vec{A}) = \psi\vec{\nabla} \times \vec{A} - \vec{A} \times \vec{\nabla}\psi$
- $\vec{\nabla} \cdot (\vec{\nabla} \times \vec{A}) = 0$
- $\vec{\nabla} \cdot (\vec{A} + \vec{B}) = \vec{\nabla} \cdot \vec{A} + \vec{\nabla} \cdot \vec{B}$
- $\vec{\nabla} \times (\vec{A} + \vec{B}) = \vec{\nabla} \times \vec{A} + \vec{\nabla} \times \vec{B}$
- $\vec{\nabla} \cdot (\vec{A} \times \vec{B}) = \vec{B} \cdot \vec{\nabla} \times \vec{A} - \vec{A} \cdot \vec{\nabla} \times \vec{B}$
- $\vec{\nabla} \times (\vec{A} \times \vec{B}) = \vec{A}\vec{\nabla} \cdot \vec{B} - \vec{B}\vec{\nabla} \cdot \vec{A} + (\vec{B} \cdot \vec{\nabla})\vec{A} - (\vec{A} \cdot \vec{\nabla})\vec{B}$
- $\vec{\nabla}(\vec{A} \cdot \vec{B}) = (\vec{A} \cdot \vec{\nabla})\vec{B} + (\vec{B} \cdot \vec{\nabla})\vec{A} + \vec{A} \times (\vec{\nabla} \times \vec{B}) + \vec{B} \times (\vec{\nabla} \times \vec{A})$

An other important operator in vector calculus is the second order scalar operator known as the Laplacian: $\Delta \equiv \vec{\nabla} \cdot \vec{\nabla} = \nabla^2$. Notice that:

- $\vec{\nabla} \cdot \vec{\nabla}\phi = \nabla^2\phi$
- $\vec{\nabla} \times (\vec{\nabla} \times \vec{A}) = \vec{\nabla}(\vec{\nabla} \cdot \vec{A}) - \nabla^2\vec{A}$
- $\psi\nabla^2\phi - \phi\nabla^2\psi = \vec{\nabla} \cdot (\psi\vec{\nabla}\phi - \phi\vec{\nabla}\psi)$

In calculus one defines the integral as being the inverse operator of the differentiation: $\int_a^b \partial_x f \, dx = f(b) - f(a)$. In vector calculus integrals can be performed along a line, a surface or a volume with distinct relations to the differentiation operator.

The inverse of the gradient operator is the line integral:

$$\int_a^b \vec{\nabla}\phi \cdot d\vec{r} = \phi(b) - \phi(a) \tag{A.4}$$

The inverse of the curl operator is the surface integral:

$$\oint_S \vec{\nabla} \times \vec{A} \cdot d\vec{S} = \oint_\Gamma \vec{A} \cdot d\vec{r} \tag{A.5}$$

where S is the surface enclosed by the closed line Γ. This equation is also known as Stokes' theorem.

The inverse of the divergence operator is the volume integral:

$$\oint_V \vec{\nabla} \cdot \vec{A} \, dV = \oint_S \vec{A} \cdot d\vec{S} = \oint_S \vec{A} \cdot \hat{n} \, dS \tag{A.6}$$

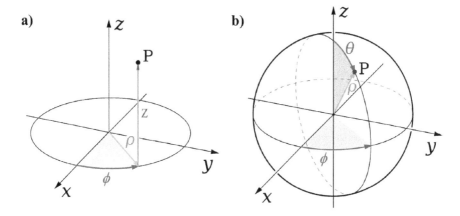

Figure A.1 (a) Cylindrical coordinate system: ρ is the radius in the xy-plane, ϕ the angle between the radius vector and the x-axis and z the vertical coordinate. (b) Spherical coordinate system: the radius ρ is the distance to the origin, θ the angle the radius vector makes with the z-axis and ϕ the angle between the x-axis and the projection of $\vec{\rho}$ unto the xy-plane.

where V is the volume enclosed by the surface S and $\hat{n}$ is the normal to that surface. This equation is also known as the divergence theorem.

Similarly:

$$\oint_V \vec{\nabla} \times \vec{A} \, dV = \oint_S \hat{n} \times \vec{A} \, dS \qquad (\text{A.7})$$

To conclude let us present the various operators (gradient, divergence, curl and Laplacian) in cartesian (x, y, z), cylindrical (ρ, ϕ, z) and spherical (r, θ, ϕ) coordinates, see Fig.A.1 for a definition of these variables.

The gradient in these coordinates is:

$$\vec{\nabla} f = \frac{\partial f}{\partial x} \hat{\mathbf{x}} + \frac{\partial f}{\partial y} \hat{\mathbf{y}} + \frac{\partial f}{\partial z} \hat{\mathbf{z}}$$
$$= \frac{\partial f}{\partial \rho} \hat{\rho} + \frac{1}{\rho} \frac{\partial f}{\partial \phi} \hat{\phi} + \frac{\partial f}{\partial z} \hat{z}$$
$$= \frac{\partial f}{\partial r} \hat{r} + \frac{1}{r} \frac{\partial f}{\partial \theta} \hat{\theta} + \frac{1}{r \sin \theta} \frac{\partial f}{\partial \phi} \hat{\phi}$$

The divergence in these coordinates is:

$$\vec{\nabla} \cdot \vec{A} = \frac{\partial A_x}{\partial x} + \frac{\partial A_y}{\partial y} + \frac{\partial A_z}{\partial z}$$

$$= \frac{1}{\rho} \frac{\partial (\rho A_\rho)}{\partial \rho} + \frac{1}{\rho} \frac{\partial A_\phi}{\partial \phi} + \frac{\partial A_z}{\partial z}$$

$$= \frac{1}{r^2} \frac{\partial (r^2 A_r)}{\partial r} + \frac{1}{r \sin \theta} \frac{\partial}{\partial \theta} (A_\theta \sin \theta) + \frac{1}{r \sin \theta} \frac{\partial A_\phi}{\partial \phi}$$

The curl in these coordinates is:

$$\vec{\nabla} \times \vec{A} = \left(\frac{\partial A_z}{\partial y} - \frac{\partial A_y}{\partial z} \right) \hat{x} + \left(\frac{\partial A_x}{\partial z} - \frac{\partial A_z}{\partial x} \right) \hat{y} + \left(\frac{\partial A_y}{\partial x} - \frac{\partial A_x}{\partial y} \right) \hat{z}$$

$$= \left(\frac{1}{\rho} \frac{\partial A_z}{\partial \phi} - \frac{\partial A_\phi}{\partial z} \right) \hat{\rho} + \left(\frac{\partial A_\rho}{\partial z} - \frac{\partial A_z}{\partial \rho} \right) \hat{\phi} + \frac{1}{\rho} \left(\frac{\partial (\rho A_\phi)}{\partial \rho} - \frac{\partial A_\rho}{\partial \phi} \right) \hat{z}$$

$$= \frac{1}{r \sin \theta} \left(\frac{\partial}{\partial \theta} (A_\phi \sin \theta) - \frac{\partial A_\theta}{\partial \phi} \right) \hat{r} + \frac{1}{r} \left(\frac{1}{\sin \theta} \frac{\partial A_r}{\partial \phi} - \frac{\partial}{\partial r} (r A_\phi) \right) \hat{\theta} + \frac{1}{r} \left(\frac{\partial}{\partial r} (r A_\theta) - \frac{\partial A_r}{\partial \theta} \right) \hat{\phi}$$

Finally, the Laplacian in these coordinates is:

$$\nabla^2 f = \frac{\partial^2 f}{\partial x^2} + \frac{\partial^2 f}{\partial y^2} + \frac{\partial^2 f}{\partial z^2}$$

$$= \frac{1}{\rho} \frac{\partial}{\partial \rho} \left(\rho \frac{\partial f}{\partial \rho} \right) + \frac{1}{\rho^2} \frac{\partial^2 f}{\partial \phi^2} + \frac{\partial^2 f}{\partial z^2}$$

$$= \frac{1}{r^2} \frac{\partial}{\partial r} \left(r^2 \frac{\partial f}{\partial r} \right) + \frac{1}{r^2 \sin \theta} \frac{\partial}{\partial \theta} \left(\sin \theta \frac{\partial f}{\partial \theta} \right) + \frac{1}{r^2 \sin^2 \theta} \frac{\partial^2 f}{\partial \phi^2}$$

A.3.1 TAYLOR EXPANSIONS

Most of the problems encountered in Science are not exactly soluble. Approximate solutions can be obtained using a number of mathematical approaches, one of which is the Taylor expansion around a point where a solution is known. Let $f(x)$ describe a perturbation to a problem soluble for $x = 0$. The idea is then to approximate $f(x)$ near $x = 0$ by its best polynomial fit:

$$f(x) = f(0) + x f^{(1)}(0) + x^2 f^{(2)}(0)/2 + x^3 f^{(3)}(0)/6 + \cdots + x^n f^{(n)}(0)/n! + \ldots$$

where $f^{(n)}(0)$ stands for the nth derivative of $f(x)$ evaluated at $x = 0$. One then looks for an approximate solution for small x. The following Taylor expansions (around

$x = 0$) are commonly used:

$$(1+x)^\alpha = 1 + \alpha x + \alpha(\alpha-1)x^2/2 + \cdots + \alpha(\alpha-1)\ldots(\alpha+1-n)x^n/n! + \ldots$$

$$\log(1+x) = x - x^2/2 + x^3/3 + \cdots + (-1)^{n+1}x^n/n + \ldots$$

$$\sin x = x - x^3/6 + x^5/5! + \cdots + (-1)^n x^{2n+1}/(2n+1)! + \ldots$$

$$\cos x = 1 - x^2/2 + x^4/4! + \cdots + (-1)^n x^{2n}/(2n)! + \ldots$$

$$e^x = 1 + x + x^2/2 + x^3/6 + \cdots + x^n/n! + \ldots$$

The binomial expansion of $(a+b)^n$ can be viewed as a particular example of an exact Taylor expansion:

$$(a+b)^n = a^n + na^{n-1}b + \cdots + \frac{m!(n-m)!}{n!}a^m b^{n-m} + \cdots + nab^{n-1} + b^n$$

A.4 FOURIER TRANSFORMS

In mathematics, the Fourier transform is an operation that decomposes a signal into its constituent frequencies. If the original signal depends on space or time, its representation is called the (space-) time domain representation of the signal, whereas its Fourier transform is called the (spatial or temporal) frequency domain representation of the signal. The Fourier transform is used in all domains of Science and Engineering, since for linear systems, knowing the response in the frequency domain allows one to deduce the response to any space or time dependent signal.

For a function $f(u)$ which is square integrable: $\int_{-\infty}^{\infty} du|f(u)|^2 < \infty$, the Fourier transform always exists (essentially all physical functions). For time varying signals $f(t)$ the Fourier transform is defined as:

$$\mathcal{F}_t(f) \equiv \tilde{f}(\omega) = \frac{1}{\sqrt{2\pi}} \int_{-\infty}^{\infty} dt f(t)e^{i\omega t} \tag{A.8}$$

where ω is the frequency. The inverse Fourier transform is:

$$f(t) = \frac{1}{\sqrt{2\pi}} \int_{-\infty}^{\infty} d\omega \tilde{f}(\omega)e^{-i\omega t} \tag{A.9}$$

Notice that for space varying signals $f(x)$ the convention adopted here is the opposite[1]

$$\mathcal{F}_x(f) \equiv \tilde{f}(k) = \frac{1}{\sqrt{2\pi}} \int_{-\infty}^{\infty} dx f(x)e^{-ikx} \tag{A.10}$$

where k is the wavenumber. The inverse Fourier transform is:

$$f(x) = \frac{1}{\sqrt{2\pi}} \int_{-\infty}^{\infty} dk \tilde{f}(k)e^{ikx} \tag{A.11}$$

---

[1] Propagating plane waves are written as: $E(x,t) = (1/2\pi)\int dk d\omega \tilde{E}(k,\omega)e^{ikx-i\omega t}$.

For example the Fourier transform of the window function $f_w(x) = 1$ when $|x| < 1$ and zero otherwise is:

$$\tilde{f}_w(k) = \frac{1}{\sqrt{2\pi}} \int_{-1}^{1} dx \, e^{-ikx} = \sqrt{\frac{2}{\pi}} \frac{\sin k}{k} \equiv \sqrt{\frac{2}{\pi}} \text{sinc } k$$

Similarly the Fourier transform of Dirac's delta function, $\delta(x)$ is $\tilde{\delta}(k) = 1/\sqrt{2\pi}$ so that:

$$\delta(x) = \frac{1}{2\pi} \int_{-\infty}^{\infty} dk e^{ikx}$$

A.4.1 PROPERTIES OF THE FOURIER TRANSFORM

The Fourier transforms has the following properties:

- Linearity: for any complex numbers a and b, if $h(x) = af(x) + bg(x)$, then $\tilde{h}(k) = a\tilde{f}(k) + b\tilde{g}(k)$.
- Translation: for any real number x_0, if $h(x) = f(x - x_0)$, then: $\tilde{h}(k) = e^{-ikx_0}\tilde{f}(k)$.
- Scaling: for a non-zero real number a, if $h(x) = f(ax)$, then:

$$\tilde{h}(k) = \frac{1}{|a|}\tilde{f}(k/a)$$

- Convolution: the Fourier transform of the product of two functions is the convolution of their Fourier transforms.

$$\widetilde{fg}(k) = \frac{1}{\sqrt{2\pi}} \int_{-\infty}^{\infty} dx f(x)g(x)e^{-ikx} = \frac{1}{\sqrt{2\pi}} \int_{-\infty}^{\infty} dk' \, \tilde{f}(k')\tilde{g}(k - k') \equiv \tilde{f} \star \tilde{g}$$

Conversely, the product of the Fourier transforms is equal to the convolution of the functions. This is a very important property used in linear signal analysis. Let $h(x)$ be the response of a linear system to a delta function input $\delta(x)$. Since any function $f(x)$ can be decomposed as a sum over delta-functions: $f(x) = \int_{-\infty}^{\infty} dx' f(x')\delta(x - x')$, the response of the linear system to an input $f(x)$ will be: $g(x) = \int dx' f(x')h(x - x')$ or in Fourier space: $\tilde{g}(k) = \tilde{f}(k) \cdot \tilde{h}(k)$. Since the Fourier transform of a delta-function is a constant, this result implies that by measuring the response of a linear system to all frequencies, one can deduce the response to any spatial (or temporal) input.
- Parseval's theorem: The integral over the function squared is equal to the integral over the square of its Fourier transform:

$$\int_{-\infty}^{\infty} dx |f(x)|^2 = \int_{-\infty}^{\infty} dk |\tilde{f}(k)|^2$$

In a physical context, this property expresses the different ways of computing the total energy of a system. For the example mentioned previously of the window function, $f_w(x)$, using Parseval's theorem:

$$\int_{-\infty}^{\infty} dk |f_w(k)|^2 = \frac{2}{\pi} \int_{-\infty}^{\infty} dk \, \text{sinc}^2 k = \int_{-\infty}^{\infty} dx |f_w(x)|^2 = 2$$

and therefore: $\int_{-\infty}^{\infty} dk \, \text{sinc}^2 k = \pi$.

- Fourier transform of space or time derivative. If $f'(x)$ is the spatial derivative of $f(x)$, from Eq.A.11 its Fourier transform is $\mathcal{F}_x(f') = ik\tilde{f}(k)$. Similarly, in the time domain: $\mathcal{F}_t(f') = -i\omega\tilde{f}(\omega)$.
- Fourier transform of space or time integral. The Fourier transform of the spatial integration: $F(x) = \int_{-\infty}^{x} f(x')dx'$ is from Eq.A.11: $\mathcal{F}_x(F) = \tilde{f}(k)/ik$. Similarly, in the time domain: $\mathcal{F}_t(F) = -\tilde{f}(\omega)/i\omega$

A.5   LAPLACE AND HELMHOLTZ EQUATIONS

In the following we shall present the solutions of the Helmholtz equation:

$$\nabla^2 \Phi = \epsilon \Phi \tag{A.12}$$

in some well defined geometries, such as a box, a cylinder or a sphere (the Laplace equation, $\nabla^2 \Phi = 0$, is obtained as the particular case $\epsilon = 0$ of the Helmholtz equation). In these cases it is often worthwhile to look for a solution Φ as a product of functions of only one coordinate. Because the Hemholtz or Laplace equations are encountered in a great variety of problems (in electrostatics, communications, quantum mechanics, heat diffusion, etc.), the functions that solve these equations in various geometries have been extensively studied[110].

A.5.1   SOLUTIONS FOR A RECTANGULAR GEOMETRY

In a rectangular box where the boundary conditions are defined at positions $x = 0, a$; $y = 0, b$; $z = 0, c$ one may try a solution of the form: $\Phi(x,y,z) = X(x)Y(y)Z(z)$ from which one obtains:

$$\nabla^2 \Phi = X_{xx}YZ + XY_{yy}Z + XYZ_{zz} = \epsilon XYZ \tag{A.13}$$

$$\text{or :} \quad \frac{X_{xx}}{X} + \frac{Y_{yy}}{Y} + \frac{Z_{zz}}{Z} = \epsilon \tag{A.14}$$

Since each of the terms on the left depends on only one variable, to be valid $\forall\{x,y,z\}$, each term separately must be equal to a constant (C_1, C_2 or C_3) and the sum of these constants: $C_1 + C_2 + C_3 = \epsilon$. For example:

$$\frac{X_{xx}}{X} = C_1 \tag{A.15}$$

Whose solution if C_1 is negative ($C_1 = -\alpha^2$) is: $X(x) \sim \sin\alpha x$ or $\cos\alpha x$. If C_1 is positive ($C_1 = \alpha^2$) the solution is: $X(x) \sim \sinh\alpha x$ or $\cosh\alpha x$ (or $e^{\pm\alpha x}$). For each value of C_1, which of the two possibilities to choose depends on the boundary conditions[2].

For example if the boundary conditions on $x = 0$ and $x = a$ are $\Phi = 0$ (i.e. Dirichlet boundary conditions, such as for a grounded conducting surface in electrostatic

---

[2] And similarly for $Y(y)$ and $Z(z)$ with $C_2 = \pm\beta^2$ and $C_3 = \pm\gamma^2$.

problems) then only a solution $X(x) = A_n \sin \alpha_n x$ will satisfy both conditions when $\alpha_n = n\pi/a$, with $n = 1, 2, \cdots$.

If on the other hand, the boundary condition on $x = 0, a$ is $\partial\Phi/\partial x = 0$ (i.e. Neumann boundary conditions, which is the case when the charge density $\sigma = 0$ on the boundary in electrostatic problems), then only a solution $X(x) = A_n \cos \alpha_n x$ will satisfy these conditions when $\alpha_n = n\pi/a$, with $n = 1, 2, \cdots$.

To be explicit let us consider the case where the electrostatic potential is defined on the box sides as: $\Phi\big|_{x=0,a} = 0$, $\Phi\big|_{y=0,b} = 0$, $\Phi\big|_{z=0} = 0$ and $\Phi\big|_{z=c} = V(x,y)$. In that case we have to solve Laplace equation, Eq.A.12 with $\epsilon = 0$. According to our previous considerations, the general solution can be written as:

$$\Phi(x,y,z) = \sum A_{n,m} \sin \alpha_n x \, \sin \beta_m y \, \sinh \gamma_{n,m} z \tag{A.16}$$

where $\alpha_n = n\pi/a$, $\beta_m = m\pi/b$ and $\gamma_{n,m}^2 = \alpha_n^2 + \beta_m^2$ (in electrostatic problems we solve the Laplace equation, i.e. set $\epsilon = 0$). The boundary condition at $z = c$ then implies:

$$V(x,y) = \sum A_{n,m} \sinh \gamma_{n,m} c \sin \alpha_n x \sin \beta_m y \tag{A.17}$$

This is a double Fourier series (see Appendix A.4) which solution is:

$$A_{n,m} = \frac{4}{ab \sinh \gamma_{n,m} c} \int_0^a dx \int_0^b dy V(x,y) \sin n\pi x/a \sin m\pi y/b \tag{A.18}$$

When studying electromagnetic waves one may want to compute the eigenmodes of a given cavity (for example in the design of a laser cavity). Similarly in quantum mechanics one may seek to determine the energy levels of a free particle in a box. In both cases the computation implies solving Eq.A.12 and determining the possible values of ϵ for which a solution for the wavefunction Φ satisfying the boundary condition $(\Phi\big|_{walls} = 0)$ exists. According to the previous analysis, a general solution is:

$$\Phi(x,y,z) = \sum A_{n,m,l} \sin \alpha_n x \, \sin \beta_m y \, \sin \gamma_l z \tag{A.19}$$

where $\alpha_n = n\pi/a$, $\beta_m = m\pi/b$, $\gamma_l = l\pi/c$ (with n, m, l integer numbers) and the eigenvalues (the frequencies of the eigenmodes of a laser cavity, a waveguide or the energy levels of a particle) are: $\epsilon_{nml} = -(\alpha_n^2 + \beta_m^2 + \gamma_l^2)$.

A.5.2   SOLUTIONS FOR A CYLINDRICAL GEOMETRY

In cylindrical coordinates (ρ, ϕ, z) the Helmholtz equation, Eq.A.12 is:

$$\nabla^2 \Phi = \frac{\partial^2 \Phi}{\partial \rho^2} + \frac{1}{\rho} \frac{\partial \Phi}{\partial \rho} + \frac{1}{\rho^2} \frac{\partial^2 \Phi}{\partial \phi^2} + \frac{\partial^2 \Phi}{\partial z^2} = \epsilon \Phi \tag{A.20}$$

It is often useful to seek for a solution as a product of functions of the individual coordinates ρ, ϕ or z:

$$\Phi = R(\rho)Q(\phi)Z(z) \tag{A.21}$$

Eq.A.20 then becomes:

$$\frac{1}{R}(R_{\rho\rho} + \frac{1}{\rho}R_\rho) + \frac{1}{\rho^2}\frac{Q_{\phi\phi}}{Q} + \frac{Z_{zz}}{Z} = \epsilon \tag{A.22}$$

The solution for the azimuthal coordinate is $Q(\phi) = e^{\pm im\phi}$ (where due to the periodicity of ϕ, $m = 0, 1, 2, \cdots$). The equation for $Z(z)$ is:

$$Z_{zz} = \pm k^2 Z \tag{A.23}$$

which solutions (for the positive sign) are $\sinh kz$ or $\cosh kz$ (i.e. $e^{\pm kz}$) or $\sin kz$ or $\cos kz$ (for the negative sign), where the value of k is determined by the boundary conditions.

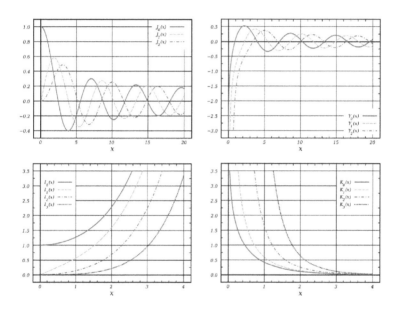

Figure A.2 Bessel functions $J_n(x)$, $Y_n(x)$ ($n = 0, 1, 2$) and modified Bessel functions $I_m(x)$ and $K_m(x)$ ($m = 0, 1, 2, 3$).

Finally the radial equation (the equation for $R(\rho)$) is:

$$\rho^2 R_{\rho\rho} + \rho R_\rho + (\kappa^2 \rho^2 - m^2)R = 0 \tag{A.24}$$

where $\kappa^2 = \pm k^2 - \epsilon$ is a number to be determined by the boundary conditions on $R(\rho)$. This equation is known as the Bessel differential equation and its solutions are known as Bessel functions of order m (when $\kappa^2 > 0$): $J_m(\kappa\rho)$ and $Y_m(\kappa\rho)$, or (when $\kappa^2 < 0$) modified Bessel functions of order m: $I_m(\kappa\rho)$ and $K_m(\kappa\rho)$, see Fig.A.2. In the limit $\rho \to 0$, Eq.A.24 reduces to: $\rho^2 R_{\rho\rho} + \rho R_\rho - m^2 R = 0$ which solution is $R \sim \rho^{\pm m}$ (the solution $R \sim \rho^{-m}$ can be discarded if the function is defined at $\rho = 0$). This defines the small ρ limit of the Bessel functions. In the limit $\rho \to \infty$ the solutions are $R \sim \sin \kappa \rho$ or $\cos \kappa \rho$ (if $\kappa^2 > 0$) or $e^{\pm \kappa \rho}$ (if $\kappa^2 < 0$). This defines the large ρ limit of the Bessel functions.

A.5.3 SOLUTIONS FOR A SPHERICAL GEOMETRY

In spherical coordinates (r, θ, ϕ) the Helmholtz equation is:

$$\nabla^2 \Phi = \frac{\partial^2}{\partial r^2} \Phi + \frac{2}{r} \frac{\partial}{\partial r} \Phi + \frac{1}{r^2 \sin \theta} \frac{\partial}{\partial \theta} \sin \theta \frac{\partial \Phi}{\partial \theta} + \frac{1}{r^2 \sin^2 \theta} \frac{\partial^2 \Phi}{\partial \phi^2} = \epsilon \Phi \qquad (A.25)$$

As previously done, we will seek a solution as a product of functions of the individual coordinates r, θ or ϕ:

$$\Phi = R(r)P(\theta)Q(\phi) \qquad (A.26)$$

Eq.A.25 then becomes:

$$PQ \frac{d^2 R}{dr^2} + PQ \frac{2}{r} \frac{dR}{dr} + \frac{RQ}{r^2 \sin \theta} \frac{d}{d\theta} \sin \theta \frac{dP}{d\theta} + \frac{RP}{r^2 \sin^2 \theta} \frac{d^2 Q}{d\phi^2} = \epsilon RPQ \qquad (A.27)$$

Multiplying both sides of this equation by $r^2 \sin^2 \theta / RPQ$ yields:

$$r^2 \sin^2 \theta \left[\frac{1}{R} \left(\frac{d^2 R}{dr^2} + \frac{2}{r} \frac{dR}{dr} \right) - \epsilon \right] + \frac{\sin \theta}{P} \frac{d}{d\theta} \sin \theta \frac{dP}{d\theta} + \frac{1}{Q} \frac{d^2 Q}{d\phi^2} = 0 \qquad (A.28)$$

Since the last term on the left depends only upon ϕ, for the equation to be valid for all ϕ this term must be constant:

$$\frac{1}{Q} \frac{d^2 Q}{d\phi^2} = -m^2 \qquad (A.29)$$

which solution is $Q = Q_m e^{\pm im\phi}$ (where due to the periodicity of ϕ, $m = 0, 1, 2, \cdots$). The equation for $R(r)$ and $P(\theta)$ then becomes:

$$\frac{r^2}{R} \left(\frac{d^2 R}{dr^2} + \frac{2}{r} \frac{dR}{dr} \right) - \epsilon r^2 + \frac{1}{P \sin \theta} \frac{d}{d\theta} \sin \theta \frac{dP}{d\theta} - \frac{m^2}{\sin \theta^2} = 0 \qquad (A.30)$$

Since the last two terms on the left depends only on θ they must be constant:

$$\frac{1}{P \sin \theta} \frac{d}{d\theta} \sin \theta \frac{dP}{d\theta} - \frac{m^2}{\sin \theta^2} = -l(l+1) \qquad (A.31)$$

where l is an integer number (see below). And the radial equation (the equation for $R(r)$) reads now:

$$\frac{d^2 R}{dr^2} + \frac{2}{r} \frac{dR}{dr} - (\epsilon + \frac{l(l+1)}{r^2})R = 0 \qquad (A.32)$$

In QM problems where $\epsilon < 0$, one defines : $x = kr = \sqrt{-\epsilon} r$ to obtain the following equation:

$$\frac{d^2 R}{dx^2} + \frac{2}{x} \frac{dR}{dx} + (1 - \frac{l(l+1)}{x^2})R = 0 \qquad (A.33)$$

A.5.3.1   Solutions of the Angular Equation

By setting $x = \cos\theta$ the equation for the function $P(\theta)$ of the angular variable θ can be recast as:

$$\frac{d}{dx}(1-x^2)\frac{dP_l}{dx} + (l(l+1) - \frac{m^2}{1-x^2})P_l = 0 \tag{A.34}$$

If the problem is azimuthally symmetric then $m = 0$ and the solutions are so-called Legendre polynomials of order $l = 0, 1, \cdots$ of which the first ones are:

$$P_0(x) = 1 \tag{A.35}$$

$$P_1(x) = x \tag{A.36}$$

$$P_2(x) = \frac{3x^2 - 1}{2} \tag{A.37}$$

$$P_3(x) = \frac{5x^3 - 3x}{2}$$

The Legendre polynomials can be obtained from Rodrigues' formula:

$$P_l(x) = \frac{1}{2^l l!}\frac{d^l}{dx^l}(x^2 - 1)^l \tag{A.38}$$

If $m \neq 0$, the solutions of Eq.A.31 are the so-called associated Legendre polynomials (with $-l \leq m \leq l$):

$$P_l^{-m}(x) = P_l^m(x) = (-1)^m (1-x^2)^{\frac{m}{2}} \frac{d^m}{dx^m} P_l(x) \tag{A.39}$$

The spherical harmonics $Y_{lm}(\theta, \phi)$ are defined as a properly normalized products of the angular functions $P(\theta)$ and $Q(\phi)$:

$$Y_{lm}(\theta, \phi) = \sqrt{\frac{(2l+1)(l-m)!}{4\pi(l+m)!}} P_l^m(\cos\theta)e^{im\phi} \tag{A.40}$$

They are encountered in many problems arising from a solution involving the Laplacian operator (∇^2, such as Laplace equation or Schrödinger's equation for the Hydrogen atom). The first spherical harmonics are (see Fig.A.3):

$$Y_{00}(\theta, \phi) = \frac{1}{2\sqrt{\pi}} \tag{A.41}$$

$$Y_{10}(\theta, \phi) = \sqrt{\frac{3}{4\pi}}\cos\theta \tag{A.42}$$

$$Y_{1,\pm 1}(\theta, \phi) = \mp\sqrt{\frac{3}{8\pi}}\sin\theta e^{\pm i\phi} \tag{A.43}$$

$$Y_{20}(\theta, \phi) = \sqrt{\frac{5}{16\pi}}(3\cos^2\theta - 1)$$

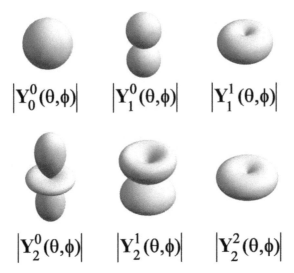

Figure A.3 The norm of the first spherical harmonics : $Y_{00}(\theta,\phi)$, $Y_{10}(\theta,\phi)$ and $Y_{1,1}(\theta,\phi)$, $Y_{20}(\theta,\phi)$, $Y_{21}(\theta,\phi)$, $Y_{22}(\theta,\phi)$. Computational graphics herein were generated with Wolfram Mathematica.

The spherical harmonics define an orthonormal basis in the space of angular functions, i.e.:

$$\int \sin\theta d\theta d\phi \; Y^*_{l'm'}(\theta,\phi) \; Y_{lm}(\theta,\phi) = \delta_{ll'}\delta_{mm'} \tag{A.44}$$

where $Y^*_{l'm'}$ stands for the complex conjugate of $Y_{l'm'}$ and the Kroenecker delta function $\delta_{ll'} = 0$ unless $l = l'$ in which case $\delta_{ll} = 1$. From Eq.A.31, it is clear that spherical harmonics satisfy the equation:

$$\mathcal{L}^2 Y_{lm} = [\frac{1}{\sin\theta}\frac{\partial}{\partial\theta}\sin\theta\frac{\partial}{\partial\theta} + \frac{1}{\sin\theta^2}\frac{\partial^2}{\partial\phi^2}]Y_{lm} = -l(l+1)Y_{lm} \tag{A.45}$$

In Quantum Mechanics the operator $\mathcal{L}^2$ is related to the angular momentum operator: $L^2 = -\hbar^2\mathcal{L}^2$ (where $\hbar$ is Planck's constant, h divided by 2π). Any square integrable function $\Psi(\theta,\phi)$ can be expanded as a sum of spherical harmonics:

$$\Psi(\theta,\phi) = \sum A_{lm}Y_{lm}(\theta,\phi) \tag{A.46}$$

where the coefficient of that expansion are:

$$A_{lm} = \int \sin\theta d\theta d\phi \; Y^*_{lm}(\theta,\phi) \; \Psi(\theta,\phi) \tag{A.47}$$

For example, if the potential $\Phi(r_0,\theta)$ on a sphere of radius r_0 is: V on one hemisphere ($\theta < \pi/2$) and $-V$ on the other ($\pi/2 < \theta < \pi$).

$$A_{lm} = 2\pi\delta_{m0} \sqrt{\frac{(2l+1)}{4\pi}} \int_0^\pi \sin\theta d\theta \, \Phi(r_0,\theta) \, P_l(\cos\theta) \tag{A.48}$$

Since the potential is odd with respect to the variable $\cos\theta = x \to -x$ the integral is non-zero only for odd values of l for which $P_l(x)$ is also odd. Performing the integrals yields:

$$\Phi(r_0,\theta) = V[\frac{3}{2}P_1(\cos\theta) - \frac{7}{8}P_3(\cos\theta) + ...] \tag{A.49}$$

A.5.3.2   Solutions of the Radial Equation

When $\epsilon = 0$, i.e. for electrostatic problems, the solutions of the radial equation, Eq.A.32 are $R = r^l$ and $R = r^{-l-1}$. With boundary conditions specified on a sphere of radius r_0, the requirement that $\Phi \sim R(r)$ does not diverge as $r \to 0$, implies that inside the sphere: $R(r) = Ar^l$ ($r < r_0$). Similarly outside the sphere, since $|\Phi_{r\to\infty}| < \infty$: $R(r) = Br^{-l-1}$ ($r > r_0$).

Thus with respect to the previous example the potential inside the sphere is:

$$\Phi(r,\theta) = V[\frac{3}{2}(\frac{r}{r_0})P_1(\cos\theta) - \frac{7}{8}(\frac{r}{r_0})^3 P_3(\cos\theta) + ...] \tag{A.50}$$

Outside the sphere the solution is obtained by replacing the terms in $(r/r_0)^l$ by $(r_0/r)^{l+1}$.

For electromagnetic radiation, quantum mechanical free particles (and for other problems) where $\epsilon \neq 0$, the solution of the radial equation, Eq.A.33 is obtained by looking for a solution:

$$R_l(x) = e^{ix} \sum_{j=0}^{\infty} a_j x^{-j-1}$$

Inserting that ansatz into Eq.A.32 yields:

$$e^{ix} \sum_{j=0}^{\infty} [j(j+1) - l(l+1)]a_j r^{-j-3} - 2ie^{ix} \sum_{j=0}^{\infty} ja_j r^{-j-2} = 0 \tag{A.51}$$

from which by equating the coefficients of r^{-j-3} we derive the relation:

$$a_{j+1} = \frac{[j(j+1) - l(l+1)]}{2i(j+1)} a_j \tag{A.52}$$

For any value l, the coefficients of the series are determined recursively starting from $a_0 = 1$ and ending at $j = l$. The solution can be cast as: $R_l(x) = y_l(x) + i j_l(x)$ where $j_l(x)$ and $y_l(x)$ are known as spherical Bessel functions of respectively the first and second kind. They can be computed from the following equation:

$$j_l(x) = (-x)^l(\frac{1}{x}\frac{d}{dx})^l\frac{\sin x}{x}$$

(A.53)

$$y_l(x) = (-x)^l(\frac{1}{x}\frac{d}{dx})^l\frac{\cos x}{x}$$

(A.54)

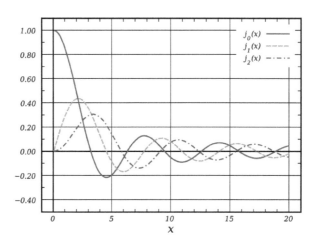

Figure A.4   The first spherical Bessel functions of the first kind: $j_0(x)$, $j_1(x)$ and $j_2(x)$.

The first spherical Bessel functions of the first kind are (see Fig.A.4) :

$$j_0(x) = \frac{\sin x}{x}$$

(A.55)

$$j_1(x) = \frac{\sin x}{x^2} - \frac{\cos x}{x}$$

(A.56)

$$j_2(x) = (\frac{3}{x^2} - 1)\frac{\sin x}{x} - \frac{3\cos x}{x^2}$$

(A.57)

The first spherical Bessel functions of the second kind (also known as spherical Neumann functions) are (see Fig.A.5):

$$y_0(x) = \frac{\cos x}{x}$$

(A.58)

$$y_1(x) = \frac{\cos x}{x^2} - \frac{\sin x}{x}$$

(A.59)

$$y_2(x) = (\frac{3}{x^2} - 1)\frac{\cos x}{x} + \frac{3\sin x}{x^2}$$

(A.60)

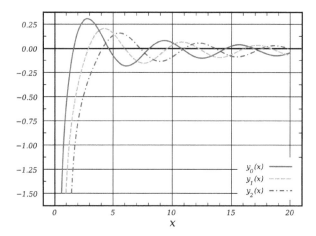

Figure A.5 The first spherical Bessel functions of the second kind: $y_0(x)$, $y_1(x)$ and $y_2(x)$.

For imaginary values of x, i.e. when $\epsilon > 0$ the solution of Eq.A.33 are known as modified spherical-Bessel functions. The modified spherical Bessel functions of the first kind $i_l(x)$ can be obtained from the analytical continuation of the spherical Bessel functions $j_l(x)$, i.e. by replacing x by ix (notice that: $\sin ix = i \sinh x$ and $\cos ix = \cosh x$). The first functions are:

$$i_0(x) = \frac{\sinh x}{x} \tag{A.61}$$

$$i_1(x) = -\frac{\sinh x}{x^2} + \frac{\cosh x}{x} \tag{A.62}$$

$$i_2(x) = \left(\frac{3}{x^2} + 1\right)\frac{\sinh x}{x} - \frac{3\cosh x}{x^2} \tag{A.63}$$

In agreement with our previous asymptotic analysis, as $x \to \infty$ these functions diverge as e^x/x. The modified spherical Bessel functions of the second kind $k_l(x)$ which behave asymptotically as e^{-x}/x are obtained as a sum of the analytical continuation of $j_l(ix)$ and $y_l(ix)$:

$$k_0(x) = \frac{e^{-x}}{x} \tag{A.64}$$

$$k_1(x) = \frac{x+1}{x^2}e^{-x} \tag{A.65}$$

$$k_2(x) = \frac{x^2 + 3x + 3}{x^3}e^{-x} \tag{A.66}$$

A.6 ◆ SPECIAL RELATIVITY

In Maxwell's equations the velocity of electromagnetic waves is a constant: the velocity of light. This result is at odds with our common experience that velocities are relative and depends on the motion of the observer: one sitting in a train and seeing a nearby train pass by will often not know which is actually moving.

That is true unless the velocity of electromagnetic waves depends (like sound waves) on the physical properties of the medium in which they propagate. Therefore to resolve the apparent contradiction between electromagnetism and our common experience, 19th century physicists introduced the concept of the luminous aether, a putative medium that served as a support for the propagation of electromagnetic waves at velocity c. In this hypothesis an observer moving with respect to the aether should experience an aether wind. Experiments attempting to observe this wind were inconclusive. The most famous of those was the Michelson-Morley experiment[15], which consisted in an interferometer with two equal and perpendicular arms of length L, see Fig.A.6.

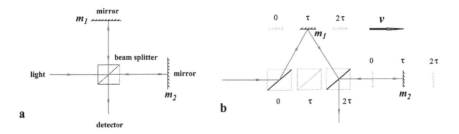

Figure A.6 a) The Michelson-Morley experiment in the rest frame of the set-up. b) The same experiment in the frame of presumed aether with respect to which the set-up (the Earth) is moving at speed v. The transit time between the beam-splitter and mirror m_2 is $\Delta t_2 = (2L/c)/(1 - v^2/c^2)$, while the transit time $\Delta t_1 = 2\tau$ between the beam-splitter and mirror m_1 satisfies: $c^2\tau^2 = L^2 + c^2\tau^2$, i.e. : $\Delta t_1 = (2L/c)/\sqrt{1 - v^2/c^2}$.

In the reference frame of the aether the Earth is moving at velocity $\vec{v} \sim 30$ km/s. If the arm ending with mirror m_2 is parallel to $\vec{v}$ the transit time Δt_2, it takes a light pulse to travel from the beam-splitter to the mirror and back is:

$$\Delta t_2 = \frac{2L/c}{1 - v^2/c^2}$$

while the transit time $\Delta t_1 = 2\tau$ in the perpendicular arm satisfies: $c^2\tau^2 = L^2 + v^2\tau^2$, i.e. :

$$\Delta t_1 = \frac{2L/c}{\sqrt{1 - v^2/c^2}}$$

The transit time difference between the two arms is accompanied by a phase shift between the two beams which value is:

$$\phi = \omega(\Delta t_2 - \Delta t_1) = 2\pi c(\Delta t_2 - \Delta t_1)/\lambda$$

For a wavelength $\lambda = 0.5\mu m$ and an interferometer length $L = 10m$ this should correspond to a phase shift $\phi \sim 0.8\pi$. However no phase shift was detected between the two arms no matter how they were positioned with respect to the direction of the Earth movement $\vec{v}$. One possible explanation to this null result was that the aether is dragged by the Earth like water is by the hull of a ship, but this led to other contradictions. An other explanation was that the arm parallel to the Earth's motion undergoes a contraction:

$$L' = L\sqrt{1 - v^2/c^2} \tag{A.67}$$

that exactly cancels the difference in transit times between the two arms of the interferometer. At that time there was no explanation for such a contraction. It was Einstein who showed that such a contraction in length is a natural result of his theory of relativity, which we discuss next[111, 112].

In 1905, Einstein decided to drop the hypothesis of the luminous aether. Instead he postulated that:

- The laws of Physics are independent of the inertial (constant velocity) reference frame in which they are observed. A postulate motivated by Faraday's observations, see section 3.3.
- The speed of light is a constant independent of the speed of the emitter or the observer. A postulate based on Michelson-Morley's experiments.

Consider a light source S at position $(0,0,0)$. After a time t the emitted light has propagated to a position (x, y, z) that satisfies: $x^2 + y^2 + z^2 = c^2 t^2$. Now consider the same light source observed by a person moving with velocity v along the x-axis. In this reference frame after a time t' the light has moved to position (x', y', z') that obeys: $x'^2 + y'^2 + z'^2 = c^2 t'^2$. The problem is that this result is inconsistent with the Galilean transformation which relates the coordinate systems among moving reference frames:

$$x' = x - vt$$
$$y' = y$$
$$z' = z$$
$$t' = t$$

That transformation is consistent with our daily experience that the velocity of an object is relative. It implies that the light speed c depends on the reference frame (along the x-axis: $c' = c - v$), but is inconsistent with Einstein's second postulate that $c' = c$. Since we know that Galilean transformations are an accurate description when $v \ll c$, Einstein looked for a transformation that would yield the classical result at low speeds yet would satisfy his second postulate. He looked for a linear transformation:

$$x' = \gamma(x - vt)$$
$$y' = y$$
$$z' = z$$
$$t' = \gamma(t - \alpha x) \qquad \text{(A.68)}$$

Inserting that ansatz into the equation for the position of the emitted light in the moving frame:

$$x'^2 + y'^2 + z'^2 = c^2 t'^2$$
$$\gamma^2(x - vt)^2 + y^2 + z^2 = c^2 \gamma^2 (t - \alpha x)^2$$
$$\gamma^2(1 - c^2 \alpha^2)x^2 + y^2 + z^2 = \gamma^2(1 - \frac{v^2}{c^2})c^2 t^2 + 2\gamma^2(v - \alpha c^2)xt$$

For this equation to be the same as the equation in the rest frame of the emitter, one must set:

$$\alpha = v/c^2$$

and

$$\gamma = 1/\sqrt{1 - v^2/c^2}$$

.

The so-called Lorentz transformations, Eq.A.68, which are the relativistic extension of the classical Galilean transformations, have some non-intuitive consequences.

- If a clock emits light at intervals δt then an observer moving at a relative velocity v will observe it to tick at intervals $\delta t' = \gamma \delta t > \delta t$, i.e. the clock appears to tick at a slower rate than in its rest frame. This is commonly verified in particle accelerators: particles moving at a velocity close to that of light are observed to decay much slower than when they are at rest. It is also commonly experienced by the orbiting atomic clocks used in the GPS system: they have to be periodically resynchronized with the ground clocks in order to reduce localization errors.
- A ruler of rest length L appears shorter for an observer moving with respect to it. Indeed to measure its length the moving observer has to observe its extremities simultaneously, i.e. for this observer the extremity at $x' = 0$ and the extremity at distance $x' = L'$ have to be measured at the same time: $t' = 0$. From the equation of the Lorentz transformation, Eq.A.68, we deduce that in the frame where the ruler is at rest: $t = \alpha L$ and thus: $L' = \gamma(L - v\alpha L) = L\sqrt{1 - v^2/c^2} < L$, which explains the results of the Michelson-Morley experiment.
- The speed of light cannot be exceeded. If a body moves with speed $u = dx/dt$ in a frame S then in a frame S' moving at speed $\vec{v} \parallel \vec{u}$ with respect to S, its velocity is:

$$u' = \frac{dx'}{dt'} = \frac{dx \pm vdt}{dt \pm \alpha dx} = \frac{u \pm v}{1 \pm vu/c^2} \qquad \text{(A.69)}$$

with + for anti-parallel motion ($\hat{v} = -\hat{u}$) and − for parallel motion ($\hat{v} = \hat{u}$) [3]. As long as $v, u < c$, $u' < c$. If the speed of light could be exceeded, then there would be a phenomenon (e.g. a rocket) for which $u > c$. In such a case in a reference frame moving at velocity $v < c$ such that $1 - vu/c^2 < 0$ (i.e. $c/u < v/c < 1$): $u' < 0$, i.e. effects will be seen before their cause (the rocket would appear to move backwards)!

A.6.1 LORENTZ INVARIANCE AND RELATIVITY

Classically the distance in space $\Delta x = x_2 - x_1$ and the time-interval $\Delta t = t_2 - t_1$ between two events are independent of the reference frame. That is not so in the theory of relativity. We have seen that the measured length of a ruler depends on the reference frame (it is shorter than in its rest frame). We have also seen that the time between events is longer in a moving frame of reference than in the frame where the observed events are at rest (e.g. the times shown by a clock). The Lorentz transformation Eq.A.68 imply that the 4-dimensional distance element:

$$\Delta s^2 = \Delta x^2 + \Delta y^2 + \Delta z^2 - c^2 \Delta t^2 \equiv \sum_{v=1}^{4} \Delta x_v \Delta x_v$$

is conserved by such a transformation (notice $\Delta x_4 = ic\Delta t$). In 3-D the group of transformations that conserve distances are translations and rotations. The Lorentz transformation is a particular case of rotation in 4-D space[113]. One may indeed write it as:

$$\begin{pmatrix} \Delta x_1' \\ \Delta x_2' \\ \Delta x_3' \\ \Delta x_4' \end{pmatrix} = \begin{pmatrix} \gamma & 0 & 0 & i\gamma v/c \\ 0 & 1 & 0 & 0 \\ 0 & 0 & 1 & 0 \\ -i\gamma v/c & 0 & 0 & \gamma \end{pmatrix} \begin{pmatrix} \Delta x_1 \\ \Delta x_2 \\ \Delta x_3 \\ \Delta x_4 \end{pmatrix}$$

$$= \begin{pmatrix} \cos\phi & 0 & 0 & \sin\phi \\ 0 & 1 & 0 & 0 \\ 0 & 0 & 1 & 0 \\ -\sin\phi & 0 & 0 & \cos\phi \end{pmatrix} \begin{pmatrix} \Delta x_1 \\ \Delta x_2 \\ \Delta x_3 \\ \Delta x_4 \end{pmatrix} \tag{A.70}$$

with $\tan\phi = iv/c$ (which can also be written as $\phi = i\,\mathrm{atanh}\,v/c$). It is a rotation (compare with Eq.1.2) but by an imaginary angle due to the fact that the time axis is imaginary. Einstein relativity postulate may then be restated differently:

---

[3]Eq.A.69 is verified in the measurement (first done by Fizeau in 1851) of the velocity c' of light propagating in water (index of refraction n) flowing at velocity u:

$$c' = \frac{c/n \pm u}{1 \pm u/nc} \approx c/n \pm u(1 - 1/n^2)$$

- For a physical law to be independent of the reference frame it must be invariant under a Lorentz transformation.

For that reason it is worthwhile to express these laws in terms of variables that are manifestly invariant under a Lorentz transformation, i.e. to express them in so-called Lorentz covariant form. For example time derivatives should be expressed with respect to the manifestly invariant distance element ds^2 also known as proper time $d\tau$ (the time in the reference frame of the particle: $ds^2 = -c^2 d\tau^2$):

$$d\tau = dt \sqrt{1 - \frac{1}{c^2}(\dot{x}^2 + \dot{y}^2 + \dot{z}^2)} = dt/\gamma$$

Hence the Lorentz covariant 4-velocity vector can be written as: $u_\nu = dx_\nu/d\tau$ ($\nu = 1...4$). It is related to the 3-D velocity vector v_i by $u_i = \gamma v_i$ and its forth component is $u_4 = ic\gamma$. The magnitude of the 4-velocity vector is constant[4]:

$$u_\nu u_\nu = \gamma^2(\vec{v}^2 - c^2) = -c^2$$

and is thus invariant under a Lorentz transformation. Similarly the 4-momentum vector: $p_\nu = mu_\nu = \{\gamma m\vec{v}, i\gamma mc\} \equiv \{\vec{p}, p_4\}$ is manifestly invariant ($p_\nu p_\nu = m^2 u_\nu u_\nu = -m^2 c^2$).

In special relativity Newton's second law becomes:

$$F_i = \frac{dp_i}{dt} = \frac{d(\gamma m v_i)}{dt} \qquad i = 1, 2, 3 \quad \text{or} \quad x, y, z \tag{A.71}$$

By computing the work performed on the particle we shall now relate the last component of the 4-momentum vector $p_4 = i\gamma mc$ to the energy E of the particle:

$$
\begin{aligned}
E &= \int dt \vec{F} \cdot \vec{v} = \int dt \frac{d(\gamma m v_i)}{dt} v_i \\
&= \int dt [\gamma m v_i v_i + mv^2 \frac{v_i v_i}{c^2} \gamma^3] = \int dt \gamma^3 m v_i v_i \\
&= \int mc^2 \frac{d\gamma}{dt} dt = \gamma mc^2 = -ip_4 c \tag{A.72}
\end{aligned}
$$

hence: $p_4 = iE/c$ from which one derives the relativistic energy-momentum relation: $-m^2 c^2 = p_\nu p_\nu = \vec{p}^2 - E^2/c^2$, or:

$$E^2 = m^2 c^4 + \vec{p}^2 c^2 \tag{A.73}$$

When $\vec{p} = 0$, we obtain the famous relation between energy and mass: $E = mc^2$.

As we have defined the Lorentz covariant 4-velocity vector, we can define the Lorentz covariant 4-force vector: $f_\nu = dp_\nu/d\tau$. Newton's Second Law, Eq. A.71,

---

[4] As customary, we adopt the summation convention: any index appearing more than once is summed over, namely: $u_\nu u_\nu \equiv \sum_\nu u_\nu u_\nu$

implies that: $f_v = (\gamma F_i, i\gamma \vec{F} \cdot \vec{v}/c)$. The equation for the fourth component f_4: $f_4 = dp_4/d\tau = idE/cd\tau = i\gamma dE/cdt$, can thus be recast as: $dE/dt = \vec{F} \cdot \vec{v}$ which is the known result relating the change in energy of the particle to the driving power.

Notice that upon a change of reference frame (a Lorentz transformation) the 4-momentum $\{p_x, p_y, p_z, iE/c\}$ and 4-force $\{f_1, f_2, f_3, f_4\}$ transform in the same way as the 4-coordinates $\{x_1, x_2, x_3, x_4\} \equiv \{x, y, z, ict\}$, Eq.A.70:

$$
\begin{aligned}
p'_x &= \gamma(p_x - vE/c^2) & f'_1 &= \gamma(f_1 + ivf_4/c) \\
p'_y &= p_y & f'_2 &= f_2 \\
p'_z &= p_z & f'_3 &= f_3 \\
E' &= \gamma(E - vp_x) & f'_4 &= \gamma(f_4 - ivf_1/c)
\end{aligned}
\tag{A.74}
$$

If a particle experiences a force $F_y = dp_y/dt$ in a frame S in which the particle moves at velocity $\vec{u} = u\hat{x}$ along the x-axis: $f_2 = \gamma_u F_y$ (with $\gamma_u = 1/\sqrt{1-u^2/c^2}$). In a frame S' moving at velocity $\vec{v} = v\hat{x}$ with respect to S, from Eq.A.69 the particle is observed to move at velocity $u' = (u-v)/(1-uv/c^2)$. In that reference frame $f'_2 = \gamma_{u'} F'_y$ (with $\gamma_{u'} = 1/\sqrt{1-u'^2/c^2}$). From the 4-force Lorentz transformation, Eq.A.74, $f'_2 = f_2$. Hence the transverse force F'_y observed in frame S' will be[5]:

$$
F'_y = \gamma_u F_y/\gamma_{u'} = \frac{F_y}{\gamma_v(1 - vu/c^2)}
\tag{A.75}
$$

If the particle is instantaneously at rest in S ($u = 0$), namely if S stands for the particle's frame of reference and the force $\vec{F}$ the force measured in that reference frame:

$$
F'_y = F_y/\gamma_v = \sqrt{1 - v^2/c^2} F_y
\tag{A.76}
$$

a result we used in section 3.2.2.

---

[5]The same equation can be derived from the definition:

$$
F'_y = dp'_y/dt' = dp_y/\gamma_v(dt - vdx/c^2) = F_y/\gamma(1 - vu/c^2)
$$

A.7   ADVANCED TOPICS IN ELECTROMAGNETISM

A.7.1   SOLUTION OF LAPLACE EQUATION IN TWO DIMENSIONS

If the boundary conditions are uniform along the vertical coordinate h, the electrostatic potential Φ will depend only on the transverse coordinates x and y. There are then powerful methods to solve the Laplace equation based on the knowledge of analytical functions[114]. These are functions $f(z)$ of the complex variable $z = x + iy$, also written as $f(z) \equiv u(x,y) + iv(x,y)$, such as $e^z = e^{x+iy} = e^x \cos y + ie^x \sin y$. Notice that here z is a complex number and does not refer to the vertical coordinate h. Since $df/dz = \partial_x f = \partial_x u + i\partial_x v$ and $df/dz = -i\partial_y f = -i\partial_y u + \partial_y v$, the real and imaginary parts[6] of $f(z)$ satisfy the Cauchy-Riemann equations:

$$\partial_x u = \partial_y v$$

$$\partial_y u = -\partial_x v$$

From which one readily shows that both $u(x,y)$ and $v(x,y)$ satisfy the 2D-Laplace equation: $\nabla^2 u = u_{xx} + u_{yy} = 0$ and $v_{xx} + v_{yy} = 0$. Therefore all analytical functions (all functions $f(z)$) are solutions of Laplace equation in 2D. One simply(!) needs to find the one satisfying the given boundary condition.

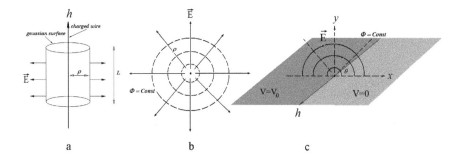

Figure A.7   (a) The electric field $\vec{E}$ generated by a wire carrying a uniform charge density per unit length λ can be computed by Gauss's theorem on a co-axial cylinder of radius r. (b) In the plane perpendicular to the wire and in cylindrical coordinates (ρ, θ, h), the electric field lines are directed along the radius ρ and the equipotential surfaces along the angular variable θ (i.e. $\rho = Const.$). (c) The electric field lines and equipotential surfaces ($\theta = Const.$) for a problem where the potential is null for $\theta = 0$ (i.e. on the half plane $x > 0, y = 0, h$) and fixed at V_0 for $\theta = \pi$ (i.e. on the half plane $x < 0, y = 0, h$). This problem is known as the conjugated problem to the uniformly charged wire, namely its solution is the imaginary part of the solution to the uniformly charged wire.

---

[6]The real and imaginary parts of $f(z)$ are denoted respectively as $\text{Re}\{f\}(\equiv u)$ and $\text{Im}\{f\}(\equiv v)$.

Consider for example the case of a very long straight wire (or cylinder) of radius $r_0 \ll L$, where L is its length, with uniform charge density per unit length: λ, see Fig.A.7. This is clearly a 2D problem as the potential does not depend on the position along the wire but on the distance from it. Gauss's theorem on a coaxial cylinder of radius r yields:

$$\int \vec{E} \cdot d\vec{S} = 2\pi r L E = 4\pi \lambda L \tag{A.77}$$

Therefore $E = 2\lambda/r$ and the potential away from the wire is $\Phi(r) = 2\lambda \log(r/r_0)$. The potential Φ is a component (here the real part: $\Phi \equiv \mathrm{Re}\{f(z)\}$) of an analytical function $f(z)$ which in this case is: $f(z) = 2\lambda \log(z/r_0) = 2\lambda \log(r/r_0) + i2\lambda\theta$ (where in polar coordinates: $z = re^{i\theta}$).

Notice that the imaginary part of $f(z)$, i.e. $v(\theta) = 2\lambda\theta$, also satisfies the Laplace equation and is thus the electric field of an electrostatic problem (known as the conjugated problem to the charged infinite wire). It is the electric field of two very long (semi-infinite) plates lying in the xh-plane: the $x > 0$ plate being grounded ($\Phi(x > 0, 0, h) = 0$) while the $x < 0$ plate is held at a fixed potential: $\Phi(x < 0, 0, h) = V_0 \equiv 2\lambda\pi$, see Fig.A.7. Due to the translation invariance in the vertical (h) direction this problem doesn't depend on h and is effectively 2D with solution: $v(\theta) = 2\lambda\theta$. Notice that while the equipotential surfaces of this problem obey $\theta = const$, the field lines are given by $\rho = const$. As expected they are orthogonal to each other. Consequently this solution also describes the field inside a wedge superposed on the equipotential lines, namely one defined by two planes intersecting along the h-axis at an angle β, one side of which is grounded ($\Phi(\rho, \theta = 0, h) = 0$) the other side being held at potential V ($\Phi(\rho, \theta = \beta, h) = V$ (i.e. constant along an equipotential line of the previous problem). The solution is: $\Phi(\rho, \theta, h) = V\theta/\beta$.

An extension of the problem of the two semi-infinite plates in the xh-plane is the case where there are many successive plates along the x-axis held at different potentials: Φ_1 for $x < \xi_1$; Φ_2 for $\xi_1 < x < \xi_2$ and Φ_3 for $x > \xi_2$. By superposition the potential at point (x, y, h) is:

$$\Phi = A_1\theta_1 + A_2\theta_2 + A_3 \tag{A.78}$$

Where θ_i is the angle between points $(x, y, 0)$, $(\xi_i, 0, 0)$ and the x-axis (due to vertical translation invariance h can be set to zero): $\tan\theta_i = y/(x - \xi_i)$. The coefficients A_i are determined by the boundary conditions:

$$\Phi_3 = A_3 \tag{A.79}$$
$$\Phi_2 = \pi A_2 + A_3 \tag{A.80}$$
$$\Phi_1 = \pi(A_2 + A_1) + A_3 \tag{A.81}$$

So that $A_i = (\Phi_i - \Phi_{i+1})/\pi$. By extension for a boundary condition on the x-axis given by a potential $\Phi_b(\xi)$ the potential Φ at a point (x,y,h) is (with $\theta(\xi) = \arctan[y/(x-\xi)]$):

$$\Phi(x,y,h) = \Phi(x,y,0) = \int A(\xi)\theta(\xi)d\xi = \frac{1}{\pi}\int\left(-\frac{d\Phi_b}{d\xi}\right)\theta(\xi)d\xi = \frac{1}{\pi}\int\Phi_b(\xi)\frac{d\theta}{d\xi}d\xi$$

$$= \frac{y}{\pi}\int_{-\infty}^{\infty}\frac{\Phi_b(\xi)}{(x-\xi)^2 + y^2}d\xi$$

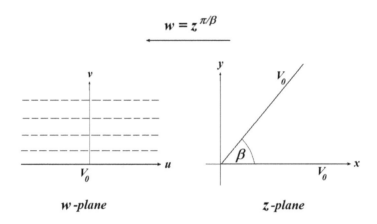

Figure A.8 Example of the solution of Laplace equation in 2D using conformal maps. The solution of Laplace equation in the plane $w = u + iv$ where the potential along the horizontal axis u is given: $\Phi(v = 0) = V_0$ is $\Phi(u,v) = \text{Im}(iV_0 + Aw)$ (the equipotential lines are shown as dashed lines). In the z-plane the potential is V_0 along a wedge of angle β. The problem in this plane can be solved by mapping the z-plane and its boundary conditions into w: $w = z^{\pi/\beta}$. Notice that by this transformation the wedge boundary is mapped into the u-axis and its interior into the upper w-plane. The solution of the potential is therefore: $\Phi(\rho,\theta) = \text{Im}(iV_0 + Az^{\pi/\beta}) = V_0 + A\rho^{\pi/\beta}\sin\pi\theta/\beta$.

The wedge we considered previously had one side held at potential V while the other was grounded. Let us now consider the potential inside a wedge both sides of which are held at constant voltage V_0, see Fig.A.8. Let us first consider the case where the wedge is flat, i.e. the case where in the complex $w = u + iv$ plane the $v = 0$ axis is held at constant voltage V_0 (i.e. $\Phi(u,0,h) = V_0$). The potential throughout is then: $\Phi = V_0 + Av$.

The solution for a wedge of angle β held at constant potential V_0 can then be found by mapping the wedge boundary into the u-axis by a so-called conformal map: $w \equiv u + iv = z^{\pi/\beta} = \rho^{\pi/\beta}e^{i\pi\theta/\beta}$. Notice that the wedge side corresponding to the positive x-axis ($\theta = 0$) in the z-plane is mapped into the positive u-axis in the w-plane, while the other side ($\theta = \beta$) is mapped into the negative u-axis. The space between the wedge is mapped into the $v > 0$ plane. The solution of Laplace equation between the wedge is then obtained from the solution of Laplace equation in the w-plane:

$$\Phi(\rho,\theta,h) = \Phi(\rho,\theta,0) = \text{Im}\{iV_0 + Aw\} = V_0 + \text{Im}\{Az^{\pi/\beta}\} = V_0 + A\rho^{\pi/\beta}\sin\pi\theta/\beta$$

This field satisfies the boundary conditions on the wedge, at $\theta = 0$, β and being the imaginary part of an analytical function it satisfies Laplace equation in the z-plane. Notice that for $\beta > \pi$ as $\rho \to 0$ the field (and thus the electrostatic force) diverges $|\vec{E}| = |\vec{\nabla}\Phi| \propto \rho^{\pi/\beta-1} \to -\infty$. In particular for a plate held at constant voltage, i.e. $\beta = 2\pi$, the electric field (and thus the force applied on charges) diverges near the edge:

$$E_r(\rho \to 0, \pi, h) = -\partial_\rho\Phi \approx -A\rho^{-1/2}/2$$

This result sets the theoretical foundation for the working of a lightning rod: its tip attracts very strongly charges of opposite sign present in the air (notice the inverse dependence on the distance ρ to the edge of the plate: $1/\sqrt{\rho}$). Even though at the time of its invention (1752) Coulomb's law was unknown to Franklin, he did show that lightning was an electrical phenomena by using it to charge a Leyden jar!

The use of conformal mapping ($w = f(z)$) to transform the Laplace equation in a given geometry (in the complex z-plane) into a simpler geometry (in the complex w-plane) where the solution is known is a very powerful tool for solving the Laplace equation in 2D. This is however beyond the scope of this book, where the purpose of the previous exercises was simply to suggest the power of these analytical methods.

A.7.2 PHYSICAL OPTICS

The field that studies the propagation of light waves in an optical system is known as physical optics[34]. As we have seen in section 3.5.3, the propagation of a light beam can be understood as a sum over spherical waves originating from an equiphase plane of the beam. In a transparent media of refractive index n the propagation of light is described by Eq.3.142 with wavenumber $k = nk_0$ (where $k_0 = 2\pi/\lambda_0$ is the wavenumber in vacuum):

$$E_{diff}(x,y,z) \approx \frac{ne^{ink_0z}}{i\lambda_0 z} \oint dx'dy' E_i(x',y')e^{ink_0((x-x')^2+(y-y')^2)/2z} \tag{A.82}$$

where $E_i(x',y')$ is the amplitude of a the electric field at position (x',y'). In the following we shall use that approach to compute the limit on the resolution of a lens (e.g. a microscope objective) and demonstrate its importance in the treatment (e.g. filtering) of optical images.

A.7.2.1 The Resolution Limit

In the following we shall use Eq.A.82 to compute the size of the image of a point source, as when observing a star with a telescope. A point source at O in a medium of refractive index n_1 (usually air or vacuum with $n_1 = 1$), see Fig.3.30(b), emits a spherical wave $E(r) = E_0 e^{in_1 k_0 r}/n_1 k_0 r$. The field impinging on a lens a distance s from the source is thus:

$$E_{lens}^{in}(u,v) \approx \frac{E_0 \exp\left[in_1 k_0 \sqrt{s^2 + u^2 + v^2}\right]}{n_1 k_0 s} \approx \frac{E_0 e^{in_1 k_0 s}}{n_1 k_0 s} \exp\left[in_1 k_0 (u^2 + v^2)/2s\right]$$

where we expanded: $\sqrt{s^2 + u^2 + v^2} \approx s + (u^2 + v^2)/2s$.

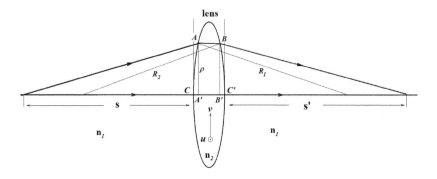

Figure A.9 Description of the passage of a light beam in a thin lens of index of refraction n_2 and radii R_1 and R_2; see text for the computation of the phase difference between the light in the input plane of the lens (passing through C) and that in its output plane (passing through C').

Between the input plane of the lens at C and the output plane of the lens at C', see Fig.A.9, the beam incident on the lens experiences a phase delay $\delta\Phi(u,v)$ that varies with its distance $\rho = \sqrt{u^2 + v^2}$ from the optical axis CC'.

$$\delta\Phi(u,v) = [n_1 k_0 (\overline{CA'} + \overline{B'C'}) + n_2 k_0 \overline{A'B'}] - n_2 k_0 \overline{CC'}$$
$$= (n_1 - n_2) k_0 (\overline{CA'} + \overline{B'C'})$$
$$\approx (n_1 - n_2) k_0 (\frac{1}{R_1} + \frac{1}{R_2}) \frac{\rho^2}{2} = -\frac{n_1 k_0 \rho^2}{2f} = -\frac{n_1 k_0 (u^2 + v^2)}{2f}$$

where we used[7] Eq.3.140. This phase delay is a result of the smaller velocity of light in the lens medium than outside (since $n_2 > n_1$), hence rays that are off the optical axis will pass through a thinner lens section and will exit the lens earlier than rays passing at the center of the lens along the path $\overline{CC'}$. Thus the electric field at the output plane of the lens is:

$$F_{lens}^{out}(u,v) = F_{lens}^{in} e^{i(\delta\Phi(u,v) + \Phi_0)} = F_{lens}^{in}(u,v) e^{i\Phi_0} e^{-in_1 k_0 (u^2 + v^2)/2f} \qquad (A.83)$$

---

[7] Notice: $\overline{CA'} = R_1 - \sqrt{R_1^2 - \rho^2} = R_1 - R_1(1 - \rho^2/2R_1^2 + \cdots) \approx \rho^2/2R_1$. Similarly: $\overline{B'C'} \approx \rho^2/2R_2$.

where Φ_0 is a constant phase (independent of the coordinates u, v). Since the field intensity ($I \sim |E|^2$) is observed, we can without loss of generality set $\Phi_0 = 0$.

From Eq.A.82, the field in the image plane a distance s' from the lens is thus:

$$E_{image}(x,y) \approx \frac{n_1 k_0 e^{in_1 k_0 s'}}{2\pi i s'} \oint du\, dv\, E_{lens}^{out}(u,v) e^{in_1 k_0((x-u)^2+(y-v)^2)/2s'}$$

$$= E_0 \frac{e^{in_1 k_0(s+s')}}{2\pi i s s'} \oint du\, dv\, \exp\{in_1 k_0[\frac{u^2+v^2}{2s} - \frac{u^2+v^2}{2f} + \frac{(x-u)^2+(y-v)^2}{2s'}]\}$$

$$= E_0 \frac{e^{in_1 k_0[s+s'+(x^2+y^2)/2s']}}{2\pi i s s'} \oint du\, dv\, \exp\{in_1 k_0[\frac{u^2+v^2}{2}(\frac{1}{s}+\frac{1}{s'}-\frac{1}{f}) - \frac{ux+vy}{s'}]\}$$

Let us define: $k_x = n_1 k_0 x/s' = n_1 k_0 \sin\theta\cos\phi$, $k_y = n_1 k_0 y/s' = n_1 k_0 \sin\theta\sin\phi$ (where the coordinates in the image plane x, y are defined with respect to the distance s' and polar angles θ, ϕ from the lens). Using the lens formula, Eq.3.140, the amplitude of the field $|E_{image}(x,y)|$ becomes[8]:

$$|E_{image}(x,y)| = \frac{|E_0|}{2\pi s s'} \left| \oint du\, dv\, e^{-ik_x u - ik_y v} \right|$$

$$= \frac{|E_0|}{2\pi s s'} \left| \int_0^{D/2} \int_0^{2\pi} \rho\, d\rho\, d\phi' \, e^{-in_1 k_0 \rho \sin\theta\cos(\phi'-\phi)} \right|$$

where D is the diameter of the lens. We have encountered the last integral when computing the diffraction pattern of a circular aperture, see Eqs.3.145 and 3.146.

Using those equations we deduce that the light intensity in the image plane is related to the intensity of the point source I_0:

$$I_{image} = I_0(\frac{D^2}{2ss'})^2 \left| \frac{J_1(k_1 D\sin\theta/2)}{k_1 D\sin\theta} \right|^2$$

The light intensity is maximal when $\theta = 0$ and vanishes at the zeros of $J_1(k_1 D\sin\theta/2)$, the first of which occurs when $k_1 D\sin\theta = 7.66$, i.e. when $\sin\theta = 1.22\lambda_1/D$. The image of a point source (such as a star observed with a telescope or a single fluorophore imaged on a microscope) is thus a spot of radius[9]:

$$r_{image} \approx s'\sin\theta = 1.22\lambda_1 s'/D > 0.61\lambda_1/N.A$$

where the last inequality is obtained since $s' > f$. Two point sources (e.g. two stars) whose images are formed at a lateral distance $\Delta l_{image} < r_{image}$ will not be discernible as individual spots since their images overlap. In other words their angular separation $\Delta\theta \approx \Delta l_{object}/s = \Delta l_{image}/s'$ must be larger than $\approx 1.22\lambda_1/D$.

---

[8] Switching the integration variables from Cartesian coordinates (u, v) to polar coordinates (ρ, ϕ'): $u = \rho\cos\phi'$; $v = \rho\sin\phi'$.

[9] The numerical aperture of a lens is defined as: $N.A \equiv D/2f$.

This criterion was first established in the second half of the 19th century by Lord Rayleigh and independently by Ernest Abbe (one of the founders with Otto Schott and Carl Zeiss of the eponymous optical company). One hundred years later it was shown by S.Hell, E.Betzig and others that this resolution limit could actually be overcome (i.e. arbitrary close fluorescent sources could be resolved on a microscope) if one uses the non-linear optical properties of the fluorescent sources to selectively induce them to emit[77, 80, 81], see section 4.12.3.1.

A.7.2.2   Optical Image Processing

Let us now consider a situation of great significance for the processing of images: the relation between the light intensity coming out from a semi-transparent sample (a microscope slide for example) positioned in one focal plane (the front plane) a distance f from a lens and that measured in the back focal plane (at same distance f on the other side of the lens), see Fig.A.10. We shall see below that the field intensity in this back focal plane is the Fourier transform of the field out of the sample. By inserting appropriate masks in the back-focal plane the Fourier components can be filtered and a processed image of the sample obtained with a similar lens configuration, performing the inverse Fourier transform.

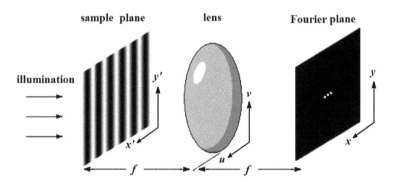

Figure A.10   A semi-transparent sample positioned in one focal plane (the front plane) of a lens is illuminated by a parallel light beam. The EM field is modulated by the sample transmission coefficient (here a sinusoidal modulation along the x-coordinate: $t(x) = a + b\sin(2\pi x/d)$). The field propagates to the lens and is focused by it to its second (back) focal plane. The image obtained in that plane i the Fourier transform of the sample image, i.e. three spots[11] whose distance is inversely proportional to the period d of the sample transmission.

Let a uniform beam of light E_0 impinge on a semi-transparent sample that alters the intensity and phase of the electric field via its transmission coefficient, $t(x',y')$: $E_s(x',y') = t(x',y')E_0$. From the sample the transmitted field propagates a distance $z = f$ in a media of refractive index n_1 (usually air with $n_1 = 1$) to a lens (of refractive index n_2) where its amplitude is by virtue of Eq.A.82:

$$E_{lens}^{in}(u,v) \approx \frac{n_1 e^{in_1 k_0 f}}{i\lambda_0 f} \oint dx' dy' E_s(x',y') e^{in_1 k_0 [(u-x')^2+(v-y')^2]/2f} \qquad (A.84)$$

From Eq.A.83, the field emerging from the lens is:

$$E_{lens}^{out}(u,v) = E_{lens}^{in}(u,v) e^{-in_1 k_0 (u^2+v^2)/2f}$$

Therefore the electric field at the back focal plane is:

$$E_f(x,y) \approx \frac{n_1 e^{in_1 k_0 f}}{i\lambda_0 f} \oint du dv E_{lens}^{out}(u,v) e^{in_1 k_0 [(x-u)^2+(y-v)^2]/2f}$$

$$= [\frac{n_1 e^{in_1 k_0 f}}{i\lambda_0 f}]^2 \oint dx' dy' E_s(x',y')$$

$$\cdot \oint du dv \exp\{i\frac{n_1 k_0}{2f}[(u-x')^2 + (v-y')^2 - (u^2+v^2) + (x-u)^2 + (y-v)^2]\}$$

$$= -\frac{n_1^2 e^{2in_1 k_0 f}}{\lambda_0^2 f^2} \oint dx' dy' E_s(x',y') e^{in_1 k_0 (xx'+yy')/f}$$

$$\cdot \int_{-\infty}^{\infty} du dv \exp\{i\frac{n_1 k_0}{2f}[(u-(x+x'))^2 + (v-(y+y')^2)]\}$$

$$= \frac{in_1 e^{2in_1 k_0 f}}{\lambda_0 f} \oint dx' dy' E_s(x',y') e^{in_1 k_0 (xx'+yy')/f}$$

$$= \frac{in_1 e^{2in_1 k_0 f}}{\lambda_0 f} \oint dx' dy' E_s(x',y') e^{ik_x x'+ik_y y'} \qquad (A.85)$$

where[12] $k_x = n_1 k_0 x/f$; $k_y = n_1 k_0 y/f$. The field $E_f(x,y)$ in the back focal plane is therefore the Fourier transform of the field transmitted by the plane.

The condition[12] $x < D$ implies that all the Fourier modes of the sample satisfy: $k_x < n_1 k_0 D/f$, i.e. that its smallest details of size $\delta x' = 2\pi/k_x^{max}$ obey: $\delta x' > \lambda_1 f/D \equiv \lambda_1/2N.A$ where $\lambda_1 = \lambda_0/n_1$ is the wavelength of light in the medium with refractive index n_1 (usually air). This condition sets a limit on the resolving power of a lens (namely on the size of smallest details in the sample that can be discerned) similar to the resolution limit we discussed above when considering the spot size of the image of a point light source.

Eq.A.85 has further important practical applications: by inserting appropriate masks in the Fourier plane, the image of a sample can be processed (filtered, modulated, etc.). Using a second lens whose front focal plane coincides with the Fourier

---

[11] The three spots correspond to the Fourier transform of $t(x)$, with spatial frequencies: $k_x = (-2\pi/d, 0, 2\pi/d)$, see Appendix A.4.

[12] If $\sqrt{\lambda_1 f} \ll D$ then in the integration over u only values for which the term $(u-(x+x'))^2$ is small will contribute. Assuming that $x, x' < D$, we can thus set the limits of integration to $\pm\infty$. Using the equality $\int_{-\infty}^{\infty} du e^{-(u-u_0)^2/2\sigma^2} = \sqrt{2\pi}\sigma$, the integration over u and v (for which the same arguments hold) yields: $i\lambda_0 f/n_1$.

plane (the back focal plane of the first lens) a processed (e.g. filtered) and inverted image of the sample is obtained in the image plane of the second lens, see Fig.A.11. If the focal distance of the second lens is f_2 and that of the first one is f_1, the image obtained is a filtered image of the sample amplified by a factor $M = f_2/f_1$.

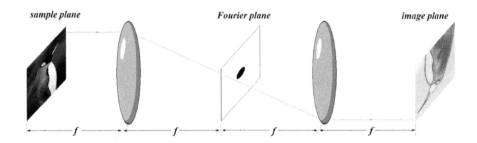

Figure A.11   The sample (an image of a penguin) is illuminated from the left. Its Fourier transform is obtained in the back focal plane of the first lens, where a small adsorbing disk eliminates its low frequency spatial Fourier components. A second lens —the front focal plane of which coincides with the mask —is used to perform the inverse Fourier transform, generating in its back focal plane an inverted high-pass filtered image of the sample, i.e. an image which highlights the contours.

A.8   ADVANCED TOPICS IN QUANTUM MECHANICS

A.8.1   QUANTUM TUNNELING

Consider a free particle moving with momentum p (wavenumber $k = p/\hbar$) along the x-axis and encountering a wall (an energy barrier of height $V > p^2/2m = E$ and thickness $x = d$, see Fig.4.14(a)). On the left side of the wall the particle's wavefunction is the sum of an incident and a reflected wave:

$$\Psi_l(x) = Ae^{ikx} + Be^{-ikx}$$

with $k^2 = 2mE/\hbar^2$, while on its right side the transmitted wavefunction is:

$$\Psi_r(x) = Ce^{ikx}$$

In the forbidden region (i.e. within the wall: $0 < x < d$) Schrödinger's equation, Eq.4.54

$$-\frac{\hbar^2}{2m}\nabla^2\Psi + V\Psi = E\Psi$$

has solutions of the form:

$$\Psi_o(x) = De^{-\kappa x} + Ee^{\kappa x}$$

with $\kappa = \sqrt{2m(V-E)}/\hbar$. At the interface with the wall (at $x = 0$ and $x = d$) the continuity of the wavefunction and its first derivative require that:

$$(1) \qquad A + B = D + E$$
$$(2) \qquad ik(A - B) = \kappa(-D + E)$$
$$(3) \qquad Ce^{ikd} = De^{-\kappa d} + Ee^{\kappa d}$$
$$(4) \qquad ikCe^{ikd} = \kappa(-De^{-\kappa d} + Ee^{\kappa d})$$

We are interested in the transmission probability, i.e. in the ratio $|C/A|^2$. Let us eliminate D by multiplying Eq.(1) by κ and adding Eq.(2) to obtain:

$$(5) \qquad (\kappa + ik)A + (\kappa - ik)B = 2\kappa E$$

Similarly let us multiply Eq.(3) by κ and add Eq.(4) to obtain:

$$(6) \qquad (\kappa + ik)e^{ikd}C = 2\kappa e^{\kappa d}E$$

and subtract Eq.(3) from Eq.(1) multiplied by $e^{-\kappa d}$:

$$(7) \qquad e^{-\kappa d}A + e^{-\kappa d}B - e^{ikd}C = E(e^{-\kappa d} - e^{\kappa d})$$

Let us know substitute E from Eq.(6) into Eqs.(5) and (7):

$$(8) \qquad (\kappa + ik)A + (\kappa - ik)B = (\kappa + ik)e^{-\kappa d + ikd}C$$
$$(9) \qquad e^{-\kappa d}A + e^{-\kappa d}B - e^{ikd}C = \frac{(\kappa + ik)}{2\kappa}(e^{-2\kappa d} - 1)e^{ikd}C$$

Eliminate B by multiplying Eq.(9) by $(\kappa - ik)e^{\kappa d}$ and subtracting it from Eq.(8):

$$4ik\kappa A = [2\kappa(\kappa + ik)e^{-\kappa d} + (\kappa^2 + k^2)(e^{\kappa d} - e^{-\kappa d}) - 2\kappa(\kappa - ik)e^{\kappa d}]e^{ikd}C$$
$$= [(k^2 - \kappa^2)(e^{\kappa d} - e^{-\kappa d}) + 2ik\kappa(e^{\kappa d} + e^{-\kappa d})]e^{ikd}C$$
$$= [2(k^2 - \kappa^2)\sinh\kappa d + 4ik\kappa\cosh\kappa d]e^{ikd}C$$

Thus the transmission probability becomes[13]:

$$
\begin{aligned}
T = \left|\frac{C}{A}\right|^2 &= \frac{4k^2\kappa^2}{(k^2 - \kappa^2)^2 \sinh^2\kappa d + 4k^2\kappa^2 \cosh^2\kappa d} \\
&= \frac{4k^2\kappa^2}{4k^2\kappa^2 + (k^2 + \kappa^2)^2 \sinh^2\kappa d} \\
&= \frac{4E(V - E)}{4E(V - E) + V^2 \sinh^2\kappa d} \rightarrow \frac{16E}{V}e^{-d/\lambda_t} \quad \text{when:} \quad E \ll V \text{ and } \kappa d > 1
\end{aligned}
\tag{A.86}
$$

Where $\lambda_t = 1/2\kappa \approx \hbar/2\sqrt{2mV}$ is the typical tunneling length. For an energy barrier $V = 1\text{eV}$, the tunneling length for an electron: $\lambda_t \simeq 1\text{Å}$ which is the typical size of an atom. The transmission probability thus decreases exponentially with the size d of the gap which can be but a few Å wide.

A.8.2 GAUGE INVARIANCE: THE AHARONOV-BOHM EFFECT

The Hamiltonian of a charged particle in a magnetic field, Eq.4.62, appears to violate the principle of gauge invariance. This principle implies that Nature is unaffected by a change in the potential fields $(\Phi, \vec{A})$ that leaves the real electromagnetic fields $(\vec{E}, \vec{B})$ unchanged. We have seen that $\vec{A}$ is defined up to the addition of the gradient of an arbitrary function $f: \vec{A} \rightarrow \vec{A} + \vec{\nabla}f$ which does not affect the real field $\vec{B} = \vec{\nabla} \times \vec{A}$, see Eq.3.57. But such a gauge transformation does modify the Hamiltonian:

$$\hat{H}' = \frac{(\hat{p} - (q/c)\vec{A} - (q/c)\vec{\nabla}f)^2}{2m} = \frac{(-i\hbar\vec{\nabla} - (q/c)\vec{A} - (q/c)\vec{\nabla}f)^2}{2m}$$

However this transformation can be compensated by a similar gauge transformation on the wavefunction:

$$\Psi'(\vec{r}) = e^{iqf/\hbar c}\Psi(\vec{r})$$

In that case:

$$
\begin{aligned}
\hat{H}'\Psi'(\vec{r}) &= \frac{1}{2m}(\hat{p} - (q/c)\vec{A} - (q/c)\vec{\nabla}f)^2 e^{iqf/\hbar c}\Psi(\vec{r}) \\
&= \frac{1}{2m}(-i\hbar\vec{\nabla} - (q/c)\vec{A} - (q/c)\vec{\nabla}f)e^{iqf/\hbar c}(-i\hbar\vec{\nabla} - (q/c)\hat{A})\Psi(\vec{r}) \\
&= \frac{1}{2m}(-i\hbar\vec{\nabla} - (q/c)\vec{A})^2\Psi(\vec{r}) = \frac{1}{2m}(\hat{p} - (q/c)\hat{A})^2\Psi(\vec{r}) = \hat{H}\Psi(\vec{r})
\end{aligned}
$$

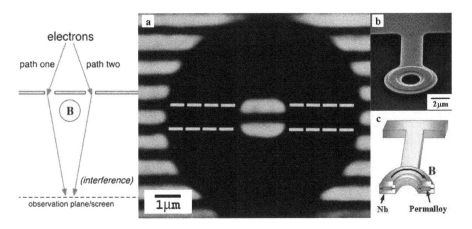

Figure A.12 The double-slit or Young's experiment performed in presence of a magnetic field confined in a region isolated from the interfering electrons. An electron passes through two slits and interferes on a far-field screen. On its way it is incident on a microscopic toroidal ring of ferromagnetic material (with azimuthal magnetization) enclosed by a superconducting Niobium (Nb) shield to prevent any flux leakage. Notice the phase shift $\delta\phi = \pi$ in the interference pattern of the electrons that have passed in the center of the ring as compared with those that have passed outside (bright stripes appear instead of dark ones). This phase shift is due to the flux $\Delta\Phi = hc/2e$ (= $2\ 10^{-7}$ Gauss cm^2 = $2\ 10^{-15}$ Tesla m^2) enclosed by the paths surrounding the ring (the value of this flux quantum is itself a result of the Aharonov-Bohm effect on the superconducting current with charge 2e flowing and interfering constructively in the Niobium shield). Reprinted with permission from ref.[115], copyright (1986) by the American Physical Society.

In general the extra phase $qf/\hbar c$ has no effect on the probability of detecting a particle at a given position, since that probability depends on the absolute value of $\Psi(x)$. There is however a situation first pointed out by Aharonov and Bohm in 1959 where this gauge transformation leads to detectable effects. This happens when the probability of detecting an electron in presence of a magnetic field results from the interference between different paths the electron can follow, as in the diffraction experiment shown in Figs.4.12 and A.12. Based on our study of particle diffraction, see section 4.5.1, the wave-function of a particle diffracted by two slits in presence

---

[13]Using the identity: $\cosh^2 x - \sinh^2 x = 1$.

of a magnetic field[14] on a screen far away is:

$$|\Psi\rangle = |I\rangle + |II\rangle = e^{ikl_1 + i(q/\hbar c)\int_1 \vec{A}\cdot d\vec{r}}|O\rangle + e^{ikl_2 + (q/\hbar c)\int_2 \vec{A}\cdot d\vec{r}}|O\rangle$$

$$= (1 + e^{ik(l_2 - l_1) + i(q/\hbar c)\oint \vec{A}\cdot d\vec{r}})|O\rangle$$

$$= (1 + e^{ik(l_2 - l_1) + i\delta\phi})|O\rangle$$

where the phase difference $\delta\phi$ is related to the magnetic flux Φ enclosed by the interfering paths:

$$\delta\phi = (q/\hbar c)\oint \vec{A}\cdot d\vec{r} = (q/\hbar c)\oint \vec{\nabla}\times\vec{A}\cdot d\vec{S} = (q/\hbar c)\oint \vec{B}\cdot d\vec{S} = q\Phi/\hbar c$$

Hence the enclosed magnetic flux can change the pattern of interference of the electron wavefunction, even if the electron does not (and classically cannot) pass through the region containing the magnetic field and no force is exerted on it! As shown in Fig.A.12, this effect has been clearly evidenced using superconducting materials to prevent leakage of the magnetic field[115]. Notice that the interference pattern is periodic in Φ with a period $\Delta\Phi = hc/q$ a property which is used to measure the magnetic field in extremely sensitive magnetic sensors known as Superconducting Quantum Interference Devices (SQUID).

A.8.3 ANGULAR MOMENTUM REPRESENTATION

We have seen that the eigenstates of the momentum are characterized by two quantum numbers: l and m that are eigenvalues of respectively the total angular momentum operator $\hat{L}^2$ and the momentum along a chosen axis $\hat{L}_z$. While the operators $\hat{L}_x, \hat{L}_y$ commute with $\hat{L}^2$ they do not commute with $\hat{L}_z$. Therefore their representation is not diagonal in the eigenstates of $\hat{L}_z$: $|l, m\rangle$. To find that representation, it is useful to define the raising and lowering operators:

$$\hat{L}_\pm = \hat{L}_x \pm i\hat{L}_y = \hbar e^{\pm i\phi}\left(\pm\frac{\partial}{\partial\theta} + i\cot\theta\frac{\partial}{\partial\phi}\right)$$

which satisfy the following commutation rules:

$$[\hat{L}_+, \hat{L}_z] = -\hbar\hat{L}_+$$

$$[\hat{L}_-, \hat{L}_z] = \hbar\hat{L}_-$$

$$[\hat{L}_+, \hat{L}_-] = 2\hbar\hat{L}_z$$

---

[14]Notice that the wavefunction of a free particle in a magnetic field is:

$$\Psi(\vec{r}, t) = \exp\left(i\vec{k}\cdot\vec{r} - i\omega t + i(q/\hbar c)\int \vec{A}\cdot d\vec{r}\right)$$

as can be verified by inserting that function into Schrödinger's equation:

$$\hat{H}\Psi = \frac{1}{2m}(-i\hbar\vec{\nabla} - (q/c)\vec{A})^2\Psi = \frac{\hbar^2 k^2}{2m}\Psi = E\Psi = i\hbar\frac{\partial\Psi}{\partial t} = \hbar\omega\Psi$$

Notice that $\hat{L}^2 = \hat{L}_x^2 + \hat{L}_y^2 + \hat{L}_z^2 = \hat{L}_z^2 + \hat{L}_+\hat{L}_- - \hbar\hat{L}_z = \hat{L}_z^2 + \hat{L}_-\hat{L}_+ + \hbar\hat{L}_z$ commutes with $\hat{L}_\pm$. Using these commutation rules one can show that:

$$\hat{L}_\pm |l, m\rangle \sim |l, m \pm 1\rangle$$

Consider for example the commutation relations for $\hat{L}_+$:

$$-\hbar\hat{L}_+ |l, m\rangle = [\hat{L}_+, \hat{L}_z] |l, m\rangle = (\hat{L}_+\hat{L}_z - \hat{L}_z\hat{L}_+) |l, m\rangle$$
$$= \hbar m \hat{L}_+ |l, m\rangle - \hat{L}_z\hat{L}_+ |l, m\rangle$$

namely : $\qquad \hat{L}_z(\hat{L}_+ |l, m\rangle) = \hbar(m+1)(\hat{L}_+ |l, m\rangle)$

Hence:

$$\hat{L}_\pm |l, m\rangle = a_{lm}^\pm |l, m \pm 1\rangle = \hbar \sqrt{l(l+1) - m(m \pm 1)} |l, m \pm 1\rangle \qquad (A.87)$$

Where the coefficient a_{lm}^+ is determined by the normalization condition:

$$|a_{lm}^+|^2 = |a_{lm}^+|^2 \langle l, m+1 | l, m+1\rangle = \langle l, m | \hat{L}_+^* \hat{L}_+ | l, m\rangle = \langle l, m | \hat{L}_- \hat{L}_+ | l, m\rangle$$
$$= \langle l, m | \hat{L}^2 - \hat{L}_z^2 - \hbar\hat{L}_z | l, m\rangle = \hbar^2 (l(l+1) - m(m+1))$$

Notice that the action of $\hat{L}_+$ on a state $|l, m\rangle$ is to generate the higher m-state $|l, m+1\rangle$ whereas the action of $\hat{L}_-$ on $|l, m\rangle$ generates the lower m-state $|l, m-1\rangle$, which justifies their names as raising and lowering operators[15]. The matrix representation $L_{m'm}^\pm$ of $\hat{L}_\pm$ is then non-zero only if $m' = m \pm 1$:

$$L_{m'm}^\pm \equiv \langle m' | \hat{L}_\pm | m\rangle = \hbar \sqrt{l(l+1) - m(m \pm 1)}\delta_{m', m\pm 1}$$

For $l = 1$ we obtain:

$$L_+ = \sqrt{2}\hbar \begin{pmatrix} 0 & 1 & 0 \\ 0 & 0 & 1 \\ 0 & 0 & 0 \end{pmatrix} \quad L_- = \sqrt{2}\hbar \begin{pmatrix} 0 & 0 & 0 \\ 1 & 0 & 0 \\ 0 & 1 & 0 \end{pmatrix}$$

$$L_x = \frac{\hbar}{\sqrt{2}} \begin{pmatrix} 0 & 1 & 0 \\ 1 & 0 & 1 \\ 0 & 1 & 0 \end{pmatrix} \quad L_y = \frac{\hbar}{\sqrt{2}} \begin{pmatrix} 0 & -i & 0 \\ i & 0 & -i \\ 0 & i & 0 \end{pmatrix}$$

It is easy to verify that the eigenvalues of L_x are as expected: $\hbar, 0, -\hbar$ with corresponding eigenstates:

$$|1\rangle_x = \frac{1}{2}\begin{pmatrix} 1 \\ \sqrt{2} \\ 1 \end{pmatrix} \quad |0\rangle_x = \frac{1}{\sqrt{2}}\begin{pmatrix} 1 \\ 0 \\ -1 \end{pmatrix} \quad |-1\rangle_x = \frac{1}{2}\begin{pmatrix} 1 \\ -\sqrt{2} \\ 1 \end{pmatrix}$$

Hence if a particle is prepared in state $|1\rangle_x$ and its angular momentum along the z-axis is measured it has probability $1/4$ of yielding a value $\hbar$ or $-\hbar$ and a probability $1/2$ of yielding a value 0.

---

[15]From the fact that $|a_{lm}^\pm|^2 \geq 0$, it is clear without even solving Schrödinger's equation that $-l \leq m \leq l$.

A.8.3.1   Total Angular Momentum Eigenstates

Since the spin is also an angular momentum operator the same commutation relations exist between $\hat{S}^2$, $\hat{S}_x$, $\hat{S}_y$, $\hat{S}_z$ and $\hat{S}_\pm$. For spin one-half particles, the eigenstates $|s, s_z\rangle$ of $\hat{S}^2 = \hbar^2\sigma^2/4$ and $\hat{S}_z = \hbar\sigma_z/2$ with eigenvalues[16] $s(s+1)\hbar^2 = 3\hbar^2/4$ and $\hbar s_z = (\pm 1/2)\hbar$ are:

$$|1/2, 1/2\rangle = \begin{pmatrix} 1 \\ 0 \end{pmatrix} \qquad\qquad |1/2, -1/2\rangle = \begin{pmatrix} 0 \\ 1 \end{pmatrix}$$

We have seen that one of the results of Dirac's equation is that spin and angular momentum operators $\hat{L}_z, \hat{S}_z$ do not commute with the Hamiltonian and are therefore not good (i.e. conserved) quantum numbers when relativistic effects are important and Dirac's equation has to be used instead of Schrödinger's equation. Rather it is the total angular momentum operator: $\hat{J}_z = \hat{L}_z + \hat{S}_z$ with eigenvalue $m_J = m + s_z$ which is conserved, together with $\hat{J}^2$, $\hat{L}^2$ and $\hat{S}^2$. Thus the angular momentum eigenstate is a function of 4 quantum numbers: $|j, l, s, m_J\rangle$. Without loss of generality we may assume that $m_J = m + 1/2$ ($m_J = m - 1/2$ is equivalent to $m_J = (m-1) + 1/2$). As the eigenstates of the angular momentum are Y_{lm} we can write:

$$|j, l, s, m_J\rangle = c_1 |l, m\rangle |s, 1/2\rangle + c_2 |l, m+1\rangle |s, -1/2\rangle$$
$$= c_1 Y_{l,m} \begin{pmatrix} 1 \\ 0 \end{pmatrix} + c_2 Y_{l,m+1} \begin{pmatrix} 0 \\ 1 \end{pmatrix} \qquad\qquad \text{(A.88)}$$

For this state to be an eigenstate of $\hat{J}^2 = \hat{L}^2 + \hat{S}^2 + 2\hat{L}\cdot\hat{S}$ it has to satisfy:

$$\hat{J}^2|j, l, s, m_J\rangle = \hbar^2[l(l+1) + \frac{3}{4} + \frac{\hat{L}\sigma}{\hbar}]|j, l, s, m_J\rangle = \hbar^2\gamma|j, l, s, m_J\rangle$$

Using the relation:

$$\hat{L}\sigma = \hat{L}_x\sigma_x + \hat{L}_y\sigma_y + \hat{L}_z\sigma_z \equiv \begin{pmatrix} L_z & L_- \\ L_+ & -L_z \end{pmatrix}$$

and the previously derived equation for the raising and lowering operators, Eq.A.87, one obtains the following eigenvalue equation for γ:

$$\begin{pmatrix} l(l+1) + 3/4 + m - \gamma & \sqrt{l(l+1) - m(m+1)} \\ \sqrt{l(l+1) - m(m+1)} & l(l+1) - 1/4 - m - \gamma \end{pmatrix} \begin{pmatrix} c_1 Y_{l,m} \\ c_2 Y_{l,m+1} \end{pmatrix} = 0$$

which has a non-zero solution when the determinant of the matrix is null, i.e. when the following quadratic equation is satisfied:

$$[(l + 1/2)^2 - \gamma]^2 - (l + 1/2)^2 = 0$$

---

[16]The matrices σ_i ($i = x, y$ or z) are Pauli matrices, Eq.4.65 .

which solution is $\gamma = j(j+1)$ with $j = l \pm 1/2$. Hence the eigenvalues of the total angular momentum operator $\hat{J}^2$ are $\hbar^2 j(j+1)$ with eigenfunctions:

$$|j,l,s,m_J\rangle = |l+1/2,l,1/2,m_J\rangle = \frac{1}{\sqrt{2l+1}}\begin{pmatrix} \sqrt{l+m+1}\; Y_{l,m} \\ \sqrt{l-m}\; Y_{l,m+1} \end{pmatrix} \qquad (A.89)$$

$$|j,l,s,m_J\rangle = |l-1/2,l,1/2,m_J\rangle = \frac{1}{\sqrt{2l+1}}\begin{pmatrix} \sqrt{l-m}\; Y_{l,m} \\ -\sqrt{l+m+1}\; Y_{l,m+1} \end{pmatrix}$$

A.8.4 PERTURBATION THEORY WITH DEGENERATE EIGENSTATES

When eigenstates have the same energy the perturbation expansion described in section 4.11 diverges at second order since the denominator in Eq.4.107 is zero for degenerate states (i.e. when $\epsilon_n^0 = \epsilon_m^0$). The way to deal with degenerate eigenstates is to neglect all transitions to non-degenerate eigenstates and to solve the equations exactly taking into account only the transitions between states of the same energy. This will usually raise the degeneracy with the new eigenstates expressed as a superposition of the original unperturbed eigenstates. One may then proceed using non-degenerate perturbation theory with these new eigenstates as zeroth order approximation.

As above we assume that the Hamiltonian is

$$H = H_0 + \lambda H_1$$

And we want to solve for it taking into account only the N degenerate eigenstates $|\phi_n\rangle$ $(n = 1, \ldots, N)$ of H_0 with energy ϵ^0:

$$H_0 |\phi_n\rangle = \epsilon^0 |\phi_n\rangle$$

One seeks the new eigenstates $|\psi_n\rangle$ of H as a superposition of these degenerate eigenstates:

$$|\psi_n\rangle = \sum_m a_{nm} |\phi_m\rangle$$

which yields: $H|\psi_n\rangle = E_n|\psi_n\rangle$ or:

$$\sum_m a_{nm}((\epsilon^0 - E_n) + \lambda H_1)|\phi_m\rangle = 0$$

Defining $V_{nm} \equiv \lambda\langle\phi_n|H_1|\phi_m\rangle$ yields the following equation:

$$\begin{vmatrix} \epsilon^0 + V_{11} - E_n & V_{12} & \cdots & V_{1N} \\ V_{21} & \epsilon^0 + V_{22} - E_n & \cdots & V_{2N} \\ \vdots & \vdots & \vdots & \vdots \\ V_{N1} & \cdots & V_{N,N-1} & \epsilon^0 + V_{NN} - E_n \end{vmatrix} = 0 \qquad (A.90)$$

The solution of this N^{th} order equation for E_n will usually yield N eigenvalues and N orthogonal eigenstates $|\psi_n\rangle$ which can then be used instead of the $|\phi_n\rangle$ in a perturbation expansion together with the other non-degenerate eigenstates of H_0.

A.8.4.1 The Stark Effect in Hydrogen

When solving for the eigen-energies of the Hydrogen atom we saw that —uniquely among all other elements —its energy levels are degenerate with respect to the quantum numbers l and m: $E_{n,l,m} = E_{n,l',m'}$, see section 4.11.1. Since the Stark effect couples between eigen-states with different l ($l' = l \pm 1$, see section 4.11.2) for which the eigen-energies for a given n quantum number are identical, the second order non-degenerate perturbation expansion, Eq.4.107 breaks down.

We shall now see how the degeneracy of the electronic energy levels in hydrogen are affected by the presence of an electric field. The Hamiltonian is:

$$H = H_0 + ez\mathcal{E}$$

Consider the case of the $n = 2$ level of hydrogen for which the states: $l = 0; m = 0$ and $l = 1; m = 0, \pm 1$ all have the same energy: $E_2 = E_1/4$ (with $E_1 = -13.6\text{eV}$). Following our previous perturbation analysis for degenerate energy levels, we need to diagonalize the Hamiltonian matrix: $\langle 2, l', m'|H|2, l, m\rangle$. While the unperturbed Hamiltonian H_0 is diagonal with energy E_2, using Eq.4.110 we deduce that the dipole coupling $\langle 2, l', m'|ez\mathcal{E}|2, l, m\rangle$ is non-zero only if $m' = m$ and $l' = l \pm 1$, i.e. only when the coupled states are $|2, 0, 0\rangle$ and $|2, 1, 0\rangle$. In that case:

$$\langle 2, 1, 0|ez\mathcal{E}|2, 0, 0\rangle = e\mathcal{E} \int r^3 dr R_{21}(r) R_{20}(r) \int d\Omega Y_{10}(\theta, \phi) \cos\theta Y_{00}(\theta, \phi)$$

$$= -3er_0\mathcal{E} \equiv -\Delta E_2$$

Where $r_0 = \hbar^2/me^2 = 0.53\text{Å}$ is Bohr's radius, section 4.1.3. The eigen-energies E are thus determined by the equation:

$$\begin{vmatrix} E_2 - E & -\Delta E_2 & 0 & 0 \\ -\Delta E_2 & E_2 - E & 0 & 0 \\ 0 & 0 & E_2 - E & 0 \\ 0 & 0 & 0 & E_2 - E \end{vmatrix} = 0$$

which solutions are: $E^I = E_2 + \Delta E_2 = E_2 + 3er_0\mathcal{E}$ and $E^{II} = E_2 - \Delta E_2$ with eigenstates:

$$|2, I\rangle = \frac{|2, 0, 0\rangle - |2, 1, 0\rangle}{\sqrt{2}} \qquad |2, II\rangle = \frac{|2, 0, 0\rangle + |2, 1, 0\rangle}{\sqrt{2}}$$

and $E^{III,IV} = E_2$ with eigenstates $|2, 1, 1\rangle$ and $|2, 1, -1\rangle$. Notice that for the hydrogen atom the Stark effect, i.e. the shift ΔE_2 in the energy level, is linear with electric field and not quadratic as it is for the other atoms or molecules for which the electronic energy levels are non-degenerate, see Eq.4.111 and Fig.A.13.

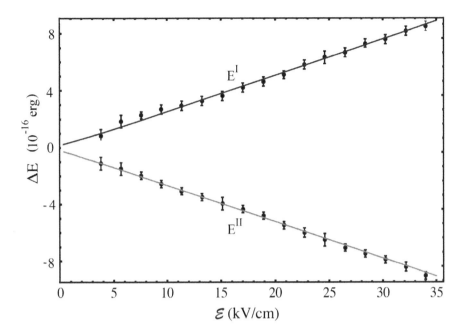

Figure A.13   The split in the energy of the $n = 2$ state upon application of an electric field in Hydrogen. The energy splits into two mixed levels with energies E^I and E^{II}, which vary linearly with the electric field. The predicted value of the slope (see text): $3er_0 = 7.63 \ 10^{-18}$ erg cm/statV is close to the measured value of about $2.5 \ 10^{-17}$ erg cm/kV $= 7.5 \ 10^{-18}$ erg cm/statV. Reprinted with permission from ref.[116], copyright (2014) by the American Physical Society.

A.9 ADVANCED TOPICS IN STATISTICAL MECHANICS

A.9.1 SPECIFIC HEAT OF SOLIDS

The atoms in a solid crystal are positioned at their equilibrium position: no net force acts on them. In such a situation if an atom is displaced by a small fluctuations of amplitude $\vec{x}$ (much smaller than the inter-atomic distance), it experiences a spring-like restoring force that brings it back to its original equilibrium position. The energy of atoms in a solid can thus be described as the energy of small particles of mass m tethered by small springs of elastic constant k_s:

$$E_i = \vec{p}_i^2/2m + k_s\vec{x}_i^2/2 \tag{A.91}$$

where $\vec{p}_i, \vec{x}_i$ are the momentum and displacement vectors of atom i from its equilibrium position. By the equipartition theorem the energy of a solid should thus be: $E = 3NkT$ (each atom has 6 degrees of freedom with mean energy $kT/2$ each). The specific heat of solids is then $\tilde{c}_V = 3k$, which is finite at zero temperature. This is a paradoxical result since it implies that the entropy change of a solid $dS = \delta Q/T = N\tilde{c}_V \delta T/T$ diverges as $T \to 0$ (this is also true for gases, but the gas phase ceases to exist (gases condense) before reaching zero temperature). This paradox (together with the divergence of the black-body radiation) was resolved by the discovery of quantum mechanics.

The quantum mechanical study of the harmonic oscillator, see section 4.10.3, reveals that its energy levels are quantized: $E(i) = (i + 1/2)\hbar\omega$ (with $\omega = \sqrt{k/m}$ and $i = 0, 1, \cdots$). In 1907, Einstein assumed that all atoms in a solid oscillated in three dimensions at the same frequency $\omega = \omega_0$. The energy of a particular configuration of vibrational excitations $\{i_n\}$ in the solid can be written as

$$E_p = \sum_{n=1}^{3N} E_n \quad \text{with}: \quad E_n(i_n) = (i_n + 1/2)\hbar\omega_0$$

and the sum is over the $3N$ oscillatory modes. Following our study of paramagnetic systems, polymers or ideal gases, see Eq.5.44, we can write the partition function for the vibrations in the solid considered by Einstein:

$$
\begin{aligned}
\mathcal{Z} = \text{Tr } e^{-\beta E_p} &= \sum_{\{i_n=1\}}^{\infty} \Pi_{n=1}^{3N} e^{-\beta E_n(i_n)} = \left(\sum_{i=1}^{\infty} e^{-\beta E(i)}\right)^{3N} \\
&= \left(\frac{e^{-\beta\hbar\omega_0/2}}{1 - e^{-\beta\hbar\omega_0}}\right)^{3N}
\end{aligned} \tag{A.92}
$$

From Eq.5.46, the mean energy of the system is then:

$$E = \langle E_p \rangle = -\frac{\partial}{\partial\beta} \log \mathcal{Z} = \frac{3N\hbar\omega_0}{2} + 3N\frac{\hbar\omega_0}{e^{\beta\hbar\omega_0} - 1} \tag{A.93}$$

The first term on the right is the ground state energy of $3N$ identical harmonic oscillators. At high temperatures $kT \gg \hbar\omega_0$, the energy $E = 3NkT$ as expected from the equipartition theorem. At low temperatures on the other hand:

$$E = \frac{3N\hbar\omega_0}{2} + 3N\hbar\omega_0 e^{-\beta\hbar\omega_0}$$

and the specific heat at low temperature behaves as:

$$\tilde{c}_V = \frac{1}{N}\frac{\partial E}{\partial T} = 3k(\beta\hbar\omega_0)^2 e^{-\beta\hbar\omega_0} \tag{A.94}$$

which goes to zero exponentially as the temperature goes to zero. Hence the entropy doesn't diverge as $T \to 0$. While the model of Einstein solved the paradox of the entropy of solids, it failed to reproduce the temperature dependence of the specific heat at low temperatures (which decreased as T^3 rather than exponentially in $1/T$).

A more realistic model was then proposed and solved by Debye in 1912. It doesn't assume that all the atoms in a solid oscillate at the same frequency. Rather at each frequency ω there are a number $n(\omega)$ of collective oscillatory modes known as phonons, so that the total number of phonons $\int d\omega\, n(\omega) = 3N$. In that case the energy Eq.A.93 becomes:

$$E = \int d\omega\, n(\omega)\hbar\omega/2 + \int d\omega\, n(\omega)\frac{\hbar\omega}{e^{\beta\hbar\omega} - 1} \tag{A.95}$$

Let us now compute the number of phonons at frequency ω: $n(\omega)$. Phonons are standing oscillatory waves (sound waves) in the solid and they behave like the free particle in the square box (of dimensions $L_x \times L_y \times L_z$) studied in quantum mechanics. Their wavefunctions characterized by wavenumbers (k_x, k_y, k_z) vanishes on the solid boundaries: $\sin k_x x \big|_{0,L_x} = 0$ (and similarly along the y and z axes), namely $k_x = n_x \pi/L_x$ ($n_x = 0, 1, \cdots, N$). Since for every wavenumber there can be three modes of vibrations —two transverse modes (akin to the vibrations of a violin string) and one longitudinal (similar to compression waves in a spring) —the maximal oscillatory wavenumber k_m must satisfy:

$$3N = 3\frac{L_x L_y L_z}{\pi^3}\int_0^{k_m} dk_x dk_y dk_z = 3\frac{V}{8\pi^3}\int_0^{k_m} 4\pi k^2 dk \tag{A.96}$$

where $V = L_x L_y L_z$ is the volume of the box. The frequency of low energy phonons (sound waves) satisfy the linear dispertion relation: $\omega = c_s k$, where c_s is the sound velocity (notice the analogy with photons, the quanta of EM radiation, see section 5.9.1). Thus we can rewrite the above equation as:

$$3N = \frac{3V}{2\pi^2 c_s^3}\int_0^{\omega_D} \omega^2 d\omega \equiv \int n(\omega) d\omega \tag{A.97}$$

where the maximal frequency of oscillation, the Debye frequency ω_D obeys: $\omega_D = c_s(6\pi^2 N/V)^{1/3}$. The number of phonons of frequency ω is thus:

$$n(\omega) = 3V\omega^2/2\pi^2 c_s^3 \equiv 9N\frac{\omega^2}{\omega_D^3}$$

and the energy Eq.A.95 is:

$$E = \frac{9N\hbar}{2\omega_D^3} \int_0^{\omega_D} d\omega\,\omega^3 + \frac{9N}{\omega_D^3}\hbar \int_0^{\omega_D} d\omega\,\frac{\omega^3}{e^{\beta\hbar\omega}-1}$$

$$= \frac{9N\hbar\omega_D}{8} + \frac{9N}{\beta^4\hbar^3\omega_D^3}\int_0^{\xi_D} d\xi\,\frac{\xi^3}{e^\xi-1} \tag{A.98}$$

where $\xi = \beta\hbar\omega$. In the low temperature limit $T \ll \Theta_D \equiv \hbar\omega_D/k$ (i.e. $\xi_D = \Theta_D/T \to \infty$), the integral on the right equals $\pi^4/15$ and therefore:

$$E = \frac{9N\hbar\omega_D}{8} + \frac{3N\pi^4 k}{5\Theta_D^3}T^4 \tag{A.99}$$

and the specific heat is:

$$\tilde{c}_V = \frac{1}{N}\frac{\partial E}{\partial T} = \frac{12\pi^4}{5}k\left(\frac{T}{\Theta_D}\right)^3 \tag{A.100}$$

which decreases as T^3 when the temperature approaches zero as indeed observed experimentally, see Fig.A.14. In the high temperature limit $T \gg \Theta_D$ (i.e. $\xi_D \to 0$) one can expand the exponential term in Eq.A.98: $e^\xi - 1 \approx \xi$. The integral over ξ yields: $\xi_D^3/3$. From the deduced value of E, we then recover the classical result: $\tilde{c}_V = 3k$.

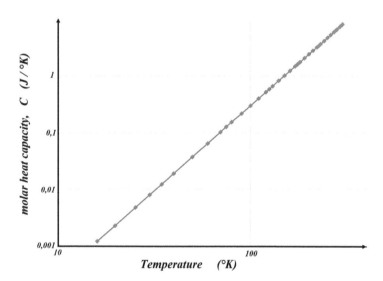

Figure A.14 The molar specific heat (see Eq.5.102) of diamond as a function of temperature. Data (blue dots) from [117, 118]. The continuous line is the theoretical prediction: $c_V = (12\pi^4/5)(T/\Theta_D)^3 R$ with a Debye temperature $\Theta_D = \hbar\omega_D/k = 1870°$K.

A.9.2 THE BIPOLAR TRANSISTOR

A bipolar transistor consists of two *pn* junctions back to back, so as to make for a *npn* (or *pnp*) sandwich. We shall here consider only the *npn* transistor (the *pnp* mode of operation is similar), see Fig.A.15(a,b). In such a transistor one of the *n*-type (the emitter) is more highly doped than the other *n*-type (the collector) and the *p*-type in the middle (the base). The base is held at a higher voltage than the emitter, thus putting the base-emitter *pn* junction under forward bias, whereas the collector is held at a higher voltage than the base, thus putting the collector-base *pn* junction under reverse bias. As we shall see such a device amplifies the current passing through the base and collected at the collector.

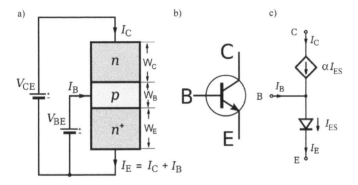

Figure A.15   (a) A *npn* bipolar transistor is formed by a sandwich of a *p*-type of width W_B held between two *n*-type semiconductors of width W_E, W_C. One of the *n*-type semiconductors (the emitter, E) has a higher concentration of impurities than either the base (B) or the collector (C). In its regular mode of function, the first *pn* junction between base and emitter is held at a forward bias $V_{BE} > 0$, while the second junction between collector and base is held at a reverse bias $V_{CB} = V_{CE} - V_{BE} > 0$. The current I_E through the emitter (hole+electron) is the sum of the base current I_B (hole injection into the emitter) and collector current I_C (electron injection from the emitter). (b) Schematics of a *npn* transistor displaying the emitter, base and collector connections. (c) The simplified Ebers-Moll model for a bipolar transistor in its habitual functional mode, see text. The first *pn* junction (between emitter and base) functions like a diode under forward bias while the current through the second junction is a result of the diffusion of the injected electrons from the emitter, through the base to the collector (details in the text).

From our previous discussion of the *pn* junction, Eq. 5.182, we know that the *p*-type side (the base) of the emitter-base junction being under a forward bias voltage V_{BE} experiences an injection of electrons from *n*-type side (the emitter) of density:

$$\delta n_B^e = n_E [e^{-\beta(V_{pn} - V_{BE})} - e^{-\beta V_{pn}}] = \frac{n_i^2}{n_B}(e^{\beta V_{BE}} - 1)$$

Where $n_E \equiv n_e^n$, $n_B \equiv n_h^p$ are the impurity densities on the emitter and base side respectively. These injected electrons diffuse in the base until they encounter the second pn junction (the base-collector junction under reverse bias) where they are drawn by the electric field across that junction to the n-type side (the collector). Since the diffusion current $\vec{J} = -De\vec{\nabla}n$ (where D is the diffusion constant, see sections 5.1.2 and 5.11.1.1) the current I_C arriving at the collector is:

$$I_C = S J_C \approx S D_B e \delta n_B^e / W_B = \frac{S D_B e n_i^2}{n_B W_B}(e^{\beta V_{BE}} - 1)$$

Where D_B is the diffusion constant of electrons in the base of length W_B and cross section S. On the emitter (n-type) side, the junction experiences an injection of holes from the p-type (the base) of density:

$$\delta n_E^h = \frac{n_i^2}{n_E}(e^{\beta V_{BE}} - 1)$$

which by the same argument as before results in a hole current from the base:

$$I_B = S J_B \approx S D_E e \delta n_E^h / W_E = \frac{S D_E e n_i^2}{n_E W_E}(e^{\beta V_{BE}} - 1)$$

The amplification factor β_e of a bipolar transistor is defined as the ratio of the collector current I_C to the current in the base I_B:

$$\beta_e \equiv \frac{I_C}{I_B} = \frac{D_B W_E n_E}{D_E W_B n_B}$$

While there is not much control on the diffusion of electrons and holes (i.e. D_B, D_E) and limited leeway with the lengths of the junctions (W_B, W_E), one can achieve large amplification factors $\beta_e > 100$ by controlling the concentrations of impurities such that $n_E \gg n_B$.

With the definition: $I_{ES} = (S D_E e n_i^2 / n_E W_E) + (S D_B e n_i^2 / n_B W_B)$ (= the saturation current in the emitter under reverse bias of both junctions) and $\alpha \equiv I_C/I_E = 1 - \beta_e^{-1}$, the previous equations can be summarized in the oft-used Ebers-Moll model for a bipolar transistor, see Fig.A.15:

$$I_E = I_C + I_B = I_{ES}(e^{\beta V_{BE}} - 1)$$
$$I_C = \alpha I_E \approx I_E \qquad\qquad (A.101)$$
$$I_B = (1 - \alpha)I_E = \frac{I_E}{\beta_e}$$

To summarize, the collector current in a bipolar transistor is determined by the current flowing in the base amplified by a factor β_e: a bipolar transistor is a current operated device. On the other hand in a Field Effect Transistor (FET) (such as a MOS-FET) described in section 5.9.2.4, the voltage controls the current flowing through the device.

A.9.3 CRITICAL PHENOMENA

The study of systems near a phase transition has been remarkably successful and has shown that the behavior of many physical system is quite independent of the details of their microscopic interactions. We shall discuss below two systems (Ising spins and gases near the gas-liquid transition) that though physically very different, nonetheless display very similar behavior at their transition.

A.9.3.1 The Ising Model Near its Critical Temperature: Critical Exponents

The study of the Ising model close to its critical temperature T_c (where it undergoes a transition from a para- to a ferro-magnetic phase) is instructive: it provides for a test ground of a critical system, one whose response diverges (as we shall see below) when $T \to T_c$.

As pointed out in section 5.10.1, when $B = 0$ a finite magnetization m exists only if the slope of the curve $m/\mu_0 \equiv y = ax$ (with $a = T/T_c$) near $m = 0$ is smaller than 1. In such case there is a non-zero solution to the equation $ax = \tanh(x)$. Expanding the function $\tanh x = x - x^3/3 + 2x^5/15 - \cdots$ for small x in Eq.5.199 yields: $ax \approx x - x^3/3$ from which we derive the spontaneous mean magnetization m_s in the vicinity of the transition temperature: $a \lesssim 1$ ($T \lesssim T_c$):

$$x_s \approx \sqrt{3(1-a)} \quad \text{i.e.:} \quad m_s \approx \mu_0 \sqrt{3(1 - T/T_c)} \sim (1 - T_R)^{\beta_f}$$

where $T_R = T/T_c$ is known as the reduced temperature and $\beta_f = 1/2$ is the so-called critical exponent associated with the mean magnetization. Hence below a critical temperature T_c a ferromagnetic system acquires a spontaneous magnetization: it is a magnet which strength (magnetization) increases as the square root of the deviation from T_c. In a magnet below T_c the internally generated magnetic field is strong enough to orient the various spins against the thermal agitation. Notice that conversely, a magnet can be demagnetized by heating it to temperatures $T > T_c$.

In presence of a small magnetic field we can compute the magnetization near T_c and from it the susceptibility, i.e. the response of the system to a magnetic field, which as we shall see below diverges as $T \to T_c$. From Eq.5.199, we can compute the variation of the susceptibility as one goes from the high temperature paramagnetic phase into the low temperature ferromagnetic one. In the high temperature phase ($T > T_c$), the spontaneous magnetization is null and by expanding $\tanh(x+b)$ for small b (and thus small x) (recall that $b = \beta\mu_0 B$):

$$ax \approx x + b \quad \text{i.e.} \quad m = \mu_0 ax = \frac{\mu_0 ab}{a - 1} = \frac{\mu_0^2 B}{k(T - T_c)}$$

from which we deduce that the magnetic susceptibility is, see Eq.5.51:

$$\chi_{T>T_c} = N\frac{\partial m}{\partial B}\bigg|_{B=0} = \frac{N\mu_0^2}{k(T - T_c)} \sim (T_R - 1)^{-1} \tag{A.102}$$

This equation is known as the Curie-Weiss law for ferromagnets. As the system gets closer to the transition the magnetic susceptibility diverges as $(T_R - 1)^{-\gamma}$ with a critical exponent $\gamma = 1$. This result reflects the fact that as the ensemble of coupled spins gets closer to the ferromagnetic transition the spins tend to align more easily with one another (they are less sensitive to the thermal agitation), thus increasing the response of the system to an external magnetic field.

At $T = T_c$, $(a = 1)$ the susceptibility can be computed from the mean magnetization in the limit $B \to 0$:

$$x = x + b - (x+b)^3/3 \quad \text{i.e.:} \quad x \approx (3b)^{1/3} \quad \text{or:} \quad m = \mu_0(3\beta_c\mu_0 B)^{1/3} \sim B^{1/\delta}$$

where $\delta = 3$ is the critical exponent associated with the magnetization at T_c. The susceptibility diverges as the magnetic field $B \to 0$:

$$\chi_{T_c} = N\frac{\partial m}{\partial B}\bigg|_{B=0} = \frac{N\mu_0^2\beta_c}{3}(\beta_c\mu_0 B)^{-2/3} \to \infty \tag{A.103}$$

Below T_c $(a < 1)$, if the magnetization change δm due to the magnetic field B is smaller than the spontaneous magnetization m_s $(m = m_s + \delta m \approx m_s$, i.e. $m/\mu_0 \equiv y = y_s + \delta y)$:

$$y = a(x_s + \delta x) = (x_s + \delta x) + b - [(x_s + \delta x) + b]^3/3$$

where the spontaneous magnetization satisfies as we have seen above: $y_s = ax_s = x_s - x_s^3/3$, i.e.: $x_s^2 = 3(1-a)$. Expanding to first order in b and δx yields when $a \approx 1$:

$$a\delta x \approx \delta x + b - x_s^2\delta x - x_s^2 b = \delta x - 3(1-a)\delta x + b - 3(1-a)b$$

$$\delta x \approx \frac{b}{2(a-1)} \quad \text{i.e.:} \quad \delta m = \frac{\mu_0^2 B}{2k(T_c - T)} \tag{A.104}$$

from which we derive the susceptibility in the low temperature phase:

$$\chi_{T<T_c} = N\frac{\partial \delta m}{\partial B}\bigg|_{B=0} = \frac{N\mu_0^2}{2k(T_c - T)} \sim (1 - T_R)^{-1} \tag{A.105}$$

The susceptibility diverges as in the high temperature phase but differs by a factor two from the result obtained for $T > T_c$. While the susceptibility is measured in the limit of zero magnetic field $B \to 0$, if the field B is large enough to induce a magnetization δm greater than the spontaneous magnetization (m_s which tends to zero as one gets closer to T_c): $\delta m > m_s$, the measured susceptibility will display a dependence on B similar to that obtained at $T = T_c$ (when $m_s = 0$). In other words at temperatures below T_c there will be a cross-over between a high magnetic field regime where the susceptibility varies as $B^{-2/3}$ and a low magnetic field regime where the susceptibility is independent of B, but diverges as $1/(T_c - T)$.

Using the results on the magnetization below T_c we may now compute the entropy in the ferromagnetic phase, near T_c, i.e. for small $y = m/\mu_0 = \sqrt{3(1 - T_R)}$. Expanding

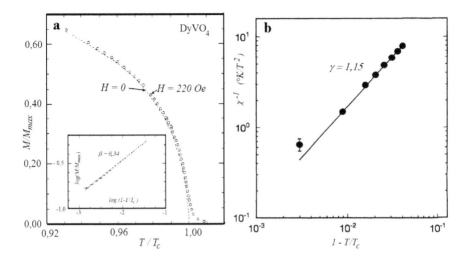

Figure A.16 a) Experimental measurement of the variation of magnetization at zero field with temperature below T_c in DyVO$_4$. Inset: the same variation in logarithmic scale near T_c displaying a critical exponent $\beta = 0.34$; reproduced with permission from [119], copyright (1975) by IOP publishing. b) Variation of the inverse of the magnetic susceptibility near T_c displaying a critical exponent $\gamma = 1.15$; reprinted with permission from ref.[120], copyright (1992) by the American Physical Society.

the logarithm[17] in Eq.5.201 one gets:

$$S^{MF} = Nk\log 2 - Nky^2/2 = Nk\log 2 - \frac{3Nk}{2}(1 - T/T_c) \quad \text{for } T < T_c \qquad (A.106)$$

As expected, the entropy in the ferromagnetic phase is smaller than in the paramagnetic one. The specific heat at constant (zero) magnetic field c_B is null above the transition and decreases with T below T_c:

$$c_B = \left.\frac{\delta Q}{\delta T}\right|_{B=0} = T\left.\frac{\partial S^{MF}}{\partial T}\right|_{B=0} = \frac{3NkT}{2T_c} \rightarrow \frac{3Nk}{2} \quad \text{as} \quad T \rightarrow T_c \qquad (A.107)$$

In other words, as $T \rightarrow T_c$ the specific heat doesn't depend on the reduced temperature difference, i.e. $c_B = (1 - T_R)^\alpha$ with $\alpha = 0$, where α is the critical exponent associated to the specific heat.

The results obtained above from the mean-field approximation are in qualitative agreement with experiments, but the measured critical exponents ($\alpha \approx 0.18$; $\beta \approx 0.34$; $\gamma \approx 1.15$; $\delta \approx 4.10$) differ from the ones computed above, see Fig.A.16.

---

[17]Using the Taylor expansion: $\log(1+y) \approx y - y^2/2 + y^3/3 - y^4/4 + \cdots$

The mean field theory which consists in approximating the interactions with neighboring spins by an interaction with the average field generated by all spins in the sample is not valid in real 3D space [18]. The reason for that break-down of mean field theory is the importance near T_c of collective and correlated spin fluctuations. These correlated fluctuations result in large variations in the magnetic field experienced by any given spin (due to the diverging magnetic susceptibility near T_c, see Eq.A.102), invalidating the approximation made in the mean field. A correct account of these effects necessitates a different approach dealing with these correlated fluctuations and known as the renormalization group approach to critical phenomena, which necessitates mathematical tools that are beyond those used in this book.

A.9.3.2 Real Gas Near its Critical Point

Just as the study of an Ising ferromagnet near its phase transition provided a quantitative test of the mean field approximation, so does the study of a gas near T_c: it allows for a quantitative test of the van der Waals approximation. At this critical point, i.e. reduced temperature $T_R = 1$ and pressure $P_R = 1$, the gas and liquid phases differ only slightly in their density (specific volume): $v_{R,1} = 1 - \epsilon$ (gas phase) and $v_{R,2} = 1 + \epsilon$ (liquid phase) (with $\epsilon \ll 1$). That small difference in specific volumes (2 ϵ) can be computed from the equality of the pressure, see Eq.5.211:

$$P_R = \frac{8T_R}{3v_{R,1} - 1} - \frac{3}{v_{R,1}^2} = \frac{8T_R}{3v_{R,2} - 1} - \frac{3}{v_{R,2}^2} \tag{A.108}$$

which can be recast as:

$$8T_R\left(\frac{1}{3v_{R,1} - 1} - \frac{1}{3v_{R,2} - 1}\right) = \frac{3}{v_{R,1}^2} - \frac{3}{v_{R,2}^2}$$

$$4T_R\left(\frac{1}{1 - 3\epsilon/2} - \frac{1}{1 + 3\epsilon/2}\right) = \frac{3}{(1 - \epsilon)^2} - \frac{3}{(1 + \epsilon)^2}$$

$$\tag{A.109}$$

Expanding to third order in ϵ yields[19]: $\epsilon = 2(1 - T_R)^{1/2} \equiv 2(1 - T_R)^\beta$ where $\beta = 1/2$ is the critical exponent associated to the difference in specific volumes (or densities). Notice that it is identical to the variation with temperature of the magnetization of an Ising ferromagnet near T_c, see section A.9.3.1.

Similarly the variation at $T = T_c$ of the specific volume with pressure can be obtained by expanding Eq.A.108 with $T_R = 1$ up to third order in ϵ. The result is: $P_R - 1 = 3\epsilon^3/2$, namely $\epsilon \sim (P_R - 1)^{1/\delta}$ with $\delta = 3$. Thus the variation at $T = T_c$ of the specific volume with pressure, i.e. the critical exponent δ, is identical to the variation

---

[18]The mean-field approximation is valid when generalizing the Ising model to higher non-physical dimensions: $D \geq 4$

[19]Using the Taylor expansion $(1 + x)^a = 1 + ax + a(a - 1)x^2/2! + a(a - 1)(a - 2)x^3/3! + \cdots$ one obtains: $12T_R(\epsilon + 9\epsilon^3/4) \approx 12(\epsilon + 2\epsilon^3)$ from which one obtains: $T_R \approx (1 + 2\epsilon^2)/(1 + 9\epsilon^2/4) \approx 1 - \epsilon^2/4$

of the magnetic susceptibility with magnetic field at $T = T_c$ in the Ising model, see section A.9.3.1.

One can also compute the equivalent for a real gas of of the magnetic susceptibility, namely the compressibility κ_R (the relative change in volume as the pressure is altered) near the critical point. When $T_R > 1$ ($V_R = 1 - \epsilon$), the reduced compressibility obeys:

$$\kappa_R = -\frac{1}{V_R}\frac{\partial V_R}{\partial P_R} \approx \frac{\partial \epsilon}{\partial P_R} = \frac{1}{6(T_R - 1)} \sim (T_R - 1)^{-\gamma}$$

with $\gamma = 1$ just like the magnetic susceptibility in the Curie-Weiss law, Eq.A.102.

Finally, notice that the specific heat of a real gas: $c_V = 3Nk/2$ does not depend on temperature (i.e. $c_V \sim (T_R - 1)^\alpha$ with $\alpha = 0$) as for the Ising ferromagnet discussed above, section A.9.3.1. The measured values of these various exponents $(\alpha, \beta, \gamma, \delta)$ differ from these predictions but are very close to the ones measured for an Ising ferromagnet (see section A.9.3.1 above): $\alpha \approx 0.1$, $\beta \approx 0.33$, $\gamma \approx 1.35$, $\delta \approx 4.2$. In fact near their critical point both models fall within the same so-called "universality class": both are binary systems (spins up or down, gas or liquid phase) whose behavior near their critical point is independent of the precise nature and form of their microscopic interactions! This universality of critical phenomena is an amazing and profound discovery of the last half of the 20th century which is valid for many different physical systems near their critical point (spanning different universality classes). The successful description and unification of critical phenomena (which is beyond the scope of this book) is a reflection of both the unity of Science and the stunning success of Mathematics in the description of Nature.

A.9.4   MONTE-CARLO METHODS

Many (most) problems in Statistical Mechanics cannot be solved analytically as was the case for the ideal gas. To find the equilibrium behavior of a system (its energy, specific heat (response to temperature change), compressibility (response to pressure change), magnetic susceptibility (response to magnetic field), etc.) one often resort to stochastic numerical simulation[5], known as Monte-Carlo simulations. The goal of these methods is to reproduce the equilibrium distribution of states of the system even if the way that distribution is reached is unrealistic. These simulations proceed from a given state i of the system with energy E_i. One then randomly chooses a new state j and compute its energy E_j (much ingenuity goes into that procedure which must sample the space of possible states in a random, unbiased way). If $E_j < E_i$ one accepts state j as the new state of the system and repeats the procedure. If $E_j > E_i$, one may accept state j as the new system state albeit with probability $P_{ij} = \exp\beta(E_i - E_j)$ (with $\beta = 1/kT$). It is easy to verify that after a large number of such random choices the program reaches a steady state where the probability of being in state i is given by its equilibrium Boltzman weight:

$$P_i = \frac{e^{-\beta E_i}}{\sum_j e^{-\beta E_j}}$$

Indeed in that steady state the probability of moving out of state i:

$$P_{out} = \sum_{j,E_j<E_i} P_i + \sum_{j,E_j>E_i} P_i e^{-\beta(E_j-E_i)}$$

is equal to the probability of moving into state i:

$$P_{in} = \sum_{j,E_j<E_i} P_j e^{-\beta(E_i-E_j)} + \sum_{j,E_j>E_i} P_j$$

as can be verified by plugging in $P_i \sim e^{-\beta E_i}$ and $P_j \sim e^{-\beta E_j}$.

References

1. E.Wigner. The Unreasonable Effectiveness of Mathematics in the Natural Sciences. *Communications on Pure and Applied Mathematics*, 13:1–14, 1960.
2. R.W. Hamming. The Unreasonable Effectiveness of Mathematics. *The American Mathematical Monthly*, 87:81–90, 1980.
3. F. Byron Jr. and R.Fuller. *Mathematics of Classical and Quantum Physics*. Dover Publications, 1970.
4. M.Kline. *Calculus: an Intuitive and Physical Approach*. Dover Publications, 1998.
5. W.H. Press, S.A. Teukolsky, W.T. Vetterling, and B.P. Flannery. *Numerical Recipes in C: The Art of Scientific Computing, 2nd ed.* Cambridge University Press, 1992.
6. K. Singh. *Linear Algebra, Step by Step*. Oxford University Press, 2014.
7. P.M. Morse and H.Feshbach. *Methods of Theoretical Physics, part I*. McGraw-Hill, 1953.
8. E.T.Jaynes. *Probability Theory: The Logic of Science*. Cambridge University Press, 2003.
9. H. Tijms. *Understanding Probability: Chance Rules in Everyday Life*. Cambridge University Press, 2004.
10. I.Newton. *Newton's Principia: The Mathematical Principles of Natural Philosophy*. Kessinger Publishing Co, 2003.
11. C.Pask. *Magnificent Principia: Exploring Isaac Newton's Masterpiece*. Prometheus books, 2019.
12. H.Goldstein. *Classical Mechanics*. Addison-Wesley., 1981.
13. Lucretius. *On the Nature of the Universe*. Penguin Classics, 1951.
14. Fluidworkshop. Shutterstock.
15. R.P.Feynmann, R.B.Leighton, and M.Sands. *The Feynman Lectures on Physics, Vol.I.* Addison-Wesley., 1966. Classical and Statistical Mechanics.
16. L.D.Landau and E.M.Lifchitz. *Course of Theoretical Physics, Vol.1*. Pergammon, 1960. Classical Mechanics.
17. L.D.Landau and E.M.Lifchitz. *Course of Theoretical Physics, Vol.2*. Pergammon, 1960. Electromagnetism.
18. R.P.Feynmann, R.B.Leighton, and M.Sands. *The Feynman Lectures on Physics, Vol.II.* Addison-Wesley., 1966. Electromagnetism.
19. J.D. Jackson. *Classical Electrodynamics*. Wiley, 1962.
20. D.J.Griffiths. *Introduction to Electrodynamics*. Cambridge University Press, Cambridge, 2017.
21. J. Israelachvili. *Intermolecular and Surface Forces*. Academic Press, 1992.
22. R.A. Millikan. The Isolation of an Ion, a Precision Measurement of its Charge, and the Correction of Stoke's Law. *The Physical Review*, 32:349–397, 1911.
23. F. Strubbe, F. Beunis, and K. Neyts. Detection of Elementary Charges on Colloidal Particles. *Physical Review Letters*, 100:218301, 2008.
24. Ojibberish, 2007. https://commons.wikimedia.org/wiki/File:Triode_tube_schematic.svg.
25. Grebbe, 2010. https://commons.wikimedia.org/wiki/File:RS242_triode_2.png.
26. Rehua, 2012. https://commons.wikimedia.org/wiki/File:Galvanic_ cell_ labeled.svg.
27. S.Okamoto. Shutterstock.

28. Akriesh, 2006. https://commons.wikimedia.org/wiki/File:Penning_ Trap.svg.
29. Fred the Oyster, 2011. https://commons.wikimedia.org/wiki/File:Galvanometer_scheme. svg.
30. BillC, 2006. https://commons.wikimedia.org/wiki/File:Transformer3d_ col3.svg.
31. S.Cymro. Shutterstock.
32. G.Wiora, 2011. https://commons.wikimedia.org/wiki/File:Redshift.svg.
33. NASA, 2007. https://commons.wikimedia.org/wiki/File:EM_ Spectrum_ Properties_ edit.svg.
34. M.Born and E.Wolf. *Principles of Optics, 5th ed.* Pergammon, 1975.
35. Lookang, 2011. https://commons.wikimedia.org/wiki/File:Huygens_Fresnel_ Principle.gif.
36. Travel-Master. Shutterstock.
37. Ikarus, 2005. https://commons.wikimedia.org/wiki/File:RainbowFormation_Droplet Primary.png.
38. N.D. Mermin. Is the Moon there when nobody looks? Reality and the Quantum Theory. *Physics Today*, pages 38–47, 1985.
39. P.A.M.Dirac. *The Principles of Quantum Mechanics, 4th ed.* Oxford University Press, 1958.
40. R.P.Feynmann, R.B.Leighton, and M.Sands. *The Feynman Lectures on Physics, Vol.III.* Addison-Wesley., 1966. Quantum Mechanics.
41. L.D.Landau and E.M.Lifchitz. *Course of Theoretical Physics, Vol.3.* Pergammon, 1960. Quantum Mechanics.
42. L.Pauling and E.B.Wilson. *Introduction to Quantum Mechanics.* McGraw-Hill, 1935.
43. S.Gasiorowicz. *Quantum Physics.* Wiley, 2003.
44. L.Susskind. *Quantum Mechanics: the Theoretical Minimum.* Penguin books, 2014.
45. D.J.Griffiths and D.F. Schroeter. *Introduction to Quantum Mechanics, 3rd ed.* Cambridge Univ. Press, Cambridge, 2018.
46. MoFarouk. Shutterstock.
47. NASA. https://asd.gsfc.nasa.gov/archive/arcade/images/cmb_ intensity.gif.
48. Physics Today, 1992. https://aether.lbl.gov/www/projects/cobe/COBE_ Home/phys_ today_ cover_ big.gif.
49. W.M.Haynes. *CRC Handbook of Chemistry and Physics, 96th edition.* CRC Press, 2016.
50. A.Einstein. The Quantum Theory of Radiation. *Physikalische Zeitschrift*, 18:121–136, 1917.
51. J.W.M. Bush. Pilot-wave Hydrodynamics. *Annual Review in Fluid Mechanics*, 47:269–292, 2015.
52. A.J. MacDermott and R.A. Hegstrom. A proposed experiment to measure the parity-violating energy difference between enantiomers from the optical rotation of chiral ammonia-like "cat" molecules. *Chemical Physics*, 305:55–68, 2004.
53. Dhatfield, 2008. https://commons.wikimedia.org/wiki/File:Schrodingers_ cat.svg.
54. A.J. Leggett and A. Garg. Quantum mechanics versus macroscopic realism: is the flux there when nobody looks? *Physical Review Letters*, 54:857–860, 1985.
55. G.C. Knee, S.Simmons, E.M. Gauger, J.J.L. Morton, H. Riemann, N.V. Abrosimov, P. Becker, H.-J. Pohl, K.M. Itoh, M.L.W. Thewalt, G.A.D. Briggs, and S.C. Benjamin. Violation of a Leggett–Garg inequality with ideal non-invasive measurements. *Nature Communications*, 3:606, 2012.

100. E. A. Guggenheim. The Principle of Corresponding States. *Journal of Chemical Physics*, 13:253, 1945.

101. J.F. Marko and E.D. Siggia. Stretching DNA. *Macromolecules*, 28:8759–8770, 1995.

102. G. Charvin, J.-F. Allemand, T.R.Strick, D. Bensimon, and V. Croquette. Twisting DNA: single molecule studies. *Contemporary Physics*, 45:383–403, 2004.

103. A. Einstein. *Investigation of the Brownian Theory of Mouvement*. Dover Publication, 1956.

104. J. Monod. *Le Hasard et la Nécessité*. Edition du Seuil, 1970.

105. T. Dobzhansky. Nothing in Biology makes sense except in the light of evolution. *American Biology Teacher*, 35:125–129, 1973.

106. J.D. Watson, N.H. Hopkins, J.W. Roberts, J.A. Steitz, and A.M. Weiner. *Molecular Biology of the Gene, 4th edition*. Benjamin/Cummings Publishing Co., 1987.

107. B. Alberts, D. Bray, J. Lewis, M. Raff, K. Roberts, and J.D. Watson. *Molecular Biology of the Cell, 3rd ed.* Garland Publishing Inc., New-York, 1994.

108. R. Phillips, J.Kondev, J.Theriot, and H.G. Garcia. *Physical Biology of the Cell, 2nd ed.* Garland Science, New-York, 2013.

109. E.Jonas and K.P. Kording. Could a neuroscientist understand a microprocessor? *PLOS Computational Biology*, 13:e100526, 2017.

110. M. Abramowitz and I. Stegun. *Handbook of Mathematical Functions, with Formulas, Graphs, and Mathematical Tables*. Dover Publications, 1972.

111. H.A.Lorentz, A.Einstein, H.Minkowski, and H.Weyl. *The Principle of Relativity*. Dover, 1952.

112. R. K. Pathria. *The Theory of Relativity*. Dover Publications, 1974.

113. W.Pauli. *Theory of Relativity*. Dover, 1981.

114. W.K.H. Panofsky and M.Phillips. *Classical Electricity and Magnetism*. Dover, 2005.

115. A.Tonomura, N. Osakabe, T. Matsuda, T. Kawasaki, J. Endo, S. Yano, and H. Yamada. Evidence for Aharonov-Bohm effect with magnetic field completely shielded from electron wave. *Physical Review Letters*, 56:792–795, 1986.

116. L. Bougas, G. E. Katsoprinakis, D. Sofikitis, T. P. Rakitzis, P. C. Samartzis, T. N. Kitsopoulos, J. Sapirstein, D. Budker, V. A. Dzuba, V. V. Flambaum, and M. G. Kozlov. Stark shift and parity nonconservation for near-degenerate states of xenon. *Physical Review A*, 89:042513, 2014.

117. W. deSorbo. Specific Heat of Diamond at Low Temperatures. *Journal of Chemical Physics , 876 (1953)*, 21:876–880, 1953.

118. J.E. Desnoyehs and J.A. Morrison. The heat capacity of diamond between 12.8°K and 277°K. *Philosophical Magazine*, 3:42–48, 1958.

119. R.T. Harley and R.M. Macfarlane. A determination of the critical exponent β in TbVO$_4$ and DyVO$_4$ using linear birefringence. *Journal of Physics C: Solid State Physics*, 8:L451–L455, 1975.

120. K.A. Reza and D.R. Taylor. Measurement of static critical exponents for a structural Ising-model phase transition with random strains: Dy(As$_x$V$_{1-x}$)O$_4$. *Physical Review B*, 46:11425–11431, 1992.

Index

Milton Keynes UK
Ingram Content Group UK Ltd.
UKHW051535141024
449569UK00001B/38

9 781032 112404

.